APPROVED LISTS OF BACTERIAL NAMES

APPROVED LISTS OF BACTERIAL NAMES

edited by

V. B. D. SKERMAN
Department of Microbiology
University of Queensland
St. Lucia 4062, Queensland, Australia

VICKI McGOWAN
Department of Microbiology
University of Queensland
St. Lucia 4062, Queensland, Australia

P. H. A. SNEATH
MRC Microbial Systematics Unit
University of Leicester
Leicester LEI 7RH, England

on behalf of

The Ad Hoc Committee of the Judicial Commission
of the
International Committee on Systematic Bacteriology

Reprinted from *International Journal of Systematic Bacteriology*,
Volume 30:225–420, 1980, with corrections

American Society for Microbiology
Washington, D.C.
1989

PREFACE

The **Approved Lists of Bacterial Names** was published in the **International Journal of Systematic Bacteriology** (IJSB **30**: 225-420, 1980) and reprinted in book form to provide for the requirements of the Bacteriological Code (1976 Revision) in initiating a new starting date for bacterial nomenclature, 1 January 1980. Despite every care, errors in the text have been detected and these errors have now been corrected in this Amended Edition. The status of the names listed and the limitations applied to them are explained in the original in the Section dealing with "Specific Information relating to the Approved Lists of Bacterial Names" (IJSB **30**:228, 1980) where emphasis was placed on 1) the nomenclatural status of all names in the Lists (Item 6, p. 229), 2) the reuse and revival of names (Item 8, p. 229-230) and 3) the correction of errors in information on names included in the **Approved Lists** (Item 9, p. 230). Item 9 (p. 230) also emphasised that no name would be added to or removed from the **Approved Lists of Bacterial Names**.

Action on Names Included in the Approved Lists

Subsequent to the publication of the Approved Lists of Bacterial Names actions related to the names in the Approved Lists have fallen into two categories.

1. Correction of errors in the information. Errors communicated to the ICSB have been assessed by Professor P.H.A. Sneath and his associates and most of these have been published in the IJSB **34**: 508-511, 1984.

2. Changes in Nomenclatural Status of names in the **Approved Lists**. As indicated in Item 6 (IJSB **30**:229, 1980) no names included in the Lists were initially conserved. However these names could be later "conserved, rejected, merged with other names or subjected to any valid nomenclatural act".

Proposals for conservation or rejection of names included in the **Approved Lists** are subject to decisions of the Judicial Commission. While some names may be conserved and others rejected the decisions of the Judicial Commission will not affect the original listing in the Approved Lists. They will simply define the status of the names for future reference in nomenclature.

New combinations (or new subdivisions) of names in the **Approved Lists** will constitute new names with new authorities as described below.

Action on Names Not in the Approved Lists

Under the Code of Nomenclature (Rule 28a) authors who choose to revive or reuse a name not included in the **Approved Lists** become the authorities for the revived names or reused names and, in this sense, each revived name or reused name is a new name and is not reinstated in the Approved Lists.

Validation of New Names.

New names here refers to names arising from revision of status of names **in** the **Approved Lists**, revival or reuse of names not included in the Approved Lists and names of newly created taxa.

The **Bacteriological Code** provides for the future validation of new names of bacteria. If papers are accepted for publication in the IJSB the names contained therein are automatically validated from the date of publication. If published elsewhere copies of the papers have to be submitted to the ICSB. If they meet the requirements of the rules of nomenclature the new names are validated by publication in Validation Lists in the IJSB. The valid date of publication is that upon which the names appear in the IJSB and not the date upon which they appeared in the original publication.

The underlying reason for this action is that there are over 800 microbiological publications in which papers dealing with nomenclatural matters may appear and it is economically impossible for any institution to acquire them all. The listing of all new names and combinations in one publication is a service to all bacteriologists and seems now to be generally accepted. The onus however remains with the authors to ensure that copies of their papers, published outside the IJSB, are promptly submitted to the IJSB for validation of the names.

New names are not 'approved' by the ICSB. They are validated as new names as a result of conformity with the Rules of Nomenclature. The ICSB did not 'approve' the names in the **Approved Lists**. The ICSB simply approved of the **listing** of the names in lists with a new starting date for the operation of the rules of priority in naming. The names accepted in the lists already had valid standing in nomenclature. Hence we have the "Approved Lists of Bacterial Names" and not "Lists of Approved Bacterial Names" as the title. Other names which were valid before 1980 were invalidated by omission, by consensus, from the lists mainly because it proved impossible to locate 'Type' 'Neotype' or authentic cultures.

The direction, under Item 9 (IJSB **30**:230, 1980) that "no names will be added to or removed from the **Approved Lists of Bacterial Names**" was adopted by the ICSB in order to preserve the identity of the Lists as the starting point in a revised era of bacterial nomenclature, in much the same way as Linnaeus' "Species Plantarum" serves as the starting point of Botanical nomenclature. For this reason the Approved Lists of Bacterial Names will remain an independent publication.

As a further service to bacteriologists the editor of the IJSB has planned to produce and continually update lists of new names and new combinations, which will combine the information contained in the lists entitled "Validation of the Publication of New Names and New Combinations Previously Effectively Published Outside the IJSB", which appears in the issues of the IJSB. The first list of this kind appeared in the IJSB **32**:146-149, 1982.

The ICSB is pleased to acknowledge the sterling services of Ms Vicki McGowan in the preparation of the Approved Lists and the amendments thereto and to thank the Department of Microbiology of the University of Queensland for its role in the generation of the Lists.

V.B.D. Skerman
Brisbane, Australia
April 1989

INTRODUCTION

At the meeting of the Judicial Commission of the ICSB held in Jerusalem on the 29th March, 1973 an Ad Hoc Committee was appointed (Minute 22) to organize a review of the currently valid names of bacteria with the object of retaining only names for those taxa which were adequately described and, if cultivable, for which there was a Type, Neotype or Reference strain available; to compile these names under the title of **Approved Lists of Bacterial Names** and to publish the lists in the International Journal of Systematic Bacteriology, to become effective on January 1, 1980. This date would then replace May 1, 1753 (International Code of Nomenclature of Bacteria and Viruses. Rule 10) as the new date for determining priorities for names of new taxa.

The members of the Ad Hoc Committee originally appointed to oversee the task were S.P. Lapage, H.P.R. Seeliger and V.B.D. Skerman (Chairman). Following his election as President-Elect of IAMS, H.P.R. Seeliger resigned from the Committee and was replaced by J.G. Holt. P.H.A. Sneath, as Chairman of the Judicial Commission, was coopted to the Committee. This Committee was responsible for the editing of the first draft of the Approved Lists of Bacterial Names published in the IJSB (26, 1976, 563-599).

The Ad Hoc Committee was reconstituted by the Judicial Commission during its meetings at the International Congress of the IAMS in Munich in September, 1978, with V.B.D. Skerman (Chairman), P.H.A. Sneath (newly elected Chairman of the ICSB) and L.G. Wayne (newly elected Chairman of the Judicial Commission) as Committee members. This Committee was assigned wide powers to complete the work on the Lists and arrange publication.

The initial Committee agreed that, as a basis for inquiry, all names which had been included in the eighth edition of Bergey's Manual of Determinative Bacteriology, whether listed as recognised species, synonyms, *species incertae sedis,* or under other headings should be circulated; that all Subcommittees on Taxonomy of the ICSB should be asked for advice on the retention of taxa for which the Subcommittees were responsible and that specialists be approached for advice on other taxa - only a small list of taxa remained to be considered by the Ad Hoc Committee itself. Advice was also sought on additional names of taxa which had been validly published since the publication of the Manual. Special provision had been made in the Code of Nomenclature for the inclusion of names validly published after the 1st January, 1978 (Rule 24a).

A list of specialists was drawn up and approved by the Ad Hoc Committee. Each specialist was asked to associate two others in reaching a decision on name retention. The Chairman accepted the responsibility for coordination of work associated with the selection of generic names and specific and subspecific epithets and P.H.A. Sneath for the names of higher taxa.

Names of Genera, Species and Subspecies:

The Lists of names of species and subspecies were circulated by airmail in April, 1976 and a draft list of names recommended for retention published in the IJSB (26, 1976, 563-599) together with the list of those people associated with that stage of the project. Reference may be made to the draft list for details of the procedures adopted in processing the information.

The object of publishing the draft list was to enable microbiologists, in general, to submit opinions to the Ad Hoc Committee. Several were received. Some of these resulted in modification to lists submitted by Subcommittees on Taxonomy and others.

Rule 24a of the Code of Nomenclature provided, *inter alia,* for the publication with each bacterial name the name(s) of the author(s) who originally proposed it, a reference to an effectively published description of each species and its nomenclatural type whenever possible, and a Type, Neotype or Reference strain by its designation.

To conserve space in publication, it was agreed that for all references to the 8th edition of Bergey's Manual (Buchanan, R.E. and N.E. Gibbons (eds). 1974. Bergey's Manual of Determinative Bacteriology, 8th ed. The Williams and Wilkins Co., Baltimore) be contracted to 'Bergey 8'.

Although these requirements were formally published in the Code on 1st January, 1976 little action was taken by those responsible for various taxa and in order to assist resolution of the problem, the Chairman and his staff produced a document which listed Collections of Cultures of Microorganisms which held Type, Neotype and Reference strains for those species whose names appeared in the draft list and subsequent amendments. This list, which will be published elsewhere as a publication of the World Data Center for Microorganisms, has been used as a guide to the deletion of numerous names for which no cultures could be located.

The requirements of Rule 24a together with the list of strains, were discussed at a meeting of the ICSB with Subcommittees on Taxonomy at the International Congress for Microbiology in Munich in September, 1978 where the decision was made to list only one designation for each strain and preferably that of the American Type Culture Collection, where available. Because of difficulties associated with locating references to publications in which Type strains, Neotype strains or Reference strains had been designated, it was agreed that references would be given only to those publications concerned with the naming of the taxa and to descriptions of the taxa.

The Judicial Commission also determined that publication should be in the January (1980) issue of the IJSB, thus negating the requirement of Rule 24a that publication take place prior to 1st January, 1980. This action enabled the inclusion in the Approved Lists of names of all new taxa which had been validated by publication in the October, 1979 issue of the IJSB and so has obviated the need to make any further additions to the Approved Lists of Bacterial Names.

Formal notice of these requirements was circulated, by airmail, to all Subcommittees and specialists on 5th February, 1979 requesting return of the information by 31st August, 1979. Meanwhile, the Chairman and his staff proceeded as far as possible with the compilation of the whole of the required information, subsequently making use of such submissions as were ultimately received from others, to verify compilations already made.

Computer printouts, in the final format of assembled information, were distributed by airmail to all participants in July, 1979 for the purpose of verification and for the provision of information which was still missing.

A final decision to retain or delete some names was based on our ability to obtain the relevant information.

Names of Taxa Above the Rank of Genus:

Compilation of names of higher taxa was undertaken by Sneath and his associates. No draft list was published but a proposal was discussed at some length in an open meeting at the Munich Congress. The format of the list as submitted was modified to conform with the format adopted for the names of the genera and species. References given in this section of the lists are restricted to those relating to the higher taxa. Those to genera will be found in the list of genera and species.

Designation of Strains:

Strains have been designated by accession numbers of Culture Collections in which they have been deposited. Every effort has been made to ensure that the strains are actually available but no guarantee of this can be given. The Names, which have been omitted, may be revived if the location of suitable strains can be ascertained.

The following is a list of addresses of institutions, together with acronyms used to assign accession numbers.

AMRC	FAO-WHO International Reference Centre for Animal Mycoplasmas, Institute for Medical Microbiology, University of Aarhus, Aarhus, Denmark.
ATCC	American Type Culture Collection, Rockville, Maryland 20852, U.S.A.
BKM	All-Union Collection of Microorganisms, Institute of Microbiology, USSR Academy of Sciences, Moscow, USSR.
BKMW	Culture Collection, Institute of Microbiology, USSR Academy of Sciences, Moscow, USSR.
CBS	Centraalbureau voor Schimmelcultures, Baarn, The Netherlands.
CCM	Czechoslovak Collection of Microorganisms, J.E. Purkyne University, Brno, Czechoslovakia.
CDC	Center for Disease Control, Atlanta, Georgia, U.S.A.
CIP	Collection of the Institute Pasteur, Paris, France.

DSM	Deutsche Sammlung von Mikroorganismen, Schnittspahnstrasse, Darmstadt, Federal Republic of Germany.
IAM	Institute of Applied Microbiology, University of Tokyo, Tokyo, Japan.
IFO	Institute for Fermentation, Osaka, Japan.
IMRU	Institute of Microbiology, Rutgers - The State University, New Brunswick, New Jersey, U.S.A.
IMV	Institute of Microbiology and Virology, Academy of Sciences of the Ukranian SSR.
INA	Institute for New Antibiotics, Moscow, USSR.
INMI	Institute for Microbiology, USSR Academy of Sciences, Moscow, USSR.
IPV	Instituto de Patologia Vegetale, Milan, Italy.
KCC	Kaken Chemical Company Ltd., Tokyo, Japan.
LIA	Museum of Cultures, Leningrad Research Institute of Antibiotics, 23 Ogorodnikov Prospect, Leningrad L-20, USSR.
LSU	Louisiana State University, Baton Rouge, Louisiana, U.S.A.
LMD	Laboratorium voor Microbiologie der Technische Hogeschool, Delft, The Netherlands.
NCDO	National Collection of Dairy Organisms, National Institute for Research in Dairying, University of Reading, England, UK.
NCIB	National Collection of Industrial Bacteria, Torry Research Station, Aberdeen, Scotland, UK.
NCMB	National Collection of Marine Bacteria, Torry Research Station, Aberdeen, Scotland, UK.
NCPPB	National Collection of Plant Pathogenic Bacteria, Plant Pathology Laboratory, Harpenden, England, UK.
NCTC	National Collection of Type Cultures, Central Public Health Laboratory, Colindale, London, England, UK.
NIAID	National Institute of Allergy and Infectious Diseases, Hamilton, Montana, U.S.A.
NIHJ	National Institute of Health, Tokyo, Japan.
NRC	National Research Council, Ottawa, Canada.
NRL	Neisseria Reference Laboratory, US Public Health Service Hospital, Seattle, Washington, U.S.A.
NRRL	Northern Utilization Research and Development Division, US Department of Agriculture, Peoria, Illinois, U.S.A.
PDDCC	Culture Collection of Plant Diseases Division, New Zealand Department of Scientific and Industrial Research, Auckland, New Zealand.
TC	Thaxter Collection, Farlow Herbarium, Harvard University, Boston, Massachusetts, U.S.A.
UMH	University of Missouri Herbarium, Missouri, U.S.A.
UQM	Culture Collection, Department of Microbiology, University of Queensland, Brisbane, Australia.
VKM	Institute of Microbiology, Academy of Sciences of the USSR, Moscow, USSR.
VPI	Virginia Polytechnic Institute and State University, Blacksburg, Virginia, U.S.A.
WINDSOR	Culture Collection, University of Windsor, Windsor, Ontario, Canada.

Acknowledgements:

Whilst the task of compilation and preparation of manuscripts for publication has been of necessity concentrated in two areas, the successful completion of the task would not have been possible without the associated efforts of a large number of people whose names are recorded here.

T.V. Aristovskaya, P.K.C. Austwick, E. Baldacci, L. Barksdale, E. Barnes, P. Baumann, J.H. Becking, E.L. Biberstein, I. Bousfield, K. Bovre, W.J. Brinley-Morgan, D. Brenner, T. Brock, E. Brockman, M. Bryant, J.C. Burnham, J. Carr, W. Catlin, E.P. Cato, D. Claus, G. Colman, R.R. Colwell, M.J. Corbel, K.T. Crabtree, T. Cross, G.A. Dubinina, P. Dugan, D.W. Dye, K. Eimhjellen, G. Eldering, W.H. Ewing, S. Faine, S.M. Finegold, E.A. Freundt,

W. Frederiksen, A.L. Furniss, E. Garvie, N.E. Gibbons, M. Goodfellow, M. Gordon, B.V. Gronov, H. Hatt, A.C. Hayward, M. Hendrie, S.D. Henriksen, A. Henssen, G.J. Hermann, G. Gobbs, L.V. Holdeman, R. Hugh, D.B. Johnstone, D. Jones, L.V. Kalakoutskii, O. Kandler, K.-A. Karlsson, R.M. Keddie, M. Killian, K. Kitahara, J. Knapp, W. Knapp, M. Kocur, N.R. Krieg, S.P. Lapage, J.M. Larkin, H. Lautrop, H.A. Lechevalier, J. DeLey, R. Locci, S. Maier, N.S. Mair, P.H. Makela, M. Mandel, E. McCoy, H.D. McCurdy, K. McNiel, W.E.C. Moore, N. Nishida, H. Nonomura, R.A. Ormsbee, F. Orskov, I. Orskov, E. van Oye, O.M. Parinkina, N. Pfennig, J.E. Phillips, F. Pichinoty, M. Pittman, T. Pridham, C.I. Randles, J.W.M. LaRiviere, M. Rhodes-Roberts, D.S. Roberts, R. Rohde, B. Rowe, E.B. Rozlycky, E. Runyon, R. Sakazaki, V. Scardovi, D. Schafer, R.J. Sidler, J.M. Shewan, G.C. Simmons, I.J. Slotnick, R.M. Smibert, P.H. Smith, J.T. Staley, M. Starr, Y. Terasaki, Y.T. Tchan, E. Thal, J. Thompson, H. Truper, G. Tunevall, D.C. Turk, M. Veron, N. Walker, S.W. Watson, G. Wauters, L. Wayne, R.E. Weaver, O.B. Weeks, E. Weiss, H.J. Welshimer, S. Williams, G.A. Zavarzin, J.G. Zeikus, K. Zinneman.

Thanks are also due to Miss Annette McLennan who typed the original manuscripts and to Dr. Lindsay Sly, Dr. Horst Doelle and Mrs. Elizabeth Marden for considerable assistance in the search for Culture Collections maintaining strains of taxa in the Approved Lists and other information and to the numerous other unnamed people who have assisted in various ways in laboratories throughout the world.

The computer programs used in compiling these lists were written and controlled by Mr. Geoffrey Dengate, Systems Programmer, Prentice Computer Centre, University of Queensland. His help and expertise is sincerely acknowledged.

SPECIFIC INFORMATION RELATING TO THE APPROVED LISTS OF BACTERIAL NAMES

The names included in these Lists are those selected by Subcommittees on Taxonomy of the ICSB, specialist advisers and, where the former have not been available or have declined to advise, by the members of the Ad Hoc Committee.

With few exceptions (vide infra) names which have been included in the Lists are those which

- were validly published before the 1st January, 1978 and which have been listed by advisers.
- have been validly published since the 1st January, 1978 by publication in the IJSB or by inclusion in the lists of new names cited in the IJSB (to and including Volume 29, Part 4, 1979) as having been effectively published in other Journals.

Specific attention is drawn to the following

1. Names included in the Approved List of Bacterial Names are the only names which are nomenclaturally valid as at the 1st January, 1980. All other names which have appeared in the literature prior to 1st January, 1980 are nomenclaturally invalid. Names which have been effectively published in Journals other than the IJSB but which have not been cited in the lists of new names published in the IJSB, have not been validly published under Rule 27 of the Code of Nomenclature. Such names may be validated by publication in forthcoming issues of the IJSB and will simply constitute names of new taxa.

2. No name appearing in the Approved Lists of Bacterial Names is conserved in the sense defined under Rule 56b of the Code of Nomenclature (1976 Revision).

3. Any name previously rejected by the Judicial Commission remains a *nomen rejiciendum* and can no longer be used to name a bacterium.

4. By majority decision of the Ad Hoc Committee, supported by an overwhelming majority of the members of the ICSB, all strains of bacteria which were submitted as nomenclatural Type strains, Neotype strains, Proposed Neotype strains or Reference strains have been elevated to the status of Type strains. The few objections which were received to the elevation of reference strains to Type status were based on concern that such reference strains may not prove to be the best choice for Type strains. This objection however, applies equally to quite a number of Type strains, particularly monotypes of which there are, and will continue to be, sev-

eral. It seemed more appropriate that names appearing in the Approved Lists of Bacterial Names, which constitutes a new base for future nomenclature, should have Type status or be omitted.

5. For some names included in the Lists no strains were available. These names were retained to preserve the status of

A. some microorganisms for which habitats are well known, which have distinct morphological characters but which have never been isolated. In most cases only the Type Species has been retained to preserve the status of the genus.

B. a genus which contained well described species for which representative strains were available for all but the Type Species of the genus itself.

Exclusion of the Type Species would have necessitated the exclusion of the genus as a whole under Rule 20c of the Code of Nomenclature. There were several of these. It is hoped that early action will be taken to request the Judicial Commission to conserve these genera with new Type Species in order that each species within these genera will be represented by a Type strain in a recognised culture collection.

6. The Ad Hoc Committee has repeatedly stressed that its task was nomenclatural and not taxonomic; further that Subcommittees and others were not required to solve taxonomic problems before advising on retention of names. They were requested to omit names of doubtful species as such species could be revived (see 8 below).

Because there are some differences of opinion regarding the generic position of some species it has been necessary to include the same microorganism under two names, each with the same specific epithet and the same nomenclatural type strain. In such instances cross-reference has been made to the alternate name. This applied specifically to the species in the genus *Beneckea* and the genus *Vibrio* and to a few instances in the Enterobacteriaceae. For this reason it is necessary here to reaffirm the statement made previously that no name in the Approved Lists of Bacterial Names has been conserved. Names in the Lists may later be conserved, rejected, merged with other names or subjected to any other valid nomenclatural act.

7. The names of some genera, for which there are no available strains for the Type Species, have been omitted from the Lists in order to permit the resolution of a very difficult nomenclatural situation arising from the recent publication of provisional names which have no standing in nomenclature but which have been cited by subsequent authors as though they were valid. Addition of new species, with nomenclatural types, to an invalidly published generic name renders the new names equally invalid. This refers specifically to the methane oxidising bacteria, where the otherwise valid name *Methanomonas* has been omitted in the hope that by so doing the impasse caused by the illegitimate substitution of the name *Methylomonas* may be resolved. Specialists in this field are now urged to publish a fully validated list of appropriate species.

8. Reuse and revival of names which do not appear in the Approved Lists of Bacterial Names is authorized by Rule 28a. In addition, the Code of Nomenclature (1976 Revision) contains a number or Provisional Rules. At the time of publication of this List these Rules still have only provisional status. It will require a specific act on the part of the Judicial Commission at a formally constituted meeting to raise these Provisional Rules to Rule status. Meanwhile they continue to be inapplicable.

Revival of Names: A name which has not appeared in the Approved Lists of Bacterial Names and which has not been listed as a *nomen rejiciendum* by the Judicial Commission may

(a) be revived (*nomen revictum*) if it is to be used for the same taxon to which (in the author's opinion) it was originally applied. For citation of such names see Rule 28a and Provisional Rules B2 and B3.

As most names which have not been included in the Approved Lists of Bacterial Names have been omitted because of uncertainty of the validity of the taxa to which they apply, the revival of names should be practised very conservatively.

Reuse of names — Specific and Subspecific epithets: There never has been any rule which forbids the use of the same specific epithet for taxa *within different genera.* Many contributors to the Approved Lists of Bacterial Names have voiced considerable concern at the possibility that a specific epithet may be used *within the same genus* as that in which it was originally used but for the naming of a taxon to which it was not originally applied.

Unless the Judicial Commission takes action to prevent such usage, the Ad Hoc Committee can only urge taxonomists to avoid such reusage as may lead to considerable confusion in future nomenclature.

For the same reason taxonomists are urged not to use as a specific epithet within a genus a name which has been used for an infrasubspecific subdivision of a species within that genus. This applies particularly to several genera of predominantly plant pathogens, where previously cited species have been relegated to the status of pathovars within species listed in the Approved Lists of Bacterial Names. As such infrasubspecific pathovars may later be elevated to the status of subspecies or species, preemption of the pathovar name as a specific or subspecific epithet of a different taxon would obviously cause confusion.

The International Society for Plant Pathology is publishing, in the Review of Plant Pathology, a list of pathovars of species which appear in the Approved Lists of Bacterial Names. Taxonomists are urged to consult this list before proposing new specific epithets or reusing or reviving old epithets for new species within the relevant genera.

Reuse of generic names: The reuse of a generic name which has not appeared in the Approved Lists of Bacterial Names and which has not been listed as a *nomen rejiciendum* by the Judicial Commission does not appear to be fraught with the same difficulties as those associated with specific and subspecific epithets. Taxonomists are nevertheless asked to seriously consider the possibility of confusion arising from the use of such names before taking such action.

The correct form of citation for reused names is given in Provisional Rules B1 and B4.

9. Omissions from Approved Lists of Bacterial Names. No names will be added to or removed from the Approved Lists of Bacterial Names. Attention may be drawn to errors in nomenclature or to errors in spelling or citations of references and such communications should be directed to the Executive Secretary, ICSB for transmission to the appropriate body. Please note that the omission of diacritical signs was intentional and dictated by the limitations of the computers used in setting up the type. These omissions do not constitute 'errors' in the above sense.

Every effort has been made to obviate errors by reference of the computer printouts of the final information to the original advisers for correction and also for insertion of some missing information. The final printout prior to production of the bromide copy for printing has been checked by the Chairman but the magnitude of the task allowed only of a search for the more glaring errors of which there were fortunately few.

NAMES OF TAXA ABOVE THE RANK OF GENUS

Note: No Nomenclatural Type is required for the categories of Division, Class or Subclass.

DIVISION

Firmacutes Gibbons and Murray 1978
Gibbons, N.E. and R.G.E. Murray. 1978. International Journal of Systematic Bacteriology **28:**1-6.

Gracilicutes Gibbons and Murray 1978
Gibbons, N.E. and R.G.E. Murray. 1978. International Journal of Systematic Bacteriology **28:**1-6.

CLASS

Actinomycetes Krassilnikov 1949
Krassilnikov, N.A. 1949. Opredelitelv Bakterii i Actinomicetov. Akademii Nauk SSSR Moscow. pp. 328.

Bacteria Haeckel 1894
Haeckel, E.H. 1894. Systematische Phylogenie der Protisten und Planzen. I.G. Reimer, Berlin. p. 40.

Microtatobiotes Philip 1956
Philip, C.B. 1956. Canadian Journal of Microbiology **2:**261-270.

Mollicutes Edward and Freundt 1967
Edward, D.G. ff. and E.A. Freundt. 1967. International Journal of Systematic Bacteriology **17:**267-268.

Photobacteria Gibbons and Murray 1978
Gibbons, N.E. and R.G.E. Murray. 1978. International Journal of Systematic Bacteriology **28:**1-6.

Schizomycetes Naegeli 1857
Naegeli, C. von. 1857 in Caspari, R. Botanische Zeitung **15:**749-776.

Scotobacteria Gibbons and Murray 1978
Gibbons, N.E. and R.G.E. Murray. 1978. International Journal of Systematic CBacteriology **28:**1-6.

SUBCLASS

Anoxyphotobacteriae Gibbons and Murray 1978
Gibbons, N.E. and R.G.E. Murray. 1978. International Journal of Systematic CBacteriology **28:**1-6.

ORDER

Actinomycetales Buchanan 1917
Buchanan, R.E. 1917. Journal of Bacteriology **2:**155-164. **Type genus:** *Actinomyces* Harz 1877

Bacillales Prevot 1953
Prevot, A.R. 1953 in Hauduroy, P., G. Ehringer, G. Guillot, J. Magrou, CA.R. Prevot, Rosset and A. Urbain. Dictionnaire des bacteries Cpathogenes. 2nd ed. Masson, Paris. pp. 692. **Type genus:** *Bacillus* Cohn 1872

Beggiatoales Buchanan 1957
Buchanan, R.E. 1957 in Breed, R.S., E.G.D. Murray and N.R. Smith (eds). Bergey's Manual of Determinative Bacteriology, 7th ed. The Williams and Wilkins Co., Baltimore. pp. 837-853. **Type genus:** *Beggiatoa* Trevisan 1842

Caryophanales Peshkoff 1939
Peshkoff, M.A. 1939. Doklady Akademii Nauk SSSR **25:**239-242. **Type genus:** *Caryophanon* Peshkoff 1939

Caulobacterales Henrici and Johnson 1935
Henrici, A.T. and D. Johnson. 1935. CJournal of Bacteriology **29:**3-4. **Type genus:** *Caulobacter* Henrici and Johnson 1935

Chlamydiales Storz and Page 1971
Storz, J. and L.A. Page. 1971. International Journal of Systematic Bacteriology **21:**332-334. **Type genus:** *Chlamydia* Jones, Rake and Stearns 1945

Chlorobiales Gibbons and Murray 1978
Gibbons, N.E. and R.G.E. Murray. 1978. International Journal of Systematic CBacteriology **28:**1-6. **Type genus:** *Chlorobium* Nadson 1906

Clostridiales Prevot 1953
Prevot, A.R. 1953 in Hauduroy, P., G. Ehringer, G. Guillot, J. Magrou, A.R. Prevot, Rosset and A. Urbain. Dictionnaire des bacteries pathogenes. 2nd ed. Masson, Paris. pp. 692. **Type genus:** *Clostridium* Prazmowski 1880

Cytophagales Leadbetter 1974
Leadbetter, E.R. 1974. in Bergey 8. pp.

99-127. **Type genus:** *Cytophaga* Winogradsky 1929

Eubacteriales Buchanan 1917
Buchanan, R.E. 1917. Journal of Bacteriology **2**:347-350. **Type genus:** *Eubacterium* Prevot 1938

Hyphomicrobiales Douglas 1957
Douglas, H.C. 1957 in Breed, CR.S., E.G.D. Murray and N.R. Smith (eds). Bergey's Manual of Determinative CBacteriology, 7th ed. The Williams and Wilkins Co., Baltimore. pp. 276-280. **Type genus:** *Hyphomicrobium* Stutzer and Hartleb 1899

Lysobacterales Christensen and Cook 1978
Christensen, P. and F.D. Cook. 1978. International Journal of Systematic Bacteriology **28**:367-393. **Type genus:** *Lysobacter* Christensen and Cook 1978

Micrococcales Prevot 1940
Prevot, A.R. 1940. Manuel de classification et de determination des bacteries anaerobies. 1st ed. Masson, Paris. pp. 223. **Type genus:** *Micrococcus* Cohn 1872

Mycobacteriales Janke 1924
Janke, A. 1924. Allgemeine technische Mikrobiologie. I. Teil. Die Mikroorganismen. T. Steinkopf, Dresden und Leipzig. pp. 1-342. **Type genus:** *Mycobacterium* Lehmann and Neumann 1896

Mycoplasmatales Freundt 1955
Freundt, E.A. 1955. International Bulletin of Bacteriological Nomenclature and Taxonomy **5**:67-78. **Type genus:** *Mycoplasma* Nowak 1929

Myxococcales Tchan, Pochon and Prevot 1948
Tchan, Y.T., J. Pochon and A.R. Prevot. 1948. Annales de Cl'Institut Pasteur (Paris) **74**:394-400. **Type genus:** *Myxococcus* Thaxter 1892

Pseudomonadales Orla-Jensen 1921
Orla-Jensen, S. 1921. Journal of Bacteriology **6**:263-273. **Type genus:** *Pseudomonas* Migula 1894

Rhodospirillales Pfennig and Truper 1971
Pfennig, N. and H.G. Truper. 1971. International Journal of Systematic Bacteriology **21**:17-18. **Type genus:** *Rhodospirillum* Molisch 1904

Rickettsiales Gieszczkiewicz 1939
Gieszczkiewicz, M. 1939. Bulletin de l'Academie Polonaise des Sciences. Serie des Sciences Biologiques **1**:9-27. **Type genus:** *Rickettsia* da Rocha-Lima 1916

Spirillales Prevot 1940
Prevot, A.R. 1940. Manuel de classification et de determination des bacteries anaerobies. 1st ed. Masson, Paris. pp. 223. **Type genus:** *Spirillum* Ehrenberg 1832

Spirochaetales Buchanan 1917
Buchanan, R.E. 1917. Journal of Bacteriology **2**:155-164. **Type genus:** *Spirochaeta* Ehrenberg 1835

SUBORDER

Caulobacterineae Breed, Murray and Hitchens 1944
Breed, R.S., E.G.D. Murray and A.P. Hitchens. 1944. Bacteriological Reviews **8**: 255-260. **Type genus:** *Caulobacter* Henrici and Johnson 1935

Eubacteriineae Breed, Murray and Hitchens 1944
Breed, R.S., E.G.D. Murray and A.P. Hitchens. 1944. Bacteriological Reviews **8**: 255-260. **Type genus:** *Eubacterium* Prevot 1938

Pseudomonadineae Breed, Murray and Smith 1957
Breed, R.S., E.G.D. Murray and N.R. Smith (eds). Bergey's Manual of Determinative Bacteriology, 7th ed. The Williams and Wilkins Co., Baltimore. p. 67. **Type genus:** *Pseudomonas* Migula 1894

TRIBE

Acetobactereae Pribram 1929
Pribram, E. 1929. Journal of Bacteriology **18**:361-394. **Type genus:** *Acetobacter* Beijerinck 1898

Bacteroideae Castellani and Chalmers 1919

Castellani, A. and A.J. Chalmers. 1919. Manual of tropical medicine, 3rd ed. Williams, Wood and Co., New York. **Type genus:** *Bacteroides* Castellani and Chalmers 1919

Brevibacterieae Prevot 1961
Prevot, A.R. 1961. Traite de systematique bacterienne. Dunod, Paris. Volume 2. pp. 771. **Type genus:** *Brevibacterium* Breed 1953

Brucelleae Murray 1948
Murray, E.G.D. 1948 in Breed, R.S., E.G.D. Murray and A.P. Hitchens (eds). Bergey's Manual of Determinative Bacteriology, 6th ed. The Williams and Wilkins Co., Baltimore. pp. 545-595. **Type genus:** *Brucella* Meyer and Shaw 1920

Chromobacterieae Winslow, Broadhurst, Buchanan, Krumwiede, Rogers and Smith 1920
Winslow, C.E.A., J. Broadhurst, R.E. Buchanan, C. Krumwiede Jr., L.A. Rogers and G.H. Smith. Journal of Bacteriology **5**:191-229. **Type genus:** *Chromobacterium* Bergonzini 1881

Ehrlichieae Philip 1957
Philip, C.B. 1957 in Breed, R.S., E.G.D. Murray and N.R. Smith (eds). Bergey's Manual of Determinative Bacteriology, 7th ed. The Williams and Wilkins Co., Baltimore. p. 933. **Type genus:** *Ehrlichia* Moshkovski 1945

Erwinieae Winslow, Broadhurst, Buchanan, Krumwiede, Rogers and Smith 1920
Winslow, C.E.A., J. Broadhurst, R.E. Buchanan, C. Krumwiede Jr., L.A. Rogers and G.H. Smith. 1920. Journal of Bacteriology **5**:191-229. **Type genus:** *Erwinia* Winslow, Broadhurst, Buchanan, Krumwiede, Rogers and Smith 1920

Escherichieae Bergey, Breed, Murray and Hitchens 1939
Bergey, D.H., R.S. Breed, E.G.D. Murray and A.P. Hitchens (eds). 1939. Bergey's Manual of Determinative Bacteriology, 5th ed. The Williams and Wilkins Co., Baltimore. p. 388. **Type genus:** *Escherichia* Castellani and Chalmers 1919

Eubacterieae Prevot 1961
Prevot, A.R. 1961. Traite de systematique bacterienne. Dunod, Paris. Volume 2. pp. 771. **Type genus:** *Eubacterium* Prevot 1938

Haemophileae Castellani and Chalmers 1919
Castellani, A. and A.J. Chalmers. 1919. Manual of tropical medicine, 3rd ed. Williams, Wood and Co., New York. p. 933. **Type genus:** *Haemophilus* Winslow, Broadhurst, Buchanan, Krumwiede, Rogers and Smith 1917

Lactobacilleae Winslow, Broadhurst, Buchanan, Krumwiede, Rogers and Smith 1920
Winslow, C.E.A., J. Broadhurst, R.E. Buchanan, C. Krumwiede Jr., L.A. Rogers and G.H. Smith. 1920. Journal of Bacteriology **5**:191-229. **Type genus:** *Lactobacillus* Beijerinck 1901

Micrococceae Prevot 1961
Prevot, A.R. 1961. Traite de systematique bacterienne. Dunod, Paris. Volume 2. p. 31. **Type genus:** *Micrococcus* Cohn 1872

Pasteurelleae Castellani and Chalmers 1919
Castellani, A. and A.J. Chalmers. 1919. Manual of tropical medicine, 3rd ed. Williams, Wood and Co., New York. p. 943. **Type genus:** *Pasteurella* Trevisan 1887

Proteeae Castellani and Chalmers 1919
Castellani, A. and A.J. Chalmers. 1919. Manual of tropical medicine, 3rd ed. Williams, Wood and Co., New York. p. 932. **Type genus:** *Proteus* Hauser 1885

Pseudomonadeae Kluyver and van Niel 1936
Kluyver, A.J. and C.B. van Niel. 1936. Zentralblatt fur Bakteriologie, Parasitenkunde, Infektionskrankheiten und Hygiene. Abteilung II. **94**:369-403. **Type genus:** *Pseudomonas* Migula 1894

Rhizobieae Prevot 1948
Prevot, A.R. 1948. Manuel de classification et de determination des bacteries anaerobies. 2nd ed. Masson, Paris. pp. 290. **Type genus:** *Rhizobium* Frank 1889

Rickettsieae Philip 1957
Philip, C.B. 1957 in Breed, R.S., E.G.D. Murray and N.R. Smith (eds). Bergey's Manual of Determinative Bacteriology, 7th ed. The Williams and Wilkins Co., Baltimore. **Type genus:** *Rickettsia* da Rocha-Lima 1916

Salmonelleae Kalz 1957
Kalz, G. 1957 in Breed, R.S., E.G.D. Murray and N.R. Smith (eds). Bergey's Manual of Determinative Bacteriology, 7th ed. The Williams and Wilkins Co., Baltimore. pp. 368-384. **Type genus:** *Salmonella* Lignieres 1900

Serratieae Bergey, Breed, Murray and Hitchens 1939
Bergey, D.H., R.S. Breed, E.G.D. Murray and A.P. Hitchens (eds). 1939. Bergey's Manual of Determinative Bacteriology, 5th ed. The Williams and Wilkins Co., Baltimore. p. 359. **Type genus:** *Serratia* Bizio 1823

Spirilleae Kluyver and van Niel 1936
Kluyver, A.J. and C.B. van Niel. 1936. Zentralblatt fur Bakteriologie, Parasitenkunde, Infektionskrankheiten und Hygiene. Abteilung II. **94**:369-403. **Type genus:** *Spirillum* Ehrenberg 1832

Staphylococceae Prevot 1940
Prevot, A.R. 1940. Manuel de classification et de determination des bacteries anaerobies. 1st ed. Masson, Paris. pp. 223. **Type genus:** *Staphylococcus* Rosenbach 1884

Streptococceae Trevisan 1889
Trevisan, V. 1889. I generi e le specie delle Batteriacee. Zanaboni and Gabuzzi, Milano. p. 29. **Type genus:** *Streptococcus* Rosenbach 1884

Thiobacilleae Pribram 1929
Pribram, E. 1929. Journal of Bacteriology **18**:361-394. **Type genus:** *Thiobacillus* Beijerinck 1904

Wolbachieae Philip 1955
Philip, C.B. 1955. Bacteriological Reviews **19**:271. **Type genus:** *Wolbachia* Hertig 1936

FAMILY

Acholeplasmataceae Edward and Freundt 1970
Edward, D.G. ff. and E.A. Freundt. 1970. Journal of General Microbiology **62**:1-2. **Type genus:** *Acholeplasma* Edward and Freundt 1970

Achromatiaceae Massart 1901
Massart, J. 1901. Recueil de l'Institut Botanique Leo Errara, (Bruxelles) **5**:251-282. **Type genus:** *Achromatium* Schewiakoff 1893

Actinomycetaceae Buchanan 1918
Buchanan, R.E. 1918. CJournal of Bacteriology **3**:403-406. **Type genus:** *Actinomyces* Harz 1877

Actinoplanaceae Couch 1955
Couch, J.N. 1955. Journal of Elisha Mitchell Scientific Society **71**:269. **Type genus:** *Actinoplanes* Couch 1950

Anaplasmataceae Philip 1957
Philip, C.B. 1957 in Breed, R.S., E.G.D. Murray and N.R. Smith (eds). Bergey's Manual of Determinative Bacteriology, 7th ed. The Williams and Wilkins Co., Baltimore. pp. 980-984. **Type genus:** *Anaplasma* Theiler 1910

Archangiaceae Jahn 1924
Jahn, E. 1924. Beitrage zur botanishcen Protistologie. I. Die Polyangiden. Verlag von Gebruder Borntraeger, Leipzig. p. 66. **Type genus:** *Archangium* Jahn 1924

Azotobacteraceae Pribram 1933
Pribram, E. 1933. Klassification der Schizomyceten. F. Deuticke, Leipzig. pp. 1-143. **Type genus:** *Azotobacter* Beijerinck 1901

Bacillaceae Fischer 1895
Fischer, A. 1895. Jahrbuch fur Wissenschaftliche Botanik **27**:1-163. **Type genus:** *Bacillus* Cohn 1872

Bacteroidaceae Pribram 1933
Pribram, E. 1933. Klassification der Schi-

zomyceten. F. Deuticke, Leipzig. pp. 1-143. **Type genus:** *Bacteroides* Castellani and Chalmers 1919

Bartonellaceae Gieszczykiewicz 1939
Gieszczykiewicz, M. 1939. Bulletin de l'Academie Polonaise des Sciences. Serie des Sciences Biologiques **1**:9-27. **Type genus:** *Bartonella* Strong, Tyzzer and Sellards 1915

Beggiatoaceae Migula 1894
Migula, W. 1894. Arbeiten aus dem Bakteriologischen Institut der Technischen Hochschule zu Karlsruhe **1**:235-238. **Type genus:** *Beggiatoa* Trevisan 1842

Brevibacteriaceae Breed 1953
Breed, R.S. 1953. Riassunti delle Communicazioni, VI Congresso Internazionale Microbiologia Roma, **1**:10-15. **Type genus:** *Brevibacterium* Breed 1953

Brucellaceae Breed, Murray and Smith 1957
Breed, R.S., E.G.D. Murray and N.R. Smith (eds). 1957. Bergey's Manual of Determinative Bacteriology, 7th ed. The Williams and Wilkins Co., Baltimore. p. 394. **Type genus:** *Brucella* Meyer and Shaw 1920

Caryophanaceae Peshkoff 1939
Peshkoff, M.A. 1939. Doklady Akademii Nauk SSSR **25**:239-242. **Type genus:** *Caryophanon* Peshkoff 1939

Caulobacteraceae Henrici and Johnson 1935
Henrici, A.T. and D. Johnson. 1935. Journal of Bacteriology **29**:3-4. **Type genus:** *Caulobacter* Henrici and Johnson 1935

Chlamydiaceae Rake 1957
Rake, G. 1957 in Breed, R.S., E.G.D. Murray and N.R. Smith (eds). Bergey's Manual of Determinative Bacteriology, 7th ed. The Williams and Wilkins Co., Baltimore. pp. 957-968. **Type genus:** *Chlamydia* Jones, Rake and Stearns 1945

Chlorobiaceae Copeland 1956
Copeland, H.F. 1956. The classification of lower organisms. Pacific Books, Palo Alto, California. p. 31. **Type genus:** *Chlorobium* Nadson 1906

Chromatiaceae Bavendamm 1924
Bavendamm, W. 1924 in Kalkwitz (ed). Die farblosen und roten Schwefelbakterien des Suss-und Salzwassers. Pflanzenforschung, Berlin-Dahlem. pp. 1-156. **Type genus:** *Chromatium* Perty 1852

Clostridiaceae Pribram 1933
Pribram, E. 1933. Klassification der Schizomyceten. F. Deuticke, Leipzig. pp. 1-143. **Type genus:** *Clostridium* Prazmowski 1880

Corynebacteriaceae Lehmann and Neumann 1907
Lehmann, K.B. and R. Neumann. 1907. Lehmann's Medizin, Handatlanten X. Atlas und Grundriss der Bakteriologie und Lehrbuch der speziellen bakteriologischen Diagnostik. 4. Auflage. J.F. Lehmann, Munchen. p. 500. **Type genus:** *Corynebacterium* Lehmann and Neumann 1896

Crenotrichaceae Hansgirg 1888
Hansgirg, A. 1888. Oesterreichische Botanische Zeitschrift **38**:227-230. **Type genus:** *Crenothrix* Cohn 1870

Cystobacteraceae McCurdy 1970
McCurdy, H.D. 1970. International Journal of Systematic Bacteriology **20**:283-296. **Type genus:** *Cystobacter* Schroeter 1886

Cytophagaceae Stanier 1940
Stanier, R.Y. 1940. Journal of Bacteriology **40**:619-636. **Type genus:** *Cytophaga* Winogradsky 1929

Dermatophilaceae Austwick 1958
Austwick, P.K.C. 1958. 1 Veterinary Reviews and Annotations **4**:33-48. **Type genus:** *Dermatophilus* (Van Saceghem 1915) Gordon 1964

Ehrlichiaceae Moshkovski 1945
Moshkovski, S.D. 1945. Advances in Modern Biology (Moscow) **19**:1-44. **Type genus:** *Ehrlichia* Moshkovski 1945

Enterobacteriaceae Rahn 1937

Rahn, O. 1937. Zentralblatt fur Bakteriologie, Parasitenkunde, Infektionskrankeit und Hygiene. Abteilung II. **96**:273-286. **Type genus:** *Escherichia* Castellani and Chalmers 1919

Frankiaceae Becking 1970
Becking, J.H. 1970. International Journal of Systematic Bacteriology **20**:201-220. **Type genus:** *Frankia* Brunchorst 1886

Gallionellaceae Henrici and Johnson 1935
Henrici, A.T. and D. Johnson. 1935. Journal of Bacteriology **29**:3-4. **Type genus:** *Gallionella* Ehrenberg 1838

Halobacteriaceae Gibbons 1974
Gibbons, N.E. 1974. in Bergey 8. pp. 269-272. **Type genus:** *Halobacterium* Elazari-Volcani 1957

Hyphomicrobiaceae Babudieri 1950
Babudieri, B. 1950. R.C. 1st Supplement, Sanita, Roma **13**:580-591. **Type genus:** *Hyphomicrobium* Stutzer and Hartleb 1899

Lactobacillaceae Winslow, Broadhurst, Buchanan, Krumwiede, Rogers and Smith 1917
Winslow, C.E.A., J. Broadhurst, R.E. Buchanan, C. Krumwiede Jr., L.A. Rogers and G.H. Smith 1917. Journal of Bacteriology **2**:505-566. **Type genus:** *Lactobacillus* Beijerinck 1901

Legionellaceae Brenner, Steigerwalt and McDade 1979
Brenner, D.J., A.G. Steigerwalt and J.E. McDade. 1979. International Journal of Systematic Bacteriology **29**:436; Effective publication Brenner, D.J., A.G. Steigerwalt and J.E. McDade. 1979. Annals of Internal Medicine **90**:656-658. **Type genus:** *Legionella* Brenner, Steigerwalt and McDade 1979

Leptospiraceae Hovind-Houger 1978
Hovind-Houger, K. 1978. International Journal of Systematic Bacteriology **29**: 245-251. **Type genus:** *Leptospira* Noguchi 1917

Leucotrichaceae Buchanan 1957
Buchanan, R.E. 1957 in Breed, R.S., E.G.D. Murray and N.R. Smith (eds). Bergey's Manual of Determinative Bacteriology, 7th ed. The Williams and Wilkins Co., Baltimore. pp. 850-851. **Type genus:** *Leucothrix* Oersted 1884

Lysobacteraceae Christensen and Cook 1978
Christensen, P. and F.D. Cook. 1978. International Journal of Systematic Bacteriology **28**:367-393. **Type genus:** *Lysobacter* Christensen and Cook 1978

Methanobacteriaceae Barker 1956
Barker, H.A. 1956. Bacterial fermentations. John Wiley, New York. p. 15. **Type genus:** *Methanobacterium* Kluyver and van Niel 1936

Micrococcaceae Pribram 1929
Pribram, E. 1929. Journal of Bacteriology **18**:361-394. **Type genus:** *Micrococcus* Cohn 1872

Micromonosporaceae Krassilnikov 1938
Krassilnikov, N.A. 1938. Ray fungi and related organisms, Actinomycetales. Akademii Nauk SSSR, Moscow. p. 272. **Type genus:** *Micromonospora* Orskov 1923

Mycobacteriaceae Chester 1897
Chester, F.D. 1897. Delaware Agricultural Experiment Station Bulletin **9**:38-145. **Type genus:** *Mycobacterium* Lehmann and Neumann 1896

Mycoplasmataceae Freundt 1955
Freundt, E.A. 1955. International Bulletin of Bacteriological Nomenclature and Taxonomy **5**:67-78. **Type genus:** *Mycoplasma* Nowak 1929

Myxococcaceae Jahn 1924
Jahn, E. 1924. Beitrage zur botanischen Protistologie. I. Die Polyangiden. Verlag von Gebruder Borntraeger, Leipzig. p. 403. **Type genus:** *Myxococcus* Thaxter 1892

Neisseriaceae Prevot 1933
Prevot, A.R. 1933. Annales des Sciences Naturelles Zoologie et Biologie Animale **15**:23-260. **Type genus:** *Neisseria* Trevisan

1885

Nevskiaceae Henrici and Johnson 1935
Henrici, A.T. and D. Johnson. 1935. Journal of Bacteriology **29**:3-4. **Type genus:** *Nevskia* Famintzin 1892

Nitrobacteraceae Buchanan 1917
Buchanan, R.E. 1917. Journal of Bacteriology **2**:347-350. **Type genus:** *Nitrobacter* Winogradsky 1892

Nocardiaceae Castellani and Chalmers 1919
Castellani, A. and A.J. Chalmers. 1919. Manual of tropical medicine, 3rd ed. Williams, Wood and Co., New York. p. 1040. **Type genus:** *Nocardia* Trevisan 1889

Oscillospiraceae Peshkoff 1940
Peshkoff, M.A. 1940. Zhurnal Obshchei Biologii **1**:597-617. **Type genus:** *Oscillospira* Chatton and Perard 1913

Pasteuriaceae Laurent 1890
Laurent, E. 1890. l Comptes Rendus de l'Academie des Sciences Paris **111**:754-756. **Type genus:** *Pasteuria* Metchnikoff 1888

Peptococcaceae Rogosa 1971
Rogosa, M. 1971. International Journal of Systematic Bacteriology **21**:234-237. **Type genus:** *Peptococcus* Kluyver and van Niel 1936

Planococcaceae Krassilnikov 1949
Krassilnikov, N.A. 1949. Opredelitelv Bakterii i Actinomicetov. Akademii Nauk SSSR, Moscow. pp. 328. **Type genus:** *Planococcus* Migula 1894

Polyangiaceae Jahn 1924
Jahn, E. 1924. Beitrage zur botanischen Protistologie. I. Die Polyangiden. Verlag von Gebruder Borntraeger, Leipzig. p. 75. **Type genus:** *Polyangium* Link 1809

Propionibacteriaceae Delwiche 1957
Delwiche, E.A. 1957 in Breed, R.S., E.G.D. Murray and N.R. Smith (eds). Bergey's Manual of Determinative Bacteriology, 7th ed. The Williams and Wilkins Co., Baltimore. p. 569. **Type genus:** *Propionibacterium* Orla-Jensen 1909

Pseudomonadaceae Winslow, Broadhust, Buchanan, Krumwiede, Rogers and Smith 1917
Winslow, C.E.A., J. Broadhurst, R.E. Buchanan, C. Krumwiede Jr., L.A. Rogers and G.H. Smith. 1917. Journal of Bacteriology **2**:505-566. **Type genus:** *Pseudomonas* Migula 1894

Rhizobiaceae Conn 1938
Conn, H.J. 1938. Journal of Bacteriology **36**:320-321. **Type genus:** *Rhizobium* Frank 1889

Rhodospirillaceae Pfennig and Truper 1971
Pfennig, N. and H.G. Truper. 1971. International Journal of Systematic Bacteriology **21**:17-18. **Type genus:** *Rhodospirillum* Molisch 1907

Rickettsiaceae Pinkerton 1936
Pinkerton, H. 1936. Parasitology **28**:172-189. **Type genus:** *Rickettsia* da Rocha-Lima 1916

Simonsiellaceae Steed 1962
Steed, P.D.M. 1962. Journal of General Microbiology **29**:615-624. **Type genus:** *Simonsiella* Schmid 1922

Spirillaceae Migula 1894
Migula, N. 1894. Arbeiten aus dem Bakteriologischen Institut der Technischen Hochschule zu Karlsruhe **1**:235:238. **Type genus:** *Spirillum* Ehrenberg 1832

Spirochaetaceae Swellengrebel 1907
Swellengrebel, N.H. 1907. Annales de l'Institut Pasteur (Paris) **21**:562-586. **Type genus:** *Spirochaeta* Ehrenberg 1835

Spirosomaceae Larkin and Borrall 1978
Larkin, J.M. and R. Borrall. 1978. International Journal of Systematic Bacteriology **28**:595-596. **Type genus:** *Spirosoma* Migula 1894

Streptomycetaceae Waksman and Henrici 1943
Waksman, S.A. and A.T. Henrici. 1943. Journal of Bacteriology **46**:337-341. **Type**

genus: *Streptomyces* Waksman and Henrici 1943

Streptococcaceae Deibel and Seeley 1974
Deibel, R.H. and H.W. Seeley Jr. 1974. in Bergey 8. pp. 490-517. **Type genus:** *Streptococcus* Rosenbach 1884

Thiocapsaceae Bavendamm 1924
Bavendamm, W. 1924. Die farblosen und roten Schwefelbakterien des Suss-und Salzwassers. Pflanzenforschung, Berlin-Dahlem. pp. 1-156. **Type genus:** *Thiocapsa* Winogradsky 1888

Treponemataceae Robinson 1948
Robinson, G.H. 1948 in Breed, R.S., E.G.D. Murray and A.P. Hitchens (eds). Bergey's Manual of Determinative Bacteriology, 6th ed. The Williams and Wilkins Co., Baltimore. pp. 1051-1058. **Type genus:** *Treponema* Schaudinn 1905

Veillonellaceae Rogosa 1971
Rogosa, M. 1971. International Journal of Systematic Bacteriology **21**:231-233. **Type genus:** *Veillonella* Prevot 1933

Vibrionaceae Veron 1965
Veron, M. 1965. Comptes Rendus de l'Academie des Sciences Paris **261**:5243-5246. **Type genus:** *Vibrio* Pacini 1854

Vitreoscillaceae Pringsheim 1949
Pringsheim, E.G. 1949. Bacteriological Reviews **13**:47-98. **Type genus:** *Vitreoscilla* Pringsheim 1949

NAMES OF GENERA, SPECIES AND SUBSPECIES

Note: Abbreviations used for genera in the following lists are used for brevity only and are not intended or recommended as contractions in the nomenclatural sense.

Acetobacter Beijerinck 1898
Beijerinck, M. 1898. Zentralblatt fur Bakteriologie, Parasitenkunde, Infektionskrankheiten und Hygiene. Abteilung II. **4**:209-216. **Type species:** *A. aceti* (Pasteur 1864) Beijerinck 1898
Description: Bergey 8.

A. aceti (Pasteur 1864) Beijerinck 1898
Pasteur, L. 1864. Annales scientifiques de l'Ecole normale superieure. Paris **1**:113-158; Beijerinck, M. 1898. Zentralblatt fur Bakteriologie, Parasitenkunde, Infektionskrankheiten und Hygiene. Abteilung II. **4**:209-216. **Type strain:** NCIB 8621
Description: Bergey 8.

A. aceti subsp. *aceti* (Pasteur 1864) De Ley and Frateur 1974
Pasteur, L. 1864. Annales scientifiques de l'Ecole normale superieure. Paris **1**:113-158; De Ley, J. and J. Frateur. 1974 in Bergey 8. **Type strain:** NCIB 8621
Description: Bergey 8.

A. aceti subsp. *liquefaciens* (Asai 1935) De Ley and Frateur 1974
Asai, T. 1935. Journal of the Agricultural Chemical Society of Japan **11**:674-708; De Ley, J. and J. Frateur. 1974 in Bergey 8. **Type strain:** IAM 1834
Description: Bergey 8.

A. aceti subsp. *orleanensis* (Henneberg 1906) De Ley and Frateur 1974
Henneberg, W. 1906. Deutsche Essigindustrie **10**:106-108; De Ley, J. and J. Frateur. 1974 in Bergey 8. **Type strain:** ATCC 12876
Description: Bergey 8.

A. aceti subsp. *xylinum* (Brown 1886) De Ley and Frateur 1974
Brown, A.J. 1886. Journal of the Chemical Society (London) **49**:432-439; De Ley, J. and J. Frateur. 1974 in Bergey 8. **Type strain:** ATCC 23767
Description: Bergey 8.

A. pasteurianus (Hansen 1879) Beijerinck and Folpmers 1916
Hansen, E.C. 1879. Compte Rendu des travaux du Laboratoire Carlsberg **1**:96-100; Beijerinck, M.W. and T. Folpmers 1916. Verslagen van de gewone vergadering der wis- en natuurkundige afdeeling. Koninklijke Akademie van wetenschappen te Amsterdam **18**:1198-1200. **Type strain:** LMD 22.1
Description: Bergey 8.

A. pasteurianus subsp. *ascendens* (Henneberg 1898) De Ley and Frateur 1974
Henneberg, W. 1898. Deutsche Essigindustrie **2**:145-148; De Ley, J. and J. Frateur. 1974 in Bergey 8. **Type strain:** LMD 51.1
Description: Bergey 8.

A. pasteurianus subsp. *estunensis* (Carr 1958) De Ley and Frateur 1974
Carr, J.G. 1958. Antonie van Leeuwenhoek Journal of Microbiology and Serology **24**:157-160; De Ley, J. and J. Frateur. 1974 in Bergey 8. **Type strain:** ATCC 23753
Description: Bergey 8.

A. pasteurianus subsp. *lovaniensis* (Frateur 1950) De Ley and Frateur 1974
Frateur, J. 1950. Cellule **53**:287-392; De Ley, J. and J. Frateur. 1974 in Bergey 8. **Type strain:** ATCC 12875
Description: Bergey 8.

A. pasteurianus subsp. *paradoxus* (Frateur 1950) De Ley and Frateur 1974
Frateur, J. 1950. Cellule **53**: 287-392; De Ley, J. and J. Frateur. 1974 in Bergey 8. **Type strain:** LMD 53.6
Description: Bergey 8.

A. pasteurianus subsp. *pasteurianus* (Hansen 1879) De Ley and Frateur 1974
Hansen, E.C. 1879. Compte Rendu des Travaux du Laboratoire Carlsberg **1**:96-100; De Ley, J. and J. Frateur. 1974 in Bergey 8. **Type strain:** LMD 22.1
Description: Bergey 8.

A. peroxydans Visser't Hooft 1925
Visser't Hooft, F. 1925. Dissertation. Technical University, Delft, Meinema. pp. 1-129. **Type strain:** ATCC 12874
Description: Bergey 8.

Acetobacterium Balch, Schoberth, Tanner and Wolfe 1977
Balch, W.E., S. Schoberth, R.S. Tanner and R.S. Wolfe. 1977. International Journal of Systematic Bacteriology **27**:355-361. **Type species:** *A. woodii* Balch, Schoberth,

Tanner and Wolfe 1977
Description: Balch, W.E., S. Schoberth, R.S. Tanner and R.S. Wolfe. 1977. Ibid.

A. woodii Balch, Schoberth, Tanner and Wolfe 1977
Balch, W.E., S. Schoberth, R.S. Tanner and R.S. Wolfe. 1977. International Journal of Systematic Bacteriology **27**:355-361. **Type strain:** ATCC 29683
Description: Balch, W.E., S. Schoberth, R.S. Tanner and R.S. Wolfe. 1977. Ibid.

Acholeplasma Edward and Freundt 1970
Edward, D.G. ff. and E.A. Freundt. 1970. Journal of General Microbiology **62**:1-2. **Type species:** *A. laidlawii* (Freundt 1955) Edward and Freundt 1970
Description: Bergey 8.

A. axanthum Tully and Razin 1970
Tully, J.G. and S. Razin. 1970. Journal of Bacteriology **103**:751-754. **Type strain:** ATCC 25176
Description: Tully, J.G. 1979 in Barile, M.F., S. Razin, J.G. Tully and R.F. Whitcomb (eds). The Mycoplasmas. Academic Press, New York, Volume I. pp. 431-449.

A. equifetale Kirchoff 1974
Kirchoff, H. 1974. Zentralblatt fur Veterinarmedizin Reihe B **21**:207-210. **Type strain:** ATCC 29724
Description: Kirchoff, H. 1978. International Journal of Systematic Bacteriology **28**:76-81.

A. granularum (Switzer 1964) Edward and Freundt 1970
Switzer, W.P. 1964 in Dunne (ed). Diseases of Swine, 2nd ed. Iowa State University Press, Ames. pp. 498-507; Edward, D.G. ff. and E.A. Freundt. 1970. Journal of General Microbiology **62**:1-2. **Type strain:** ATCC 19168
Description: Tully, J.G. 1979 in Barile, M.F., S. Razin, J.G. Tully and R.F. Whitcomb (eds). The Mycoplasmas. Academic Press, New York, Volume I. pp. 431-449.

A. hippikon Kirchoff 1974
Kirchoff, H. 1974. Zentralblatt fur Veterinarmedizin Reihe B **21**:207-210. **Type strain:** ATCC 29725
Description: Kirchoff, H. 1978. International Journal of Systematic Bacteriology **28**:76-81.

A. laidlawii (Freundt 1955) Edward and Freundt 1970
Freundt, E.A. 1955. International Bulletin of Bacteriological Nomenclature and Taxonomy **5**:67-78; Edward, D.G. ff. and E.A. Freundt. 1970. Journal of General Microbiology **62**:1-2. **Type strain:** ATCC 23206
Description: Tully, J.G. 1979 in Barile, M.F., S. Razin, J.G. Tully and R.F. Whitcomb (eds). The Mycoplasmas. Academic Press, New York, Volume I. pp. 431-449.

A. modicum Leach 1973
Leach, R.H. 1973. Journal of General Microbiology **75**:135-153. **Type strain:** ATCC 29102
Description: Tully, J.G. 1979 in Barile, M.F., S. Razin, J.G. Tully and R.F. Whitcomb (eds). The Mycoplasmas. Academic Press, New York, Volume I. pp. 431-449.

A. oculi Al-Aubaidi, Dardiri, Muscoplatt and McCauley 1973
Al-Aubaidi, J.M., A.H. Dardiri, C.C. Muscoplatt and E.H. McCauley. 1973. Cornell Veterinarian **63**:117-129. **Type strain:** ATCC 27350
Description: Tully, J.G. 1979 in Barile, M.F., S. Razin, J.G. Tully R.F. Whitcomb (eds). The Mycoplasmas. Academic Press, New York, Volume I. pp. 431-449.

Achromatium Schewiakoff 1893
Schewiakoff, W. 1893. Uber einen neuen bacterienahnlichen Organismus des Susswassers. Habilitations schrift, Universitat Heidelberg C. Winter. pp. 1-36. **Type species:** *A. oxaliferum* Schewiakoff 1893
Description: Bergey 8.

A. oxaliferum Schewiakoff 1893
Schewiakoff, W. 1893. Uber einen neuen bacterienahnlichen Organismus des Susswassers. Habilitations schrift, Universitat Heidelberg C. Winter. pp. 1-36. **Type strain:** no culture isolated.
Description: West, G.S. and B.M. Griffiths. 1909. Proceedings of the Royal Society London B, Biological Sciences **81**: 398-405.

Acidaminococcus Rogosa 1969
Rogosa, M. 1969. Journal of Bacteriology **98**:756-766. **Type species:** *A. fermentans* Rogosa 1969
Description: Bergey 8.

A. fermentans Rogosa 1969
Rogosa, M. 1969. Journal of Bacteriology **98**:756-766. **Type strain:** ATCC 25085
Description: Holdeman, L.V., E.P. Cato and W.E.C. Moore (eds). 1977. Anaerobe Laboratory Manual, 4th ed. Virginia Polytechnic Institute and State University, Blacksburg, Virginia. pp. 1-156.

Acinetobacter Brisou and Prevot 1954
Brisou, J. and A.R. Prevot. 1954. Annales de l'Institut Pasteur (Paris) **86**:722-728. **Type species:** *A. calcoaceticus* (Beijerinck 1911) Baumann, Doudoroff and Stanier 1968
Description: Bergey 8.

A. calcoaceticus (Beijerinck 1911) Baumann, Doudoroff and Stanier 1968
Beijerinck, M.W. 1911. Proceedings of the Royal Academy of Sciences (Amsterdam) **13**:1066-1077; Baumann, P., M. Doudoroff and R.Y. Stanier. 1968. Journal of Bacteriology **95**:1520-1541. **Type strain** ATCC 23055
Description: Henriksen, S.D. 1973. Bacteriological Reviews **37**:522-561.

A. lwoffii (Audureau 1940) Brisou and Prevot 1954
Audureau, A. 1940. Annales de l'Institut Pasteur (Paris) **64**:126-166; Brisou, J. and A.R. Prevot. 1954. Annales de l'Institut Pasteur (Paris) **86**:722-728. **Type strain:** ATCC 15309
Description: Henriksen, S.D. 1973. Bacteriological Reviews **37**:522-561.

Actinobacillus Brumpt 1910
Brumpt, E. 1910. Precis de Parasitologie. 1st ed., Masson et Cie, Paris. **Type species:** *A. lignieresii* Brumpt 1910
Description: Bergey 8.

A. actinomycetemcomitans (Klinger 1912) Topley and Wilson 1929
Klinger, R. 1912. Zentralblatt fur Bakteriologie, Parasitenkunde, Infektionskrankheiten und Hygiene. Abteilung I. **62**:191-200; Topley, W.W.C. and G.S. Wilson. 1929. The principles of bacteriology and immunity. 1st ed. Edward Arnold and Co., London. pp. 1-587. **Type strain:** NCTC 9710
Description: King, E.O. and H.W. Tatum. 1962. Journal of Infectious Diseases **111**:85-94.

A. capsulatus Arseculeratne 1962
Arseculeratne, S.N. 1962. Journal of Comparative Pathology and Therapeutics **72**:33-39. **Type strain:** Frederiksen P243
Description: Arseculeratne, S.N. 1962. Ibid.

A. equuli (van Straaten 1918) Haupt 1934
van Straaten, H. 1918. Bacteriologische Bevindingen bij eenigo gevallen van Pyosepticaemie (Lahme) der veulens. Verslag van den Werksaamheden der Rijksserum-inrichting voor 1916-1917, Rotterdam. pp. 71-76; Haupt, H. 1934. Archiv fur wissenschaftliche und praktische Tierheilkunde **67**:513-524. **Type strain:** ATCC 19392
Description: Bergey 8.

A. lignieresii Brumpt 1910
Brumpt, E. 1910. Precis de Parasitologie. 1st ed., Masson et Cie, Paris. p. 849. **Type strain:** NCTC 4189
Description: Bergey 8.

A. suis van Dorssen and Jaartsveld 1962
van Dorssen, C.A. and F.H.J. Jaartsveld. 1962. Tijdschrift voor Diergeneeskunde **87**:448-450. **Type strain:** CCM 5586
Description: Mair, N.S., C.J. Randall, G.W. Thomas, J.F. Harbourne, C.T. McCrea and K.P. Cowl. 1974. Journal of Comparative Pathology **84**:113-119.

Actinomadura Lechevalier and Lechevalier 1968
Lechevalier, H.A. and M.P. Lechevalier. 1968 in Prauser, H. (ed). The Actinomycetales. Sept. 1968. VEB Gustav Fischer Verlag, Jena. pp. 393-405. **Type species:** *A. madurae* (Vincent 1894) Lechevalier and Lechevalier 1968
Description: Lechevalier, H.A. and M.P. Lechevalier. 1968. Ibid.

A. africana Preobrazhenskaya and Sveshnikova 1974
Preobrazhenskaya, T.P. and M.A. Sveshnikova. 1974. Mikrobiologiya **43**:735-738. **Type strain:** INA 1839
Description: Preobrazhenskaya, T.P. and M.A. Sveshnikova. 1974. Ibid.

A. aurantiaca Lavrova and Preobrazhenskaya 1975
Lavrova, N.V. and T.P. Preobrazhenskaya. 1975. Antibiotiki **20**:483-488. **Type strain:** INA 1933
Description: Lavrova, N.V. and T.P. Preobrazhenskaya. 1975. Ibid.

A. carminata Gauze, Sveshnikova, Ukholina, Gavrilina, Filicheva and Gladkikh 1973
Gauze, G.F., M.A. Sveshnikova, R.S. Ukholina, G.V. Gavrilina, V.A. Filicheva and E.G. Gladkikh. 1973. Antibiotiki **18**: 675-678. **Type strain:** INA 4281
Description: Gauze, G.F., M.A. Sveshnikova, R.S. Ukholina, G.V. Gavrilina, V.A. Filicheva and E.G. Gladkikh. 1973. Ibid.

A. citrea Lavrova, Preobrazhenskaya and Sveshnikova 1972
Lavrova, N.V., T.P. Preobrazhenskaya and M.A. Sveshnikova. 1972. Antibiotiki **17**:965-970. **Type strain:** ATCC 27887
Description: Lavrova, N.V., T.P. Preobrazhenskaya and M.A. Sveshnikova. 1972. Ibid.

A. coerulea Preobrazhenskaya, Lavrova, Ukholina and Nechaeva 1975
Preobrazhenskaya, T.P., N.V. Lavrova, R.S. Ukholina and N.P. Nechaeva. 1975. Antibiotiki **20**:404-409. **Type strain:** INA 765
Description: Preobrazhenskaya, T.P., N.V. Lavrova, R.S. Ukholina and N.P. Nechaeva. 1975. Ibid.

A. coeruleofusca Preobrazhenskaya and Sveshnikova 1974
Preobrazhenskaya, T.P. and M.A. Sveshnikova. 1974. Mikrobiologiya **43**:735-738. **Type strain:** INA 1335
Description: Preobrazhenskaya, T.P. and M.A. Sveshnikova. 1974. Ibid.

A. cremea Preobrazhenskaya, Lavrova, Ukholina and Nechaeva 1975
Preobrazhenskaya, T.P., N.V. Lavrova, R.S. Ukholina and N.P. Nechaeva. 1975. Antibiotiki **20**:404-409. **Type strain:** INA 292
Description: Preobrazhenskaya, T.P., N.V. Lavrova, R.S. Ukholina and N.P. Necheva. 1975. Ibid.

A. fastidiosa Soina, Sokolov and Agre 1975
Soina, V.S., A.A. Sokolov and N.S. Agre. 1975. Mikrobiologiya **44**:883-887. **Type strain:** INMI 104
Description: Soina, V.S., A.A. Sokolov and N.S. Agre. 1975. Ibid.

A. flava Gauze, Maksimova, Ollkhovatova, Sveshnikova, Kochetkova and Ilchenko 1974
Gauze, G.F., T.S. Maksimova, O.L. Ollkhovatova, M.A. Sveshnikova, G.V. Kochetkova and G.B. Ilchenko. 1974. Antibiotiki **19**:771-775. **Type strain:** ATCC 29533
Description: Gauze, G.F., T.S. Maksimova, O.L. Ollkhovatova, M.A. Sveshnikova, G.V. Kochetkova and G.B. Ilchenko. 1974. Ibid.

A. helvata Nonomura and Ohara 1971
Nonomura, H. and Y. Ohara. 1971. Journal of Fermentation Technology **49**:904-912. **Type strain:** ATCC 27295
Description: Nonomura, H. and Y. Ohara. 1971. Ibid.

A. livida Lavrova and Preobrazhenskaya 1975
Lavrova, N.V. and T.P. Preobrazhenskaya. 1975. Antibiotiki **20**:483-488. **Type strain:** INA 1678
Description: Lavrova, N.V. and T.P. Preobrazhenskaya. 1975. Ibid.

A. longispora Preobrazhenskaya and Sveshnikova 1974
Preobrazhenskaya, T.P. and M.A. Sveshnikova. 1974. Mikrobiologiya **43**: 864-868. **Type strain:** INA 10222
Description: Preobrazhenskaya, T.P. and M.A. Sveshnikova. 1974. Ibid.

A. luteofluorescens (Shinobu 1962) Preobrazhenskaya, Lavrova and Blinov 1975
Shinobu, R. 1962. Memoirs of the Osaka University of the Liberal Arts and Education B Natural Science **11**:115-122; Preobrazhenskaya, T.P., N.V. Lavrova and N.O. Blinov. 1975. Mikrobiologiya **44**: 524-527. **Type strain:** IFO 13057
Description: Preobrazhenskaya, T.P., N.V. Lavrova and N.O. Blinov. 1975. Ibid.

A. madurae (Vincent 1894) Lechevalier and Lechevalier 1968
Vincent, H. 1894. Annales de l'Institut Pasteur (Paris) **8**:129-151; Lechevalier, H.A. and M.P. Lechevalier. 1968 in Prauser, H. (ed). The Actinomycetales. Sept. 1968. VEB Gustav Fischer Verlag, Jena. pp. 393-405. **Type strain:** NCTC

5654
Description: Lechevalier, H.A. and M.P. Lechevalier. 1968. Ibid.

A. malachitica Lavrova, Preobrazhenskaya and Sveshnikova 1972
Lavrova, N.V., T.P. Preobrazhenskaya and M.A. Sveshnikova. 1972. Antibiotiki **17**:965-970. **Type strain:** ATCC 27888
Description: Lavrova, N.V., T.P. Preobrazhenskaya and M.A. Sveshnikova. 1972. Ibid.

A. pelletieri (Laveran 1906) Lechevalier and Lechevalier 1968
Laveran, A. 1906. Comptes Rendus des Seances de la Societe de Biologie (Paris) **61**:340-341; Lechevalier, H.A. and M.P. Lechevalier. 1968 in Prauser, H. (ed). The Actinomycetales. Sept. 1968. VEB Gustav Fischer Verlag, Jena. pp. 393-405. **Type strain:** NCTC 4162
Description: Lechevalier, H.A. and M.P. Lechevalier. 1968. Ibid.

A. pusilla Nonomura and Ohara 1971
Nonomura, H. and Y. Ohara. 1971. Journal of Fermentation Technology **49**:904-912. **Type strain:** ATCC 27296
Description: Nonomura, H. and Y. Ohara. 1971. Ibid.

A. roseola Lavrova and Preobrazhenskaya 1975
Lavrova, N.V. and T.P. Preobrazhenskaya. 1975. Antibiotiki **20**:483-488. **Type strain:** INA 1671
Description: Lavrova, N.V. and T.P. Preobrazhenskaya. 1975. Ibid.

A. roseoviolacea Nonomura and Ohara 1971
Nonomura, H. and Y. Ohara. 1971. Journal of Fermentation Technology **49**:904-912. **Type strain:** ATCC 27297
Description: Nonomura, H. and Y. Ohara. 1971. Ibid.

A. rubra (Sveshnikova, Maksimova and Kudrina 1969) Meyer and Sveshnikova 1974
Sveshnikova, M.A., T.S. Maksimova and E.S. Kudrina. 1969. Mikrobiologiya **38**:883-893; Meyer, J. and M.A. Sveshnikova. 1974. Zeitschrift fur Allgemeine Mikrobiologie **14**:167-170. **Type strain:** ATCC 27031
Description: Meyer, J. and M.A. Sveshnikova. 1974. Ibid.

A. salmonea Preobrazhenskaya, Lavrova, Ukholina and Nechaeva 1975
Preobrazhenskaya, T.P., N.V. Lavrova, R.S. Ukholina and N.P. Nechaeva. 1975. Antibiotiki **20**:404-409. **Type strain:** INA 2488
Description: Preobrazhenskaya, T.P., N.V. Lavrova, R.S. Ukholina and N.P. Nechaeva. 1975. Ibid.

A. spadix Nonomura and Ohara 1971
Nonomura, H. and Y. Ohara. 1971. Journal of Fermentation Technology **49**:904-912. **Type strain:** ATCC 27298
Description: Nonomura, H. and Y. Ohara. 1971. Ibid.

A. verrucosospora Nonomura and Ohara 1971
Nonomura, H. and Y. Ohara. 1971. Journal of Fermentation Technology **49**:904-912. **Type strain:** ATCC 27299
Description: Nonomura, H. and Y. Ohara. 1971. Ibid.

A. vinacea Lavrova and Preobrazhenskaya 1975
Lavrova, N.V. and T.P. Preobrazhenskaya. 1975. Antibiotiki **20**:483-488. **Type strain:** INA 1682
Description: Lavrova, N.V. and T.P. Preobrazhenskaya. 1975. Ibid.

Actinomyces Harz 1877
Harz, C.O. 1877-1878. Deutsche Zeitschrift fur Thiermedizin **5**:125-140. **Type species:** *A. bovis* Harz 1877
Description: Bergey 8.

A. bovis Harz 1877
Harz, C.O. 1877-1878. Deutsche Zeitschrift fur Thiermedizin **5**:125-140. **Type strain:** ATCC 13683
Description: Bergey 8.

A. humiferus Gledhill and Casida 1969
Gledhill, W.E. and L.E. Casida. 1969. Applied Microbiology **18**:114-121. **Type strain:** ATCC 25174
Description: Gledhill, W.E. and L.E. Casida. 1969. Ibid.

A. israelii (Kruse 1896) Lachner-Sandoval 1898
Kruse, W. 1896 in Flugge, C. Die Mikroorganismen 3rd ed. Vol. 2, Vogel, Leipzig. pp. 48-96; Lachner-Sandoval, V. 1898. Uber Strahlenpilze. Inaugural Dissertation Strassburg. Universitats Buchdruckerei Von Carl Georgi, Bonn. **Type strain:** ATCC 12102
Description: Bergey 8.

A. naeslundii Thompson and Lovestedt 1951
Thompson, L. and S.A. Lovestedt. 1951. Proceedings of the Staff Meetings of the

Mayo Clinic **26**:169-175. **Type strain:** ATCC 12104
Description: Bergey 8.

A. odontolyticus Batty 1958
Batty, I. 1958. Journal of Pathology and Bacteriology **75**:455-459. **Type strain:** ATCC 17929
Description: Bergey 8.

A. viscosus (Howell, Jordan, Georg and Pine 1965) Georg, Pine and Gerencser 1969
Howell, A., H.V. Jordan, L.K. Georg and L. Pine. 1965. Sabouraudia **4**:65-67; Georg, L.K., L. Pine and M.A. Gerencser. 1969. International Journal of Systematic Bacteriology **19**:291-293. **Type strain:** ATCC 15987
Description: Bergey 8.

Actinoplanes Couch 1950
Couch, J.N. 1950. Journal of the Elisha Mitchell Scientific Society **66**:87-92. **Type species:** *A. philippinensis* Couch 1950
Description: Bergey 8.

A. armeniacus Kalakoutskii and Kusnetsov 1964
Kalakoutskii, L.V. and V.D. Kusnetsov. 1964. Mikrobiologiya **33**:613. **Type strain:** ATCC 15676
Description: Bergey 8.

A. brasiliensis Thiemann, Beretta, Coronelli and Pagani 1969
Thiemann, J.E., G. Beretta, C. Coronelli and H. Pagani. 1969. Journal of Antibiotics **22**:119-125. **Type strain:** ATCC 25844
Description: Thiemann, J.E., G. Beretta, C. Coronelli and H. Pagani. 1969. Ibid.

A. deccanensis Parenti, Pagani and Beretta 1975
Parenti, F., H. Pagani and G. Beretta. 1975. Journal of Antibiotics **28**:247-252. **Type strain:** ATCC 21983
Description: Parenti, F., H. Pagani and G. Beretta. 1975. Ibid.

A. ferrugineus Palleroni 1979
Palleroni, N.J. 1979. International Journal of Systematic Bacteriology **29**:51-55. **Type strain:** ATCC 29868
Description: Palleroni, N.J. 1979. Ibid.

A. italicus Beretta 1973
Beretta, G. 1973. International Journal of Systematic Bacteriology **23**:37-41. **Type strain:** ATCC 27366
Description: Beretta, G. 1973. Ibid.

A. missouriensis Couch 1963
Couch, J.N. 1963. Journal of the Elisha Mitchell Scientific Society **79**:53-70. **Type strain:** ATCC 14538
Description: Bergey 8.

A. philippinensis Couch 1950
Couch, J.N. 1950. Journal of the Elisha Mitchell Scientific Society **66**:87-92. **Type strain:** ATCC 12427
Description: Bergey 8.

A. rectilineatus Lechevalier and Lechevalier 1975
Lechevalier, M.P. and H.A. Lechevalier. 1975. International Journal of Systematic Bacteriology **25**:371-376. **Type strain:** ATCC 29234
Description: Lechevalier, M.P. and H.A. Lechevalier. 1975. Ibid.

A. utahensis Couch 1963
Couch, J.N. 1963. Journal of the Elisha Mitchell Scientific Society **79**:53-70. **Type strain:** ATCC 14539
Description: Bergey 8.

Actinopolyspora Gochnauer, Leppard, Komaratat, Kates, Novitsky and Kushner 1975
Gochnauer, M.B., G.G. Leppard, P. Komaratat, M. Kates, T. Novitsky and D.J. Kushner. 1975. Canadian Journal of Microbiology **21**:1500-1511. **Type species:** *A. halophila* Gochnauer, Leppard, Komaratat, Kates, Novitsky and Kushner 1975
Description: Gochnauer, M.B., G.G. Leppard, P. Komaratat, M. Kates, T. Novitsky and D.J. Kushner. 1975. Ibid.

A. halophila Gochnauer, Leppard, Komaratat, Kates, Novitsky and Kushner 1975
Gochnauer, M.B., G.G. Leppard, P. Komaratat, M. Kates, T. Novitsky and D.J. Kushner. 1975. Canadian Journal of Microbiology **21**:1500-1511. **Type strain:** ATCC 27976
Description: Gochnauer, M.B., G.G. Leppard, P. Komaratat, M. Kates, T. Novitsky and D.J. Kushner. 1975. Ibid.

Actinopycnidium Krassilnikov 1962
Krassilnikov, N.A. 1962. Mikrobiologiya **31**:250-253 (English Translation 204-207). **Type species:** *A. caeruleum* Krassilnikov 1962

Description: Krassilnikov, N.A. 1962. Ibid.

A. caeruleum Krassilnikov 1962
Krassilnikov, N.A. 1962. Mikrobiologiya **31**:204-207. **Type strain:** ATCC 15719
Description: Krassilnikov, N.A. 1962. Ibid.

Actinosporangium Krassilnikov and Yuan 1961
Krassilnikov, N.A. and C.S. Yuan. 1961. Izvestiya Akademii Nauk SSSR Seriya Biologicheskaya **8**:113-116. **Type species:** *A. violaceum* Krassilnikov and Yuan 1961
Description: Krassilnikov, N.A. and C.S. Yuan. 1961. Ibid.

A. violaceum Krassilnikov and Yuan 1961
Krassilnikov, N.A. and C.S. Yuan. 1961. Izvestiya Akademii Nauk SSSR Seriya Biologicheskaya **8**:113-116. **Type strain:** ATCC 15813
Description: Krassilnikov, N.A. and C.S. Yuan. 1961. Ibid.

Actinosynnema Hasegawa, Lechevalier and Lechevalier 1978
Hasegawa, T., M.P. Lechevalier and H.A. Lechevalier. 1978. International Journal of Systematic Bacteriology **28**: 304-310. **Type species:** *A. mirum* Hasegawa, Lechevalier and Lechevalier 1978
Description: Hasegawa, T., M.P. Lechevalier and H.A. Lechevalier. 1978. Ibid.

A. mirum Hasegawa, Lechevalier and Lechevalier 1978
Hasegawa, T., M.P. Lechevalier and H.A. Lechevalier. 1978. International Journal of Systematic Bacteriology **28**: 304-310. **Type strain:** IMRU 3971
Description: Hasegawa, T., M.P. Lechevalier and H.A. Lechevalier. 1978. Ibid.

Aegyptianella Carpano 1929
Carpano, M. 1929. Bulletin of the Ministry of Agriculture Egypt **86**:1-12. **Type species:** *A. pullorum* Carpano 1929
Description: Bergey 8.

A. pullorum Carpano 1929
Carpano, M. 1929. Bulletin of the Ministry of Agriculture Egypt **86**:1-12. **Type strain:** no culture isolated.
Description: Bergey 8.

Aerococcus Williams, Hirch and Cowan 1953
Williams, R.E.O., A. Hirch and S.T. Cowan. 1953. Journal of General Microbiology **8**:475-480. **Type species:** *A. viridans* Williams, Hirch and Cowan 1953
Description: Bergey 8.

A. viridans Williams, Hirch and Cowan 1953
Williams, R.E.O., A. Hirch and S.T. Cowan. 1953. Journal of General Microbiology **8**:475-480. **Type strain:** NCTC 8251
Description: Williams, R.E.O., A. Hirch and S.T. Cowan. 1953. Ibid.

Aeromonas Kluyver and van Niel 1936
Kluyver, A.J. and C.B. van Niel. 1936. Zentralblatt fur Bakteriologie, Parasitenkunde, Infektionskrankheiten und Hygiene. Abteilung II. **94**:369-403. **Type species:** *A. hydrophila* (Chester 1901) Stanier 1943
Description: Bergey 8.

A. hydrophila (Chester 1901) Stanier 1943
Chester, F.D. 1901. A Manual of Determinative Bacteriology. The MacMillan Co., New York. pp. 1-401; Stanier, R.Y. 1943. Journal of Bacteriology **46**: 213-214. **Type strain:** ATCC 7966
Description: Bergey 8.

A. hydrophila subsp. *anaerogenes* Schubert 1964
Schubert, R.H.W. 1964. Zentralblatt fur Bakteriologie, Parasitenkunde, Infektionskrankheiten und Hygiene. Abteilung I. **193**:343-352. **Type strain:** ATCC 15467
Description: Bergey 8.

A. hydrophila subsp. *hydrophila* (Chester 1901) Stanier 1943
Chester, F.D. 1901. A Manual of Determinative Bacteriology. The MacMillan Co., New York. pp. 1-401; Stanier, R.Y. 1943. Journal of Bacteriology **46**:213-214. **Type strain:** ATCC 7966
Description: Bergey 8.

A. hydrophila subsp. *proteolytica* (Merkel, Traganza, Mukherjee, Griffin and Prescott 1964) Schubert 1969
Merkel, J.R., E.D. Traganza, B.B. Mukherjee, T.B. Griffin and J.M. Prescott. 1964. Journal of Bacteriology **87**:1227-1233; Schubert, R.H.W. 1969. Zentralblatt fur Bakteriologie, Parasitenkunde, Infektionskrankheiten und Hygiene. Abteilung I. **211**:409-412. **Type strain:** ATCC

15338
Description: Bergey 8.

A. punctata (Zimmermann 1890) Snieszko 1957
Zimmermann, O.E.R. 1890. 1. Reihe, Elfter Bericht der Naturwissenschaftlichen Gesellschaft Chemnitz, pp. 54-154; Snieszko, S.F. 1957 in Breed, R.S., E.G.D. Murray and N.R. Smith (eds). Bergey's Manual of Determinative Bacteriology, 7th ed. The Williams and Wilkins Co., Baltimore. pp. 189-193. **Type strain:** ATCC 15468
Description: Bergey 8.

A. punctata subsp. *caviae* (Scherago 1936) Schubert 1964
Scherago, M. 1936. Journal of Bacteriology **31**:83; Schubert, R.H.W. 1964. Zentralblatt fur Bakteriologie, Parasitenkunde, Infektionskrankheiten und Hygiene. Abteilung I. **193**:343-352. **Type strain:** ATCC 15468
Description: Bergey 8.

A. punctata subsp. *punctata* (Zimmermann 1890) Snieszko 1957
Zimmermann, O.E.R. 1890. 1 Reihe Elfter Bericht der Naturwissenschaftlichen Gesellschaft Chemnitz, pp. 54-154; Snieszko, S.F. 1957 in Breed, R.S., E.G.D. Murray and N.R. Smith (eds). Bergey's Manual of Determinative Bacteriology, 7th ed. The Williams and Wilkins Co., Baltimore. pp. 189-193. **Type strain:** ATCC 15468
Description: Bergey 8.

A. salmonicida (Lehmann and Neumann 1896) Griffin, Snieszko and Friddle 1953
Lehmann, K.B. and R. Neumann. 1896. Atlas und Grundriss der Bakteriologie und Lehrbuch der speziellen bakteriologischen Diagnostik. 1st ed. J.F. Lehmann, Munchen; Griffin, P.J., S.F. Snieszko and S.B. Friddle. 1953. Transactions of the American Fisheries Society **82**:129-138. **Type strain:** NCMB 1102
Description: Bergey 8.

A. salmonicida subsp. *achromogenes* (Smith 1963) Schubert 1967
Smith, I.W. 1963. Journal of General Microbiology **33**:263-274; Schubert, R.H.W. 1967. International Journal of Systematic Bacteriology **17**:273-279. **Type strain:** NCMB 1110
Description: Bergey 8.

A. salmonicida subsp. *masoucida* Kimura 1969
Kimura, T. 1969. Fish Pathology (Tokyo) **3**:45-52. **Type strain:** ATCC 27013
Description: Bergey 8.

A. salmonicida subsp. *salmonicida* (Lehmann and Neumann 1896) Griffin, Snieszko and Friddle 1953
Lehmann, K.B. and R. Neumann. 1896. Atlas und Grundriss der Bakteriologie und Lehrbuch der speziellen bakteriologischen Diagnostik. 1st ed. J.F. Lehmann, Munchen; Griffin, P.J., S.F. Snieszko and S.B. Friddle. 1953. Transactions of the American Fisheries Society **82**:129-138. **Type strain:** NCMB 1102
Description: Bergey 8.

Agrobacterium Conn 1942
Conn, H.J. 1942. Journal of Bacteriology **44**:353-360. **Type species:** *A. tumefaciens* (Smith and Townsend 1907) Conn 1942
Description: Bergey 8.

A. radiobacter (Beijerinck and van Delden 1902) Conn 1942
Beijerinck, M.W. and A. van Delden. 1902. Zentralblatt fur Bakteriologie, Parasitenkunde, Infektionskrankheiten und Hygiene. Abteilung II. **9**:3-43; Conn, H.J. 1942. Journal of Bacteriology **44**:353-360. **Type strain:** ATCC 19358
Description: Bergey 8.

A. rhizogenes (Riker, Banfield, Wright, Keitt and Sagen 1930) Conn 1942
Riker, A.J., W.M. Banfield, W.H. Wright, G.W. Keitt and H.E. Sagen. 1930. Journal of Agricultural Research **41**:507-540; Conn, H.J. 1942. Journal of Bacteriology **44**:353-360. **Type strain:** ATCC 11325
Description: Burkholder, W.J. 1957 in Breed, R.S., E.G.D. Murray and N.R. Smith (eds). Bergey's Manual of Determinative Bacteriology, 7th ed. The Williams and Wilkins Co., Baltimore. pp. 288-292.

A. rubi (Hildebrand 1940) Starr and Weiss 1943
Hildebrand, E.M. 1940. Journal of Agricultural Research **61**:685-696; Starr, M.P. and J.E. Weiss. 1943. Phytopathology **33**: 314-318. **Type strain:** ATCC 13335
Description: Burkholder, W.J. 1957.

A. tumefaciens (Smith and Townsend 1907) Conn 1942
Smith, E.F. and C.O. Townsend. 1907.

Science (Washington) **25**:671-673; Conn, H.J. 1942. Journal of Bacteriology **44**: 353-360. **Type strain:** ATCC 23308
Description: Bergey 8.

Agromyces Gledhill and Casida 1969
Gledhill, W.E. and L.E. Casida. 1969. Applied Microbiology **18**:340-349. **Type species:** *A. ramosus* Gledhill and Casida 1969
Description: Gledhill, W.E. and L.E. Casida. 1969. Ibid.

A. ramosus Gledhill and Casida 1969
Gledhill, W.E. and L.E. Casida. 1969. Applied Microbiology **18**:340-349. **Type strain:** ATCC 25173
Description: Gledhill, W.E. and L.E. Casida. 1969. Ibid.

Alcaligenes Castellani and Chalmers 1919
Castellani, A. and A.J. Chalmers. 1919. Manual of Tropical Medicine, 3rd ed. Williams, Wood and Co., New York. **Type species:** *A. faecalis* Castellani and Chalmers 1919
Description: Bergey 8.

A. aestus Baumann, Baumann, Mandel and Allen 1972
Baumann, L., P. Baumann, M. Mandel and R.D. Allen. 1972. Journal of Bacteriology **110**:402-429. **Type strain:** ATCC 27128
Description: Baumann, L., P. Baumann, M. Mandel and R.D. Allen. 1972. Ibid.

A. aquamarinus (ZoBell and Upham 1944) Hendrie, Holding and Shewan 1974
ZoBell, C.E. and H.C. Upham. 1944. Bulletin Scripps Institution of Oceanography University of California (Technical Series) **5**:239-292; Hendrie, M.S., A.J. Holding and J.M. Shewan. 1974. International Journal of Systematic Bacteriology **24**:534-550. **Type strain:** ATCC 14400
Description: Zobell, C.E. and H.C. Upham. 1944. Ibid.

A. cupidus Baumann, Baumann, Mandel and Allen 1972
Baumann, L., P. Baumann, M. Mandel and R.D. Allen. 1972. Journal of Bacteriology **110**:402-429. **Type strain:** ATCC 27124
Description: Baumann, L., P. Baumann, M. Mandel and R.D. Allen. 1972. Ibid.

A. eutrophus Davis 1969
Davis, D.H. 1969 in Davis, D.H., M. Doudoroff, R.Y. Stanier and M. Mandel. International Journal of Systematic Bacteriology **19**:375-390. **Type strain:** ATCC 17697
Description: Davis, D.H. 1969. Ibid.

A. faecalis Castellani and Chalmers 1919
Castellani, A. and A.J. Chalmers. 1919. Manual of Tropical Medicine, 3rd ed. Williams, Wood and Co., New York. **Type strain:** ATCC 8750
Description: Castellani, A. and A.J. Chalmers. 1919. Ibid.

A. latus Palleroni and Palleroni 1978
Palleroni, N.J. and A.V. Palleroni. 1978. International Journal of Systematic Bacteriology **28**:416-424. **Type strain:** ATCC 29712
Description: Palleroni, N.J. and A.V. Palleroni. 1978. Ibid.

A. pacificus Baumann, Baumann, Mandel and Allen 1972
Baumann, L., P. Baumann, M. Mandel and R.D. Allen. 1972. Journal of Bacteriology **110**:402-429. **Type strain:** ATCC 27122
Description: Baumann, L., P. Baumann, M. Mandel and R.D. Allen. 1972. Ibid.

A. paradoxus Davis 1969
Davis, D.H. 1969 in Davis, D.H., M. Doudoroff, R.Y. Stanier and M. Mandel. International Journal of Systematic Bacteriology **19**:375-390. **Type strain:** ATCC 17713
Description: Davis, D.H. 1969. Ibid.

A. ruhlandii (Packer and Vishniac 1955) Aragno and Schlegel 1977
Packer, L. and W. Vishniac. 1955. Journal of Bacteriology **70**:216-233; Aragno, M. and H. Schlegel. 1977. International Journal of Systematic Bacteriology **27**: 279-281. **Type strain:** ATCC 15749
Description: Aragno, M. and H. Schlegel. 1977. Ibid.

A. venustus Baumann, Baumann, Mandel and Allen 1972
Baumann, L., P. Baumann, M. Mandel and R.D. Allen. 1972. Journal of Bacteriology **110**:402-429. **Type strain:** ATCC 27125
Description: Baumann, L., P. Baumann, M. Mandel and R.D. Allen. 1972. Ibid.

Alteromonas Baumann, Baumann, Mandel and Allen 1972
Baumann, L., P. Baumann, M. Mandel and R.D. Allen. 1972. Journal of Bacteriology **110**:402-429. **Type species:** *A. macleodii* Baumann, Baumann, Mandel and Allen 1972.
Description: Baumann, L., P. Baumann, M. Mandel and R.D. Allen. 1972. Ibid.

A. aurantia Gauthier and Breittmayer 1979
Gauthier, M.J. and V.A. Breittmayer. 1979. International Journal of Systematic Bacteriology **29**:366-372. **Type strain:** ATCC 33046
Description: Gauthier, M.J. and V.A. Breittmayer. 1979. Ibid.

A. citrea Gauthier 1977
Gauthier, M.J. 1977. International Journal of Systematic Bacteriology **27**:349-354. **Type strain:** ATCC 29719
Description: Gauthier, M.J. 1977. Ibid.

A. communis Baumann, Baumann, Mandel and Allen 1972
Baumann, L., P. Baumann, M. Mandel and R.D. Allen. 1972. Journal of Bacteriology **110**:402-429. **Type strain:** ATCC 27118
Description: Baumann, L., P. Baumann, M. Mandel and R.D. Allen. 1972. Ibid.

A. espejiana Chan, Baumann, Garza and Baumann 1978
Chan, K.Y., L. Baumann, M.M. Garza and P. Baumann. 1978. International Journal of Systematic Bacteriology **28**: 217-222. **Type strain:** ATCC 29659
Description: Chan, K.Y., L. Baumann, M.M. Garza and P. Baumann. 1978. Ibid.

A. haloplanktis (ZoBell and Upham 1944) Reichelt and Baumann 1973
ZoBell, C.E. and H.C. Upham. 1944. Bulletin of the Scripps Institution of Oceanography, University of California **5**: 239-292; Reichelt, J.L. and P. Baumann. 1973. International Journal of Systematic Bacteriology **23**:438-441. **Type strain:** ATCC 14393
Description: Reichelt, J.L. and P. Baumann. 1973. Ibid.

A. macleodii Baumann, Baumann, Mandel and Allen 1972
Baumann, L., P. Baumann, M. Mandel and R.D. Allen. 1972. Journal of Bacteriology **110**:402-429. **Type strain:** ATCC 27126
Description: Baumann, L., P. Baumann, M. Mandel and R.D. Allen. 1972. Ibid.

A. rubra Gauthier 1976
Gauthier, M.J. 1976. International Journal of Systematic Bacteriology **26**:459-466. **Type strain:** ATCC 29570
Description: Gauthier, M.J. 1976. Ibid.

A. undina Chan, Baumann, Garza and Baumann 1978
Chan, K.Y., L. Baumann, M.M. Garza and P. Baumann. 1978. International Journal of Systematic Bacteriology **28**: 217-222. **Type strain:** ATCC 29660
Description: Chan, K.Y., L. Baumann, M.M. Garza and L. Baumann. 1978. Ibid.

A. vaga Baumann, Baumann, Mandel and Allen 1972
Baumann, L., P. Baumann, M. Mandel and R.D. Allen. 1972. Journal of Bacteriology **110**: 402-429. **Type strain:** ATCC 27119
Description: Baumann, L., P. Baumann, M. Mandel and R.D. Allen. 1972. Ibid.

Alysiella Langeron 1923
Langeron, M. 1923. Annales de Parasitologie Humaine et Comparee **1**:113-123. **Type species:** *A. filiformis* (Schmid 1922) Langeron 1923
Description: Bergey 8.

A. filiformis (Schmid 1922) Langeron 1923
Schmid, G. 1922 in Simons, H. Zentralblatt fur Bakteriologie, Parasitenkunde, Infektionskrankheiten und Hygiene. Abteilung I. **88**:501-510; Langeron, M. 1923. Annales de Parasitologie Humaine et Comparee **1**:113-123. **Type strain:** ATCC 15532
Description: Steed, P.D.M. 1962. Journal of General Microbiology **29**:615-624.

Amoebobacter Winogradsky 1888
Winogradsky, S. 1888. Beitrage zur Morphologie und Physiologie der Bacterien. Heft I. Zur Morphologie und Physiologie der Schwefelbacterien. Arthur Felix, Leipzig. pp. 1-120. **Type species:** *A. roseus* Winogradsky 1888
Description: Bergey 8.

A. pendens (Molisch 1906) Pfennig and Truper

1971
Molisch, H. 1906. Botanische Zeitung. Abteilung I. **64**:223-232; Pfennig, N. and H.G. Truper. 1971. International Journal of Systematic Bacteriology **21**:11-14. **Type strain:** DSM 236
Description: Bergey 8.

A. roseus Winogradsky 1888
Winogradsky, S. 1888. Beitrage zur Morphologie und Physiologie der Bacterien. Heft I. Zur Morphologie und Physiologie der Schwefelbacterien. Arthur Felix, Leipzig. pp. 1-120. **Type strain:** DSM 235
Description: Bergey 8.

Amorphosporangium Couch 1963
Couch, J.N. 1963. Journal of the Elisha Mitchell Scientific Society **79**:53-70. **Type species:** *A. auranticolor* Couch 1963
Description: Bergey 8.

A. auranticolor Couch 1963
Couch, J.N. 1963. Journal of the Elisha Mitchell Scientific Society **79**:53-70. **Type strain:** ATCC 15330
Description: Bergey 8.

A. globisporum Thiemann 1967
Thiemann, J.E. 1967. Mycopathologia et Mycologia Applicata **33**:233-240. **Type strain:** ATCC 23056
Description: Thiemann, J.E. 1967. Ibid.

Ampullariella Couch 1964
Couch, J.N. 1964. Journal of the Elisha Mitchell Scientific Society **80**:29. **Type species:** *A. regularis* (Couch 1963) Couch 1964
Description: Bergey 8.

A. campanulata (Couch 1963) Couch 1964
Couch, J.N. 1963. Journal of the Elisha Mitchell Scientific Society **79**:53-70; Couch, J.N. 1964. Journal of the Elisha Mitchell Scientific Society **80**:29. **Type strain:** DSM 43148
Description: Bergey 8.

A. digitata (Couch 1963) Couch 1964
Couch, J.N. 1963. Journal of the Elisha Mitchell Scientific Society **79**:53-70; Couch, J.N. 1964. Journal of the Elisha Mitchell Scientific Society **80**:29. **Type strain:** ATCC 15349
Description: Bergey 8.

A. lobata (Couch 1963) Couch 1964
Couch, J.N. 1963. Journal of the Elisha Mitchell Scientific Society **79**:53-70; Couch, J.N. 1964. Journal of the Elisha Mitchell Scientific Society **80**:29. **Type strain:** ATCC 15350
Description: Bergey 8.

A. regularis (Couch 1963) Couch 1964
Couch, J.N. 1963. Journal of the Elisha Mitchell Scientific Society **79**:53-70; Couch, J.N. 1964. Journal of the Elisha Mitchell Scientific Society **80**:29. **Type strain:** DSM 43151
Description: Bergey 8.

Anaerobiospirillum Davis, Cleven, Brown and Balish 1976
Davis, C.P., D. Cleven, J. Brown and E. Balish. 1976. International Journal of Systematic Bacteriology **26**:498-504. **Type species:** *A. succiniciproducens* Davis, Cleven, Brown and Balish 1976
Description: Davis, C.P., D. Cleven, J. Brown and E. Balish. 1976. Ibid.

A. succiniciproducens Davis, Cleven, Brown and Balish 1976
Davis, C.P., D. Cleven, J. Brown and E. Balish. 1976. International Journal of Systematic Bacteriology **26**:498-504. **Type strain:** ATCC 29305
Description: Davis, C.P., D. Cleven, J. Brown and E. Balish. 1976. Ibid.

Anaeroplasma Robinson, Allison and Hartman 1975
Robinson, I.M., M.J. Allison and P.A. Hartman. 1975. International Journal of Systematic Bacteriology **25**:173-181. **Type species:** *A. abactoclasticum* Robinson, Allison and Hartman 1975
Description: Robinson, I.M., M.J. Allison and P.A. Hartman. 1975. Ibid.

A. abactoclasticum Robinson, Allison and Hartman 1975
Robinson, I.M., M.J. Allison and P.A. Hartman. 1975. International Journal of Systematic Bacteriology **25**:173-181. **Type strain:** ATCC 27879
Description: Robinson, I.M., M.J. Allison and P.A. Hartman. 1975. Ibid.

A. bactoclasticum (Robinson and Hungate 1973) Robinson and Allison 1975
Robinson, J.P. and R.E. Hungate. 1973. International Journal of Systematic Bacteriology **23**:171-181; Robinson, I.M. and

M.J. Allison. 1975. International Journal of Systematic Bacteriology **25**:182-186. **Type strain:** ATCC 27112
Description: Robinson, I.M. and M.J. Allison. 1975. Ibid.

Anaerovibrio Hungate 1966
Hungate, R.E. 1966. The rumen and its microbes. Academic Press, New York. **Type species:** *A. lipolytica* Hungate 1966
Description: Hungate, R.E. 1966. Ibid.

A. lipolytica Hungate 1966
Hungate, R.E. 1966. The rumen and its microbes. Academic Press, New York. **Type strain:** VPI 7553
Description: Holdeman, L.V., E.P. Cato and W.E.C. Moore (eds). 1977. Anaerobe Laboratory Manual, 4th ed. Virginia Polytechnic Institute and State University, Blacksburg, Virginia. pp. 1-156.

Anaplasma Theiler 1910
Theiler, A. 1910. Report to the Government, Transvaal, South Africa. Veterinary Bacteriology, Department of Agriculture **1908-9**:7-64. **Type species:** *A. marginale* Theiler 1910
Description: Bergey 8.

A. marginale Theiler 1910
Theiler, A. 1910. Transvaal S. Afr. Rep. Vet. Bacteriol. Dep. Agr. **1908-9**:7-64. **Type strain:** no culture isolated.
Description: Bergey 8.

A. ovis Lestoquard 1924
Lestoquard, F. 1924. Bulletin de la Societe de Pathologie Exotique **17**:784-787. **Type strain:** no culture isolated.
Description: Bergey 8.

Ancalochloris Gorlenko and Lebedeva 1971
Gorlenko, V.M. and E.V. Lebedeva. 1971. Mikrobiologiya **40**:1035-1039. **Type species:** *A. perfilievii* Gorlenko and Lebedeva 1971
Description: Gorlenko, V.M. and E.V. Lebedeva. 1971. Ibid.

A. perfilievii Gorlenko and Lebedeva 1971
Gorlenko, V.M. and E.V. Lebedeva. 1971. Mikrobiologiya **40**:1035-1039. **Type strain:** no pure culture.
Description: Gorlenko, V.M. and E.V. Lebedeva. 1971. Ibid.

Ancalomicrobium Staley 1968
Staley, J.T. 1968. Journal of Bacteriology **95**:1921-1942. **Type species:** *A. adetum* Staley 1968
Description: Bergey 8.

A. adetum Staley 1968
Staley, J.T. 1968. Journal of Bacteriology **95**:1921-1942. **Type strain:** ATCC 23632
Description: Staley, J.T. 1968. Ibid.

Aquaspirillum Hylemon, Wells, Krieg and Jannasch 1973
Hylemon, P.B., J.S. Wells Jr., N.R. Krieg and H.W. Jannasch. 1973. International Journal of Systematic Bacteriology **23**:340-380. **Type species:** *A. serpens* (Muller 1786) Hylemon, Wells, Krieg and Jannasch 1973
Description: Bergey 8.

A. anulus (Williams and Rittenberg 1957) Hylemon, Wells, Krieg and Jannasch 1973
Williams, M.A. and S.C. Rittenberg. 1957. International Bulletin of Bacteriological Nomenclature and Taxonomy **7**: 49-110; Hylemon, P.B., J.S. Wells Jr., N.R. Krieg and H.W. Jannasch. 1973. International Journal of Systematic Bacteriology **23**:340-380. **Type strain:** NCIB 9012
Description: Hylemon, P.B., J.S. Wells Jr., N.R. Krieg and H.W. Jannasch. 1973. Ibid.

A. aquaticum Hylemon, Wells, Krieg and Jannasch 1973
Hylemon, P.B., J.S. Wells Jr., N.R. Krieg and H.W. Jannasch. 1973. International Journal of Systematic Bacteriology **23**:340-380. **Type strain:** ATCC 11330
Description: Hylemon, P.B., J.S. Wells Jr., N.R. Krieg and H.W. Jannasch. 1973. Ibid.

A. autotrophicum Aragno and Schlegel 1978
Aragno, M. and H.G. Schlegel. 1978. International Journal of Systematic Bacteriology **28**:112-116. **Type strain:** DSM 732
Description: Aragno, M. and H.G. Schlegel. 1978. Ibid.

A. bengal Kumar, Banerjee, Bowdre, McElroy and Krieg 1974
Kumar, R., A.K. Banerjee, J.H. Bowdre, L.J. McElroy and N.R. Krieg. 1974. International Journal of Systematic Bacteriology **24**:453-458. **Type strain:** ATCC 27641

Description: Kumar, R., A.K. Banerjee, J.H. Bowdre, L.J. McElroy and N.R. Krieg. 1974. Ibid.

A. delicatum (Leifson 1962) Hylemon, Wells, Krieg and Jannasch 1973
Leifson, E. 1962. International Bulletin of Bacteriological Nomenclature and Taxonomy **12**:161-170; Hylemon, P.B., J.S. Wells Jr., N.R. Krieg and H.W. Jannasch. 1973. International Journal of Systematic Bacteriology **23**:340-380. **Type strain:** ATCC 14667
Description: Hylemon, P.B., J.S. Wells Jr., N.R. Krieg and H.W. Jannasch. 1973. Ibid.

A. dispar Hylemon, Wells, Krieg and Jannasch 1973
Hylemon, P.B., J.S. Wells Jr., N.R. Krieg and H.W. Jannasch. 1973. International Journal of Systematic Bacteriology **23**:340-380. **Type strain:** ATCC 27510
Description: Krieg, N.R. and P.B. Hylemon. 1976. Annual Review of Microbiology **30**:303-325.

A. fasciculus Strength, Isani, Linn, Williams, Vandermolen, Laughon and Krieg 1976
Strength, W.J., B. Isani, D.M. Linn, F.D. Williams, G.E. Vandermolen, B.E. Laughon and N.R. Krieg. 1976. International Journal of Systematic Bacteriology **26**:253-267. **Type strain:** ATCC 27740
Description: Strength, W.J., B. Isani, D.M. Linn, F.D. Williams, G.E. Vandermolen, B.E. Laughon and N.R. Krieg. 1976. Ibid.

A. giesbergeri (Williams and Rittenberg 1957) Hylemon, Wells, Krieg and Jannasch 1973
Williams, M.A. and S.C. Rittenberg. 1957. International Bulletin of Bacteriological Nomenclature and Taxonomy **7**: 49-110; Hylemon, P.B., J.S. Wells Jr., N.R. Krieg and H.W. Jannasch. 1973. International Journal of Systematic Bacteriology **23**:340-380. **Type strain:** ATCC 11334
Description: Hylemon, P.B., J.S. Wells Jr., N.R. Krieg and H.W. Jannasch. 1973. Ibid.

A. gracile (Canale-Parola, Rosenthal and Kupfer 1966) Hylemon, Wells, Krieg and Jannasch 1973
Canale-Parola, E., S.L. Rosenthal and D.G. Kupfer. 1966. Antonie van Leeuwenhoek Journal of Microbiology and Serology **32**:113-124; Hylemon, P.B., J.S. Wells Jr., N.R. Krieg and H.W. Jannasch. 1973. International Journal of Systematic Bacteriology **23**:340-380. **Type strain:** ATCC 19624
Description: Hylemon, P.B., J.S. Wells Jr., N.R. Krieg and H.W. Jannasch. 1973. Ibid.

A. itersonii (Giesberger 1936) Hylemon, Wells, Krieg and Jannasch 1973
Giesberger, G. 1936. Inaugural Dissertation Utrecht. pp. 1-136; Hylemon, P.B., J.S. Wells Jr., N.R. Krieg and H.W. Jannasch. 1973. International Journal of Systematic Bacteriology **23**:340-380. **Type strain:** ATCC 12639
Description: Hylemon, P.B., J.S. Wells Jr., N.R. Krieg and H.W. Jannasch. 1973. Ibid.

A. itersonii subsp. *itersonii* (Giesberger 1936) Hylemon, Wells, Krieg and Jannasch 1973
Giesberger, G. 1936. Inaugural Dissertation Utrecht. pp. 1-136; Hylemon, P.B., J.S. Wells Jr., N.R. Krieg and H.W. Jannasch. 1973. International Journal of Systematic Bacteriology **23**:340-380. **Type strain:** ATCC 12639
Description: Hylemon, P.B., J.S. Wells Jr., N.R. Krieg and H.W. Jannasch. 1973. Ibid.

A. itersonii subsp. *nipponicum* (Terasaki 1973) Terasaki 1979
Terasaki, Y. 1973. Bulletin of the Suzugamine Women's College, Natural Science **17**:1-71; Terasaki, Y. 1979. International Journal of Systematic Bacteriology **29**: 130-144. **Type strain:** IFO 13615
Description: Terasaki, Y. 1979. Ibid.

A. metamorphum (Terasaki 1961) Hylemon, Wells, Krieg and Jannasch 1973
Terasaki, Y. 1961. Botanical Magazine (Tokyo) **74**:220-227; Hylemon, P.B., J.S. Wells Jr., N.R. Krieg and H.W. Jannasch. 1973. International Journal of Systematic Bacteriology **23**:340-380. **Type strain:** ATCC 15280
Description: Hylemon, P.B., J.S. Wells Jr., N.R. Krieg and H.W. Jannasch. 1973. Ibid.

A. peregrinum (Pretorius 1963) Hylemon, Wells, Krieg and Jannasch 1973
Pretorius, W.A. 1963. Journal of General

Microbiology **32**:403-408; Hylemon, P.B., J.S. Wells Jr., N.R. Krieg and H.W. Jannasch. 1973. International Journal of Systematic Bacteriology **23**:340-380. **Type strain:** ATCC 15387
Description: Hylemon, P.B., J.S. Wells Jr., N.R. Krieg and H.W. Jannasch. 1973. Ibid.

A. peregrinum subsp. *integrum* (Terasaki 1973) Terasaki 1979
Terasaki, Y. 1973. Bulletin of the Suzugamine Women's College, Natural Science **17**:1-71; Terasaki, Y. 1979. International Journal of Systematic Bacteriology **29**: 130-144. **Type strain:** IFO 13617
Description: Terasaki, Y. 1979. Ibid.

A. peregrinum subsp. *peregrinum* (Pretorius 1963) Hylemon, Wells, Krieg and Jannasch 1973
Pretorius, W.A. 1963. Journal of General Microbiology **32**:403-408; Hylemon, P.B., J.S. Wells Jr., N.R. Krieg and H.W. Jannasch. 1973. International Journal of Systematic Bacteriology **23**:340-380. **Type strain:** ATCC 15387
Description: Hylemon, P.B., J.S. Wells Jr., N.R. Krieg and H.W. Jannasch. 1973. Ibid.

A. polymorphum (Williams and Rittenberg 1957) Hylemon, Wells, Krieg and Jannasch 1973
Williams, M.A. and S.C. Rittenberg. 1957. International Bulletin of Bacteriological Nomenclature and Taxonomy **7**: 49-110; Hylemon, P.B., J.S. Wells Jr., N.R. Krieg and H.W. Jannasch. 1973. International Journal of Systematic Bacteriology **23**:340-380. **Type strain:** ATCC 11332
Description: Hylemon, P.B., J.S. Wells Jr., N.R. Krieg and H.W. Jannasch. 1973. Ibid.

A. psychrophilum (Terasaki 1973) Terasaki 1979
Terasaki, Y. 1973. Bulletin of the Suzugamine Women's College, Natural Science **17**:1-71; Terasaki, Y. 1979. International Journal of Systematic Bacteriology **29**: 130-144. **Type strain:** IFO 13611
Description: Terasaki, Y. 1979. Ibid.

A. putridiconchylium (Terasaki 1961) Hylemon, Wells, Krieg and Jannasch 1973
Terasaki, Y. 1961. Botanical Magazine (Tokyo) **74**:79-85; Hylemon, P.B., J.S. Wells Jr., N.R. Krieg and H.W. Jannasch. 1973. International Journal of Systematic Bacteriology **23**:340-380. **Type strain:** ATCC 15279
Description: Hylemon, P.B., J.S. Wells Jr., N.R. Krieg and H.W. Jannasch. 1973. Ibid.

A. serpens (Muller 1786) Hylemon, Wells, Krieg and Jannasch 1973
Muller, O.F. 1786. Animalcula Infusoria Fluviatilia et Marina, Quae Detexit. Systematice Descripsit et Ad Vivum Delineari Curavit, pp. 1-367; Hylemon, P.B., J.S. Wells Jr., N.R. Krieg and H.W. Jannasch. 1973. International Journal of Systematic Bacteriology **23**:340-380. **Type strain:** ATCC 12638
Description: Hylemon, P.B., J.S. Wells Jr., N.R. Krieg and H.W. Jannasch. 1973. Ibid.

A. sinuosum (Williams and Rittenberg 1957) Hylemon, Wells, Krieg and Jannasch 1973
Williams, M.A. and S.C. Rittenberg. 1957. International Bulletin of Bacteriological Nomenclature and Taxonomy **7**: 49-110; Hylemon, P.B., J.S. Wells Jr., N.R. Krieg and H.W. Jannasch. 1973. International Journal of Systematic Bacteriology **23**:340-380. **Type strain:** ATCC 9786
Description: Hylemon, P.B., J.S. Wells Jr., N.R. Krieg and H.W. Jannasch. 1973. Ibid.

Arachnia Pine and Georg 1969
Pine, L. and L.K. Georg. 1969. International Journal of Systematic Bacteriology **19**:267-272. **Type species:** *A. propionica* (Buchanan and Pine 1962) Pine and Georg 1969
Description: Bergey 8.

A. propionica (Buchanan and Pine 1962) Pine and Georg 1969
Buchanan, B.B. and L. Pine. 1962. Journal of General Microbiology **28**:305-323; Pine, L. and L. Georg. 1969. International Journal of Systematic Bacteriology **19**: 267-272. **Type strain:** ATCC 14157
Description: Holdeman, L.V., E.P. Cato and W.E.C. Moore (eds). 1977. Anaerobe Laboratory Manual, 4th ed. Virginia Polytechnic Institute and State University,

Blacksburg, Virginia. pp. 1-156.

Archangium Jahn 1924
Jahn, E. 1924. Beitrage zur botanischen Protistologie. I. Die Polyangiden. Verlag von Gebruder Borntraeger, Leipzig. **Type species:** *A. gephyra* Jahn 1924
Description: Bergey 8.

A. gephyra Jahn 1924
Jahn, E. 1924. Beitrage zur botanischen Protistologie. I. Die Polyangiden. Verlag von Gebruder Borntraeger, Leipzig. **Type strain:** ATCC 25201
Description: Bergey 8.

Arthrobacter Conn and Dimmick 1947
Conn, H.J. and I. Dimmick. 1947. Journal of Bacteriology **54**:291-303. **Type species:** *A. globiformis* (Conn 1928) Conn and Dimmick 1947
Description: Bergey 8.

A. atrocyaneus Kuhn and Starr 1960
Kuhn, D.A. and M.P. Starr. 1960. Archiv fur Mikrobiologie **36**:175-181. **Type strain:** ATCC 13752
Description: Kuhn, D.A. and M.P. Starr. 1960. Ibid.

A. aurescens Phillips 1953
Phillips, H.C. 1953. Iowa State Journal of Science **27**:240-241. **Type strain:** ATCC 13344
Description: Phillips, H.C. 1953. Ibid.

A. citreus Sacks 1954
Sacks, L.E. 1954. Journal of Bacteriology **67**:342-345. **Type strain:** ATCC 11624
Description: Bergey 8.

A. crystallopoietes Ensign and Rittenberg 1963
Ensign, J.C. and S.C. Rittenberg. 1963. Archiv fur Mikrobiologie **47**:137-153. **Type strain:** ATCC 15481
Description: Ensign, J.C. and S.C. Rittenberg. 1963. Ibid.

A. duodecadis Lochhead 1958
Lochhead, A.G. 1958. Archiv fur Mikrobiologie **31**:163-170. **Type strain:** ATCC 13347
Description: Lochhead, A.G. 1958. Ibid.

A. flavescens Lochhead 1958
Lochhead, A.G. 1958. Archiv fur Mikrobiologie **31**:163-170. **Type strain:** ATCC 13348
Description: Lochhead, A.G. 1958. Ibid.

A. globiformis (Conn 1928) Conn and Dimmick 1947
Conn, H.J. 1928. New York State Agricultural Experimental Station Technical Bulletin No. **138**:3-26; Conn, H.J. and I. Dimmick. 1947. Journal of Bacteriology **54**:291-303. **Type strain:** ATCC 8010
Description: Bergey 8.

A. histidinolovorans Adams 1954
Adams, E. 1954. Journal of Biological Chemistry **209**:829-846. **Type strain:** ATCC 11442
Description: Keddie, R.M. and G.L. Cure. 1977. Journal of Applied Bacteriology **42**:229-252.

A. mysorens Nand and Rao 1972
Nand, K. and D.V. Rao. 1972. Zentralblatt fur Bakteriologie, Parasitenkunde, Infektionskrankheiten und Hygiene. Abteilung II. **127**:324-331. **Type strain:** NCIB 10583
Description: Nand, K. and D.V. Rao. 1972. Ibid.

A. nicotianae Giovannozzi-Sermanni 1959
Giovannozzi-Sermanni, G. 1959. Il Tabacco (Rome) **63**:83-86. **Type strain:** ATCC 15236
Description: Giovannozzi-Sermanni, G. 1959. Ibid.

A. oxydans Sguros 1954
Sguros, P.L. 1954. Bacteriological Proceedings **1954**:21-22. **Type strain:** ATCC 14358
Description: Sguros, P.L. 1955. Journal of Bacteriology **69**:28-37.

A. pascens Lochhead and Burton 1953
Lochhead, A.G. and M.O. Burton. 1953. Canadian Journal of Botany **31**:7-22. **Type strain:** ATCC 13346
Description: Lochhead, A.G. and M.O. Burton. 1953. Ibid.

A. picolinophilus Tate and Ensign 1974
Tate, R.L. and J.C. Ensign. 1974. Canadian Journal of Microbiology **20**:691-694. **Type strain:** ATCC 27854
Description: Tate, R.L. and J.C. Ensign. 1974. Ibid.

A. polychromogenes Schippers-Lammertse, Muijsers and Klatser-Oedekerk 1963
Schippers-Lammertse, A.F., A.O. Muijsers and K.B. Klatser-Oedekerk. 1963. Antonie van Leeuwenhoek Journal of Microbiology and Serology **29**:1-15. **Type strain:** ATCC 15216

Description: Schippers-Lammertse, A.F., A.O. Muijsers and K.B. Klatser-Oedekerk. 1963. Ibid.

A. radiotolerans Yoshinaka, Yano and Yamaguchi 1973
Yoshinaka, T., K. Yano and H. Yamaguchi. 1973. Agricultural and Biological Chemistry **37**:2269-2275. **Type strain:** IAM 12072
Description: Yoshinaka, T., K. Yano and H. Yamaguchi. 1973. Ibid.

A. ramosus Jensen 1960
Jensen, V. 1960. Kongelige Veterinaer-og Landbohojskole Arsskrift pp. 123-132. **Type strain:** ATCC 13727
Description: Jensen, V. 1960. Ibid.

A. siderocapsulatus Dubinina and Zhdanov 1975
Dubinina, G. and A.V. Zhdanov. 1975. International Journal of Systematic Bacteriology **25**:340-350. **Type strain:** BKM-B N 1122
Description: Dubinin, G. and A.V. Zhdanov. 1975. Ibid.

A. simplex (Jensen 1934) Lochhead 1957
Jensen, H.L. 1934. Proceedings of the Linnean Society of New South Wales **59**: 19-61; Lochhead, A.G. 1957 in Breed, R.S., E.G.D. Murray and N.R. Smith (eds). Bergey's Manual of Determinative Bacteriology, 7th ed. The Williams and Wilkins Co., Baltimore. pp. 605-612. **Type strain:** ATCC 6946
Description: Bergey 8.

A. terregens Lochhead and Burton 1953
Lochhead, A.G. and M.O. Burton. 1953. Canadian Journal of Botany **31**:7-22. **Type strain:** ATCC 13345
Description: Bergey 8.

A. tumescens (Jensen 1934) Conn and Dimmick 1947
Jensen, H.L. 1934. Proceedings of the Linnean Society of New South Wales **59**: 19-61; Conn, H.J. and I. Dimmick. 1947. Journal of Bacteriology **54**:291-303. **Type strain:** ATCC 6947
Description: Bergey 8.

A. ureafaciens (Krebs and Eggleston 1939) Clark 1955
Krebs, H.A. and L.V. Eggleston. 1939. Enzymologia **7**:310-320; Clark, F.E. 1955. International Bulletin of Bacteriological Nomenclature and Taxonomy **5**:111-113. **Type strain:** ATCC 7562
Description: Bergey 8.

A. variabilis Muller 1961
Muller, G. 1961. Zentralblatt fur Bakteriologie, Parasitenkunde, Infektionskrankheiten und Hygiene. Abteilung II. **114**: 520-537. **Type strain:** ATCC 15753
Description: Muller, G. 1961. Ibid.

A. viscosus Gasdorf, Benedict, Cadmus, Anderson and Jackson 1965
Gasdorf, H.J., R.G. Benedict, M.C. Cadmus, R.F. Anderson and R.W. Jackson. 1965. Journal of Bacteriology **90**:147-150. **Type strain:** ATCC 19584
Description: Gasdorf, H.J., R.G. Benedict, M.C. Cadmus, R.F. Anderson and R.W. Jackson. 1965. Ibid.

Asticcacaulis Poindexter 1964
Poindexter, J.S. 1964. Bacteriological Reviews **28**:231-295. **Type species:** *A. excentricus* Poindexter 1964
Description: Bergey 8.

A. biprosthecium Pate, Porter and Jordan 1973
Pate, J.L., J.S. Porter and T.L. Jordan. 1973. Antonie van Leeuwenhoek Journal of Microbiology and Serology **39**:569-583. **Type strain:** ATCC 27554
Description: Pate, J.L., J.S. Porter and T.L. Jordan. 1973. Ibid.

A. excentricus Poindexter 1964
Poindexter, J.S. 1964. Bacteriological Reviews **28**:231-295. **Type strain:** ATCC 15261
Description: Poindexter, J.S. 1964. Ibid.

Azomonas Winogradsky 1938
Winogradsky, S. 1938. Annales de l'Institut Pasteur (Paris) **60**:351-400. **Type species:** *A. agilis* (Beijerinck 1901) Winogradsky 1938
Description: Bergey 8.

A. agilis (Beijerinck 1901) Winogradsky 1938
Beijerinck, M.W. 1901. Zentralblatt fur Bakteriologie, Parasitenkunde, Infektionskrankheiten und Hygiene. Abteilung II. **7**:561-582; Winogradsky, S. 1938. Annales de l'Institut Pasteur (Paris) **60**:351-400. **Type strain:** ATCC 7494
Description: Thompson, J.P. and V.B.D. Skerman. 1979. Azotobacteraceae - the taxonomy and ecology of the aerobic nitrogen-fixing bacteria. Academic Press, London. pp. 290-294.

A. insignis (Derx 1951) Jensen 1955
Derx, H.G. 1951. Proceedings Koninklijke Nederlandse Akademie van Wetenschappen Series C **54**:342-350; Jensen, V. 1955. Botanisk Tidsskrift **52**:143-157. **Type strain:** UQM 1966
Description: Thompson, J.P. and V.B.D. Skerman. 1979. Azotobacteraceae - the taxonomy and ecology of the aerobic nitrogen-fixing bacteria. Academic Press, London. pp. 294-297.

Azospirillum Tarrand, Krieg and Dobereiner 1979
Tarrand, J.J., N.R. Krieg and J. Dobereiner. 1979. International Journal of Systematic Bacteriology **29**:79-80; Effective publication Tarrand, J.J., N.R. Krieg and J. Dobereiner. 1978. Canadian Journal of Microbiology **24**:967-980. **Type species:** *A. lipoferum* (Beijerinck 1925) Tarrand, Krieg and Dobereiner 1979
Description: Tarrand, J.J., N.R. Krieg and J. Dobereiner. 1978. Ibid.

A. brasilense Tarrand, Krieg and Dobereiner 1979
Tarrand, J.J., N.R. Krieg and J. Dobereiner. 1979. International Journal of Systematic Bacteriology **29**:79-80; Effective publication Tarrand, J.J., N.R. Krieg and J. Dobereiner. 1978. Canadian Journal of Microbiology **24**:967-980. **Type strain:** ATCC 29145
Description: Tarrand, J.J., N.R. Krieg and J. Dobereiner. 1978. Ibid.

A. lipoferum (Beijerinck 1925) Tarrand, Krieg and Dobereiner 1979
Beijerinck, M.W. 1925. Zentralblatt fur Bakteriologie, Parasitenkunde, Infektionskrankheiten und Hygiene. Abteilung II. **63**:353-359; Tarrand, J.J., N.R. Krieg and J. Dobereiner. 1979. International Journal of Systematic Bacteriology **29**:79-80; Effective publication Tarrand, J.J., N.R. Krieg and J. Dobereiner. 1978. Canadian Journal of Microbiology **24**:967-980. **Type strain:** ATCC 29707
Description: Tarrand, J.J., N.R. Krieg and J. Dobereiner. 1978. Ibid.

Azotobacter Beijerinck 1901
Beijerinck, M.W. 1901. Zentralblatt fur Bakteriologie, Parasitenkunde, Infektionskrankheiten und Hygiene. Abteilung II. **7**:561-582. **Type species:** *A. chroococcum* Beijerinck 1901
Description: Bergey 8.

A. beijerinckii Lipman 1904
Lipman, J.G. 1904. Report of the New Jersey Agricultural Experiment Station **25**:237-289. **Type strain:** ATCC 19360
Description: Thompson, J.P. and V.B.D. Skerman. 1979. Azotobacteraceae - the taxonomy and ecology of the aerobic nitrogen-fixing bacteria. Academic Press, London. pp. 277-281.

A. chroococcum Beijerinck 1901
Beijerinck, M.W. 1901. Zentralblatt fur Bakteriologie, Parasitenkunde, Infektionskrankheiten und Hygiene. Abteilung II. **7**:561-582. **Type strain:** ATCC 9043
Description: Thompson, J.P. and V.B.D. Skerman 1979. Azotobacteraceae - the taxonomy and the ecology of the aerobic nitrogen-fixing bacteria. Academic Press, London. pp. 263-266.

A. macrocytogenes Jensen 1955
Jensen, H. 1955. Acta Agriculturae Scandinavica **5**:280-294. **Type strain:** ATCC 12335
Description: Thompson, J.P. and V.B.D. Skerman. 1979. Azotobacteraceae - the taxonomy and ecology of the aerobic nitrogen-fixing bacteria. Academic Press, London. pp. 306-309. (as Azomonotrichon macrocytogenes)

A. nigricans Krassilnikov 1949
Krassilnikov, N.A. 1949. Guide to the bacteria and actinomycetes. Akademiya Nauk SSSR, Moscow. pp. 1-830. **Type strain:** UQM 1967
Description: Thompson, J.P. and V.B.D. Skerman. 1979. Azotobacteraceae - taxonomy and ecology of the aerobic nitrogen-fixing bacteria. Academic Press, London. pp. 267-271.

A. paspali Dobereiner 1966
Dobereiner, J. 1966. Pesquisas Agropecuar Brasil **1**:357-365. **Type strain:** ATCC 23833
Description: Thompson, J.P. and V.B.D. Skerman. 1979. Azotobacteraceae - taxonomy and ecology of the aerobic nitrogen-fixing bacteria. Academic Press, London. pp. 306-309. (as Azorhizophilus paspali)

A. vinelandii Lipman 1903

Lipman, J.G. 1903. Report of the New Jersey Agricultural Experiment Station **24**:217-285. **Type strain:** ATCC 478
Description: Thompson, J.P. and V.B.D. Skerman. 1979. Azotobacteraceae - taxonomy and ecology of the aerobic nitrogen-fixing bacteria. Academic Press, London. pp. 283-287.

Bacillus Cohn 1872
Cohn, F. 1872. Beitrage zur Biologie der Pflanzen **1**:127-224. **Type species:** *B. subtilis* (Ehrenberg 1835) Cohn 1872
Description: Bergey 8.

B. acidocaldarius Darland and Brock 1971
Darland, G. and T.D. Brock. 1971. Journal of General Microbiology **67**:9-15. **Type strain:** ATCC 27009
Description: Bergey 8.

B. alcalophilus Vedder 1934
Vedder, A. 1934. Antonie van Leeuwenhoek Journal of Microbiology and Serology **1**:141-147. **Type strain:** ATCC 27647
Description: Boyer, E.W., M.B. Ingle and G.D. Mercer. 1973. International Journal of Systematic Bacteriology **23**:238-242.

B. alvei Cheshire and Cheyne 1885
Cheshire, F.R. and W.W. Cheyne. 1885. Journal of the Royal Microscopic Society, Series II **5**:581-601. **Type strain:** ATCC 6344
Description: Bergey 8.

B. anthracis Cohn 1872
Cohn, F. 1872. Beitrage zur Biologie der Pflanzen **1**:127-224. **Type strain:** ATCC 14578
Description: Bergey 8.

B. badius Batchelor 1919
Batchelor, M.D. 1919. Journal of Bacteriology **4**:23-34. **Type strain:** ATCC 14574
Description: Bergey 8.

B. brevis Migula 1900
Migula, W. 1900. System der Bakterien, Vol. 2. Gustav Fischer, Jena. **Type strain:** ATCC 8246
Description: Bergey 8.

B. cereus Frankland and Frankland 1887
Frankland, G.C. and P.F. Frankland. 1887. Royal Society London, Philosophical Transactions, Series B, Biological Sciences **178**:257-287. **Type strain:** ATCC 14579
Description: Bergey 8.

B. circulans Jordan 1890
Jordan, E.O. 1890 in Sedgewick, W.T. in The Report of the Massachusetts Board of Health. pp. 821-844. **Type strain:** ATCC 4513
Description: Bergey 8.

B. coagulans Hammer 1915
Hammer, B.W. 1915. Iowa Agricultural Experimental Station Research Bulletin **19**:119-131. **Type strain:** ATCC 7050
Description: Bergey 8.

B. fastidiosus den Dooren de Jong 1929
den Dooren de Jong, L.E. 1929. Zentralblatt fur Bakteriologie, Parasitenkunde, Infektionskrankheiten und Hygiene. Abteilung II. **79**:344-353. **Type strain:** Delft LMD 29-14
Description: Bongaerts, G.P.A. and G.D. Vogels. 1976. Journal of Bacteriology **125**:689-697.

B. firmus Bredemann and Werner 1933
Bredemann, G. and W. Werner. 1933 in Werner, W. Zentralblatt fur Bakteriologie, Parasitenkunde, Infektionskrankheiten und Hygiene. Abteilung II. **87**: 446-475. **Type strain:** ATCC 14575
Description: Gordon, R.E., J.L. Hyde and J.A. Moore. 1977. International Journal of Systematic Bacteriology **27**: 256-262.

B. globisporus Larkin and Stokes 1967
Larkin, J.M. and J.L. Stokes. 1967. Journal of Bacteriology **94**:889-895. **Type strain:** ATCC 23301
Description: Bergey 8.

B. globisporus subsp. *globisporus* Larkin and Stokes 1967
Larkin, J.M. and J.L. Stokes. 1967. Journal of Bacteriology **94**:889-895. **Type strain:** ATCC 23301
Description: Bergey 8.

B. globisporus subsp. *marinus* Ruger and Richter 1979
Ruger, H.-J. and G. Richter. 1979. International Journal of Systematic Bacteriology **29**:196-203. **Type strain:** ATCC 29841
Description: Ruger, H.-J. and G. Richter. 1979. Ibid.

B. insolitus Larkin and Stokes 1967
Larkin, J.M. and J.L. Stokes. 1967. Journal of Bacteriology **94**:889-895. **Type strain:** ATCC 23299
Description: Larkin, J.M. and J.L.

Stokes. 1967. Ibid.

B. larvae White 1906
White, G.F. 1906. United States Department of Agriculture, Bureau of Entomology, Technical Series No. 14. **Type strain:** ATCC 9545
Description: Bergey 8.

B. laterosporus Laubach 1916
Laubach, C.A. 1916. Journal of Bacteriology **1**:505-512. **Type strain:** ATCC 64
Description: Bergey 8.

B. lentimorbus Dutky 1940
Dutky, S.R. 1940. Journal of Agricultural Research **61**:57-68. **Type strain:** ATCC 14707
Description: Bergey 8.

B. lentus Gibson 1935
Gibson, T. 1935. Zentralblatt fur Bakteriologie, Parasitenkunde, Infektionskrankheiten und Hygiene. Abteilung II. **92**: 364-380. **Type strain:** ATCC 10840
Description: Gordon, R.E., J.L. Hyde and J.A. Moore. 1977. International Journal of Systematic Bacteriology **27**: 256-262.

B. licheniformis (Weigmann 1898) Chester 1901
Weigmann, H. 1898. Zentralblatt fur Bakteriologie, Parasitenkunde, Infektionskrankheiten und Hygiene. Abteilung II. **4**:820-834; Chester, F.D. 1901. A Manual of Determinative Bacteriology. The Macmillan Co., New York, pp. 1-401. **Type strain:** ATCC 14580
Description: Bergey 8.

B. macerans Schardinger 1905
Schardinger, F. 1905. Zentralblatt fur Bakteriologie, Parasitenkunde, Infektionskrankheiten und Hygiene. Abteilung II. **14**:772-781. **Type strain:** ATCC 8244
Description: Bergey 8.

B. macquariensis Marshall and Ohye 1966
Marshall, B.J. and D.F. Ohye. 1966. Journal of General Microbiology **44**:41-46. **Type strain:** ATCC 23464
Description: Marshall, B.J. and D.F. Ohye. 1966. Ibid.

B. megaterium de Bary 1884
de Bary, A. 1884. Vergleichende Morphologie und Biologie der Pilze, Mycetozoen und Bacterien. Wilhelm Engelmann, Leipzig. **Type strain:** ATCC 14581
Description: Bergey 8.

B. mycoides Flugge 1886
Flugge, C. 1886. Die Mikroorganismen. F.C.W. Vogel, Leipzig. **Type strain:** ATCC 6462
Description: Bergey 8.

B. pantothenticus Proom and Knight 1950
Proom, H. and B.C.J.G. Knight. 1950. Journal of General Microbiology **4**:539-541. **Type strain:** ATCC 14576
Description: Bergey 8.

B. pasteurii (Miquel 1889) Chester 1898
Miquel, P. 1889. Annales Micrographie **1**: 506-519; Chester, F.D. 1898. Delaware Agricultural Experimental Station Annual Report **10**:47-137. **Type strain:** ATCC 11859
Description: Bergey 8.

B. polymyxa (Prazmowski 1880) Mace 1889
Prazmowski, A. 1880. Inaugural Dissertation. Hugo Voigt, Leipzig. pp. 1-58; Mace, E. 1889. Traite Pratique de Bacteriologie, 1st ed. J.-B. Balliere. Paris. pp. 1-711. **Type strain:** ATCC 842
Description: Bergey 8.

B. popilliae Dutky 1940
Dutky, S.R. 1940. Journal of Agricultural Research **61**:57-68. **Type strain:** ATCC 14706
Description: Bergey 8.

B. pumilus Meyer and Gottheil 1901
Meyer, A. and O. Gottheil. 1901 in Gottheil, O. Zentralblatt fur Bakteriologie, Parasitenkunde, Infektionskrankheiten und Hygiene. Abteilung II. **7**:680-691. **Type strain:** ATCC 7061
Description: Bergey 8.

B. sphaericus Meyer and Neide 1904
Meyer, A. and E. Neide. 1904 in Neide, E. Zentralblatt fur Bakteriologie, Parasitenkunde, Infektionskrankheiten und Hygiene. Abteilung II. **12**:337-352. **Type strain:** ATCC 14577
Description: Bergey 8.

B. stearothermophilus Donk 1920
Donk, P.J. 1920. Journal of Bacteriology **5**:373-374. **Type strain:** ATCC 12980
Description: Bergey 8.

B. subtilis (Ehrenberg 1835) Cohn 1872
Ehrenberg, C.G. 1835. Physikalische Abhandlungen der Koeniglichen Akademie der Wissenschaften zu Berlin aus den Jahren 1833-1835, pp. 145-336; Cohn, F. 1872. Beitrage zur Biologie der Pflanzen 1875 **1**:127-224. **Type strain:** ATCC 6051

Description: Bergey 8.

B. thuringiensis Berliner 1915
Berliner, E. 1915. Zeitschrift fur angewandte Entomologie Berlin **2**:29-56. **Type strain:** ATCC 10792
Description: Bergey 8.

Bacterionema Gilmour, Howell and Bibby 1961
Gilmour, M.N., A. Howell Jr. and B.G. Bibby. 1961. Bacteriological Reviews **25**: 131-141. **Type species:** *B. matruchotii* (Mendel 1919) Gilmour, Howell and Bibby 1961
Description: Bergey 8.

B. matruchotii (Mendel 1919) Gilmour, Howell and Bibby 1961
Mendel, J. 1919. Comptes Rendus des Seances de la Societe de Biologie (Filiales) **82**:583-586; Gilmour, M.N., A. Howell Jr. and B.G. Bibby. 1961. Bacteriological Reviews **25**:131-141. **Type strain:** ATCC 14266
Description: Bergey 8.

Bacteroides Castellani and Chalmers 1919
Castellani, A. and A.J. Chalmers. 1919. Manual of tropical medicine. 3rd ed. Williams, Wood and Co., New York. pp. 959-960. **Type species:** *B. fragilis* (Veillon and Zuber 1898) Castellani and Chalmers 1919
Description: Bergey 8.

B. amylophilus Hamlin and Hungate 1956
Hamlin, L.J. and R.E. Hungate. 1956. Journal of Bacteriology **72**:548-554. **Type strain:** ATCC 29744
Description: Cato, E.P., W.E.C. Moore and M.P. Bryant. 1978. International Journal of Systematic Bacteriology **28**: 491-495.

B. asaccharolyticus (Holdeman and Moore 1970) Finegold and Barnes 1977
Holdeman, L.V. and W.E.C. Moore. 1970 in Cato, E.P., C.S. Cummins, L.V. Holdeman, J.L. Johnson, W.E.C. Moore, R.M. Smibert and L.DS. Smith (eds). Outline of clinical methods in anaerobic bacteriology, 2nd revision. Virginia Polytechnic Institute, Anaerobe Laboratory, Blacksburg, Virginia. pp. 33-44; Finegold, S.M. and E.M. Barnes. 1977. International Journal of Systematic Bacteriology **27**: 388-391. **Type strain:** ATCC 25260
Description: Finegold, S.M. and E.M. Barnes. 1977. Ibid.

B. bivius Holdeman and Johnson 1977
Holdeman, L.V. and J.L. Johnson. 1977. International Journal of Systematic Bacteriology **27**:337-345. **Type strain:** ATCC 29303
Description: Holdeman, L.V. and J.L. Johnson. 1977. Ibid.

B. capillosus (Tissier 1908) Kelly 1957
Tissier, H. 1908. Annales de l'Institut Pasteur (Paris) **22**:189-208; Kelly, C.D. 1957 in Breed, R.S., E.G.D. Murray and N.R. Smith (eds). Bergey's Manual of Determinative Bacteriology, 7th ed. The Williams and Wilkins Co., Baltimore. pp. 424-436. **Type strain:** ATCC 29799
Description: Holdeman, L.V., E.P. Cato and W.E.C. Moore (eds). 1977. Anaerobe Laboratory Manual, 4th ed. Virginia Polytechnic Institute and State University, Blacksburg, Virginia. pp. 1-156.

B. coagulans Eggerth and Gagnon 1933
Eggerth, A.H. and B.H. Gagnon. 1933. Journal of Bacteriology **25**:389-413. **Type strain:** ATCC 29798
Description: Holdeman, L.V., E.P. Cato and W.E.C. Moore (eds). 1977. Anaerobe Laboratory Manual, 4th ed. Virginia Polytechnic Institute and State University, Blacksburg, Virginia. pp. 1-156.

B. disiens Holdeman and Johnson 1977
Holdeman, L.V. and J.L. Johnson. 1977. International Journal of Systematic Bacteriology **27**:337-345. **Type strain:** ATCC 29426
Description: Holdeman, L.V. and J.L. Johnson. 1977. Ibid.

B. distasonis Eggerth and Gagnon 1933
Eggerth, A.H. and B.H. Gagnon. 1933. Journal of Bacteriology **25**:389-413. **Type strain:** ATCC 8503
Description: Holdeman, L.V., E.P. Cato and W.E.C. Moore (eds). 1977. Anaerobe Laboratory Manual, 4th ed. Virginia Polytechnic Institute and State University, Blacksburg, Virginia. pp. 1-156.

B. eggerthii Holdeman and Moore 1974
Holdeman, L.V. and W.E.C. Moore. 1974. International Journal of Systematic Bacteriology **24**:260-277. **Type strain:** ATCC 27754
Description: Holdeman, L.V., E.P. Cato and W.E.C. Moore (eds). 1977. Anaerobe

Laboratory Manual, 4th ed. Virginia Polytechnic Institute and State University, Blacksburg, Virginia. pp. 1-156.

B. fragilis (Veillon and Zuber 1898) Castellani and Chalmers 1919
Veillon, A. and A. Zuber. 1898. Archives de Medecine Experimentale et d'Anatomie Pathologique **10**:517-545; Castellani, A. and A.J. Chalmers. 1919. Manual of tropical medicine. 3rd ed. Williams, Wood and Co., New York. **Type strain:** ATCC 25285
Description: Holdeman, L.V., E.P. Cato and W.E.C. Moore (eds). 1977. Anaerobe Laboratory Manual, 4th ed. Virginia Polytechnic Institute and State University, Blacksburg, Virginia. pp. 1-156.

B. furcosus (Veillon and Zuber 1898) Hauduroy, Ehringer, Urbain, Guillot and Magrou 1937
Veillon, A. and A. Zuber. 1898. Archives de Medecine Experimentale et d'Anatomie Pathologique **10**:517-545; Hauduroy, P., G. Ehringer, A. Urbain, G. Guillot and J. Magrou. 1937. Dictionnaire des bacteries pathogenes. Masson et Cie, Paris. **Type strain:** ATCC 25662
Description: Holdeman, L.V., E.P. Cato and W.E.C. Moore (eds). 1977. Anaerobe Laboratory Manual, 4th ed. Virginia Polytechnic Institute and State University, Blacksburg, Virginia. pp. 1-156.

B. hypermegas Harrison and Hansen 1963
Harrison, Jr., A.P. and P.A. Hansen. 1963. Antonie van Leeuwenhoek Journal of Microbiology and Serology **29**:22-28. **Type strain:** ATCC 25560
Description: Holdeman, L.V., E.P. Cato and W.E.C. Moore (eds). 1977. Anaerobe Laboratory Manual, 4th ed. Virginia Polytechnic Institute and State University, Blacksburg, Virginia. pp. 1-156.

B. melaninogenicus (Oliver and Wherry 1921) Roy and Kelly 1939
Oliver, W.W. and W.B. Wherry. 1921. Journal of Infectious Diseases **28**:341-344; Roy, T.E. and C.D. Kelly. 1939 in Bergey, D.H., R.S. Breed, E.G.D. Murray and A.P. Hitchens (eds). Bergey's Manual of Determinative Bacteriology, 5th ed. The Williams and Wilkins Co., Baltimore. pp. 556-558. **Type strain:** ATCC 25845
Description: Holdeman, L.V., E.P. Cato and W.E.C. Moore (eds). 1977. Anaerobe Laboratory Manual, 4th ed. Virginia Polytechnic Institute and State University, Blacksburg, Virginia. pp. 1-156.

B. melaninogenicus subsp. *intermedius* Holdeman and Moore 1970
Holdeman, L.V. and W.E.C. Moore. 1970 in Cato, E.P., C.S. Cummins, L.V. Holdeman, J.L. Johnson, W.E.C. Moore, R.M. Smibert and L.DS. Smith (eds). Outline of clinical methods in anaerobic bacteriology, 2nd revision. Virginia Polytechnic Institute Anaerobe Laboratory, Blacksburg, Virginia. pp. 33-44. **Type strain:** ATCC 25611
Description: Holdeman, L.V., E.P. Cato and W.E.C. Moore (eds). 1977. Anaerobe Laboratory Manual, 4th ed. Virginia Polytechnic Institute and State University, Blacksburg, Virginia. pp. 1-156.

B. melaninogenicus subsp. *melaninogenicus* (Oliver and Wherry 1921) Holdeman and Moore 1970
Oliver, W.W. and W.B. Wherry. 1921. Journal of Infectious Diseases **28**:341-344; Holdeman, L.V. and W.E.C. Moore. 1970 in Cato, E.P., C.S. Cummins, L.V. Holdeman, J.L. Johnson, W.E.C. Moore, R.M. Smibert and L.DS. Smith (eds). Outline of clinical methods in anaerobic bacteriology, 2nd revision. Virginia Polytechnic Institute Anaerobe Laboratory, Blacksburg, Virginia. pp. 33-44. **Type strain:** ATCC 25845
Description: Holdeman, L.V., E.P. Cato and W.E.C. Moore (eds). 1977. Anaerobe Laboratory Manual, 4th ed. Virginia Polytechnic Institute and State University, Blacksburg, Virginia. pp. 1-156.

B. microfusus Kaneuchi and Mitsuoka 1978
Kaneuchi, C. and T. Mitsuoka. 1978. International Journal of Systematic Bacteriology **28**:478-481. **Type strain:** ATCC 29728
Description: Kaneuchi, C. and T. Mitsuoka. 1978. Ibid.

B. multiacidus Mitsuoka, Terada, Watanabe and Uchida 1974
Mitsuoka, T., A. Terada, K. Watanabe and K. Uchida. 1974. International Journal of Systematic Bacteriology **24**:35-41. **Type strain:** ATCC 27723
Description: Holdeman, L.V., E.P. Cato

and W.E.C. Moore (eds). 1977. Anaerobe Laboratory Manual, 4th ed. Virginia Polytechnic Institute and State University, Blacksburg, Virginia. pp. 1-156.

B. nodosus (Beveridge 1941) Mraz 1963
Beveridge, W.I.B. 1941. Council of Scientific Industrial Research Australian Bulletin **140**:1-56; Mraz, O. 1963 in Mraz, O., J. Tesarcik and R. Varejka (eds). Nomina und Synonyma der pathogenen und saprophytaren Mikroben, isoliert aus den wirtschaftlich oder epidemiologisch bedeutenden Wirbeltieren und Lebensmitteen tierischer Herkunft. VEB Gustav Fischer Verlag, Jena. pp. 53-334. **Type strain:** ATCC 25549
Description: Holdeman, L.V., E.P. Cato and W.E.C. Moore (eds). 1977. Anaerobe Laboratory Manual, 4th ed. Virginia Polytechnic Institute and State University, Blacksburg, Virginia. pp. 1-156.

B. ochraceus Prevot, Joubert, Tardieux and de Cadore 1956
Prevot, A.R., P. Tardieux, L. Joubert and F. de Cadore. 1956. Annales de l'Institut Pasteur (Paris) **91**:787-798. **Type strain:** ATCC 27872
Description: Bergey 8.

B. oralis Loesche, Socransky and Gibbons 1964
Loesche, W.J., S.S. Socransky and R.J. Gibbons. 1964. Journal of Bacteriology **88**:1329-1337. **Type strain:** VPI D27B-24
Description: Holdeman, L.V., E.P. Cato and W.E.C. Moore (eds). 1977. Anaerobe Laboratory Manual, 4th ed. Virginia Polytechnic Institute and State University, Blacksburg, Virginia. pp. 1-156.

B. ovatus Eggerth and Gagnon 1933
Eggerth, A.H. and B.H. Gagnon. 1933. Journal of Bacteriology **25**:389-413. **Type strain:** ATCC 8483
Description: Holdeman, L.V., E.P. Cato and W.E.C. Moore (eds). 1977. Anaerobe Laboratory Manual, 4th ed. Virginia Polytechnic Institute and State University, Blacksburg, Virginia. pp. 1-156.

B. pneumosintes (Olitsky and Gates 1921) Holdeman and Moore 1970
Olitsky, P.K. and F.L. Gates. 1921. Journal of Experimental Medicine **33**:713-729; Holdeman, L.V. and W.E.C. Moore. 1970 in Cato, E.P., C.S. Cummins, L.V. Holdeman, J.L. Johnson, W.E.C. Moore, R.M. Smibert and L.DS. Smith (eds). Outline of clinical methods in anaerobic bacteriology. 2nd revision. Virginia Polytechnic Institute Anaerobe Laboratory, Blacksburg, Virginia. **Type strain:** ATCC 33048
Description: Holdeman, L.V., E.P. Cato and W.E.C. Moore (eds). 1977. Anaerobe Laboratory Manual, 4th ed. Virginia Polytechnic Institute and State University, Blacksburg, Virginia. pp. 1-156.

B. praeacutus (Tissier 1908) Holdeman and Moore 1970
Tissier, H. 1908. Annales de l'Institut Pasteur (Paris) **22**:189-208; Holdeman, L.V. and W.E.C. Moore. 1970 in Cato, E.P., C.S. Cummins, L.V. Holdeman, J.L. Johnson, W.E.C. Moore, R.M. Smibert and L.DS. Smith (eds). Outline of clinical methods in anaerobic bacteriology, 2nd revision. Virginia Polytechnic Institute Anaerobe Laboratory, Blacksburg, Virginia. **Type strain:** ATCC 25539
Description: Holdeman, L.V., E.P. Cato and W.E.C. Moore (eds). 1977. Anaerobe Laboratory Manual, 4th ed. Virginia Polytechnic Institute and State University, Blacksburg, Virginia. pp. 1-156.

B. putredinis (Weinberg, Nativelle and Prevot 1937) Kelly 1957
Weinberg, M., R. Nativelle and A.R. Prevot. 1937. Les Microbes Anaerobies. Masson et Cie, Paris. pp. 1-1186; Kelly, C.D. 1957 in Breed, R.S., E.G.D. Murray and N.R. Smith (eds). Bergey's Manual of Determinative Bacteriology, 7th ed. The Williams and Wilkins Co., Baltimore. pp. 424-436. **Type strain:** ATCC 29800
Description: Holdeman, L.V., E.P. Cato and W.E.C. Moore (eds). 1977. Anaerobe Laboratory Manual, 4th ed. Virginia Polytechnic Institute and State University, Blacksburg, Virginia. pp. 1-156.

B. ruminicola Bryant, Small, Bouma and Chu 1958
Bryant, M.P., N. Small, C. Bouma and H. Chu. 1958. Journal of Bacteriology **76**:15-23. **Type strain:** ATCC 19189
Description: Holdeman, L.V., E.P. Cato and W.E.C. Moore (eds). 1977. Anaerobe Laboratory Manual, 4th ed. Virginia Polytechnic Institute and State University, Blacksburg, Virginia. pp. 1-156.

B. ruminicola subsp. *brevis* Bryant, Small, Bouma and Chu 1958
Bryant, M.P., N. Small, C. Bouma and H. Chu. 1958. Journal of Bacteriology **76**:15-23. **Type strain:** ATCC 19188
Description: Holdeman, L.V., E.P. Cato and W.E.C. Moore (eds). 1977. Anaerobe Laboratory Manual, 4th ed. Virginia Polytechnic Institute and State University, Blacksburg, Virginia. pp. 1-156.

B. ruminicola subsp. *ruminicola* Bryant, Small, Bouma and Chu 1958
Bryant, M.P., N. Small, C. Bouma and H. Chu. 1958. Journal of Bacteriology **76**:15-23. **Type strain:** ATCC 19189
Description: Holdeman, L.V., E.P. Cato and W.E.C. Moore (eds). 1977. Anaerobe Laboratory Manual, 4th ed. Virginia Polytechnic Institute and State University, Blacksburg, Virginia. pp. 1-156.

B. splanchnicus Werner, Rintelen and Kunstek-Santos 1975
Werner, H., G. Rintelen and H. Kunstek-Santos. 1975. Zentralblatt fur Bakteriologie, Parasitenkunde, Infektionskrankheiten und Hygiene. Abteilung I. **231**: 133-144. **Type strain:** ATCC 29572
Description: Holdeman, L.V., E.P. Cato and W.E.C. Moore (eds). 1977. Anaerobe Laboratory Manual, 4th ed. Virginia Polytechnic Institute and State University, Blacksburg, Virginia. pp. 1-156.

B. succinogenes Hungate 1950
Hungate, R.E. 1950. Bacteriological Reviews **14**:1-49. **Type strain:** ATCC 19169
Description: Cato, E.P., W.E.C. Moore and M.P. Bryant. 1978. International Journal of Systematic Bacteriology **28**: 491-495.

B. termitidis (Sebald 1962) Holdeman and Moore 1970
Sebald, M. 1962. Etude sur les bacteries anaerobies gram-negatives asporulees. Theses de L'universite Paris, Imprimerie Barneoud S. A. Laval, France. pp. 1-171; Holdeman, L.V. and W.E.C. Moore. 1970 in Cato, E.P., C.S. Cummins, L.V. Holdeman, J.L. Johnson, W.E.C. Moore, R.M. Smibert and L.DS. Smith (eds). Outline of clinical methods in anaerobic bacteriology, 2nd revision. Virginia Polytechnic Institute Anaerobe Laboratory, Blacksburg, Virginia. **Type strain:** NCTC 11300
Description: Sebald, M. 1962. Ibid.

B. thetaiotaomicron (Distaso 1912) Castellani and Chalmers 1919
Distaso, A. 1912. Zentralblatt fur Bakteriologie, Parasitenkunde, Infektionskrankheiten und Hygiene. Abteilung I. **62**:433-468; Castellani, A. and A.J. Chalmers. 1919. Manual of Tropical Medicine, 3rd ed. Williams, Wood and Co., New York. **Type strain:** ATCC 29148
Description: Holdeman, L.V., E.P. Cato and W.E.C. Moore (eds). 1977. Anaerobe Laboratory Manual, 4th ed. Virginia Polytechnic Institute and State University, Blacksburg, Virginia. pp. 1-156.

B. uniformis Eggerth and Gagnon 1933
Eggerth, A.H. and B.H. Gagnon. 1933. Journal of Bacteriology **25**:389-413. **Type strain:** ATCC 8492
Description: Holdeman, L.V., E.P. Cato and W.E.C. Moore (eds). 1977. Anaerobe Laboratory Manual, 4th ed. Virginia Polytechnic Institute and State University, Blacksburg, Virginia. pp. 1-156.

B. ureolyticus Jackson and Goodman 1978
Jackson, F.L. and Y.E. Goodman. 1978. International Journal of Systematic Bacteriology **28**:197-200. **Type strain:** NCTC 10941
Description: Jackson, F.L. and Y.E. Goodman. 1978. Ibid.

B. vulgatus Eggerth and Gagnon 1933
Eggerth, A.H. and B.H. Gagnon. 1933. Journal of Bacteriology **25**:389-413. **Type strain:** ATCC 8482
Description: Holdeman, L.V., E.P. Cato and W.E.C. Moore (eds). 1977. Anaerobe Laboratory Manual, 4th ed. Virginia Polytechnic Institute and State University, Blacksburg, Virginia. pp. 1-156.

Bactoderma Tepper and Korshunova 1973
Tepper, S.Z. and G.F. Korshunova. 1973. Mikrobiologiya **42**:430-434. **Type species:** *B. alba* Tepper and Korshunova 1973
Description: Tepper, S.Z. and G.F. Korshunova. 1973. Ibid.

B. alba Tepper and Korshunova 1973
Tepper, S.Z. and G.F. Korshunova. 1973. Mikrobiologiya **42**:430-434. **Type strain:** no culture available.
Description: Tepper, S.Z. and G.F. Kor-

shunova. 1973. Ibid.

B. rosea (Winogradsky and Winogradsky 1933) Tepper and Korshunova 1973
Winogradsky, S.N. and H. Winogradsky. 1933. Annales de l'Institut Pasteur (Paris) **50**:350-434; Tepper, S.Z. and G.F. Korshunova. 1973. Mikrobiologiya **42**:430-434. **Type strain:** BKMW 1344
Description: Tepper, S.Z. and G.F. Korshunova. 1973. Ibid.

Bartonella Strong, Tyzzer and Sellards 1915
Strong, R.P., E.E. Tyzzer and A.W. Sellards. 1915. Journal of the American Medical Association **64**:806-808. **Type species:** *B. bacilliformis* (Strong, Tyzzer, Brues, Sellards and Gastiaburu 1913) Strong, Tyzzer and Sellards 1915
Description: Bergey 8.

B. bacilliformis (Strong, Tyzzer, Brues, Sellards and Gastiaburu 1913) Strong, Tyzzer and Sellards 1915
Strong, R.P., E.E. Tyzzer, C.T. Brues, A.W. Sellards and J.C. Gastiaburu. 1913. Journal of the American Medical Association **61**:1713-1716; Strong, R.P., E.E. Tyzzer and A.W. Sellards. 1915. Journal of the American Medical Association **64**: 806-808. **Type strain:** no culture available.
Description: Bergey 8.

Bdellovibrio Stolp and Starr 1963
Stolp, H.J. and M.P. Starr. 1963. Antonie van Leeuwenhoek Journal of Microbiology and Serology **29**:217-248. **Type species:** *B. bacteriovorus* Stolp and Starr 1963
Description: Bergey 8.

B. bacteriovorus Stolp and Starr 1963
Stolp, H.J. and M.P. Starr. 1963. Antonie van Leeuwenhoek Journal of Microbiology and Serology **29**:217-248. **Type strain:** NCIB 9529
Description: Stolp, H.J. and M.P. Starr. 1963. Ibid.

B. starrii Seidler, Mandel and Baptist 1972
Seidler, R.J., M. Mandel and J.N. Baptist. 1972. Journal of Bacteriology **109**: 209-217. **Type strain:** ATCC 15145
Description: Seidler, R.J., M. Mandel and J.N. Baptist. 1972. Ibid.

B. stolpii Seidler, Mandel and Baptist 1972
Seidler, R.J., M. Mandel and J.N. Baptist. 1972. Journal of Bacteriology **109**: 209-217. **Type strain:** DSM 50722
Description: Seidler, R.J., M. Mandel and J.N. Baptist. 1972. Ibid.

Beggiatoa Trevisan 1842
Trevisan, V. 1842. Prospetto della Flora Euganea. Coi Tipi Del Seminario, Padova. pp. 1-68. **Type species:** *B. alba* (Vaucher 1803) Trevisan 1845
Description: Bergey 8.

B. alba (Vaucher 1803) Trevisan 1845
Vaucher, J.P. 1803. Histoire des conferves d'eau douce, contenant leurs different modes de reproduction, et la description de leurs principales especes. J.J. Paschoud, Geneva. pp. 1-285; Trevisan, V. 1845. Nomenclator Algarum. Impr du seminaire, Padone. pp. 58-59. **Type strain:** LSU B18LD
Description: Bergey 8.

Beijerinckia Derx 1950
Derx, H.G. 1950. Proceedings Koninklijke Nederlandse Akademie van Wetenschappen Series C **53**:140-147. **Type species:** *B. indica* (Starkey and De 1939) Derx 1950
Description: Bergey 8.

B. derxii Tchan 1957
Tchan, Y.T. 1957. Proceedings of the Linnean Society New South Wales **82**:314-316. **Type strain:** UQM 1968
Description: Thompson, J.P. and V.B.D. Skerman. 1979. Azotobacteraceae - taxonomy and ecology of the aerobic nitrogen-fixing bacteria. Academic Press, London. pp. 347-352.

B. fluminensis Dobereiner and Ruschel 1958
Dobereiner, J. and A.P. Ruschel. 1958. Revista de Biologia **1**:261-272. **Type strain:** UQM 1685
Description: Thompson, J.P. and V.B.D. Skerman. 1979. Azotobacteraceae - taxonomy and ecology of the aerobic nitrogen-fixing bacteria. Academic Press, London. pp. 337-341.

B. indica (Starkey and De 1939) Derx 1950
Starkey, R.L. and P.K. De. 1939. Soil Science **47**:329-343; Derx, H.G. 1950. Proceedings Koninklijke Nederlandse Akademie van Wetenschappen Series C **53**: 140-147. **Type strain:** ATCC 9039

Description: Thompson, J.P. and V.B.D. Skerman. 1979. Azotobacteraceae - taxonomy and ecology of the aerobic nitrogen-fixing bacteria. Academic Press, London. pp. 328-332.

B. mobilis Derx 1950
Derx, H.G. 1950. Bogoriensis **1**:1-11. **Type strain:** UQM 1969
Description: Thompson, J.P. and V.B.D. Skerman. 1979. Azotobacteraceae - taxonomy and ecology of the aerobic nitrogen-fixing bacteria. Academic Press, London. pp. 333-336.

Beneckea Campbell 1957
Campbell, L.L. 1957 in Breed, R.S., E.G.D. Murray and N.R. Smith (eds). Bergey's Manual of Determinative Bacteriology, 7th ed. The Williams and Wilkins Co., Baltimore. pp. 328-332. **Type species:** *B. campbellii* Baumann, Baumann and Mandel 1971
Description: Campbell, L.L. 1957. Ibid.

B. alginolytica (Miyamoto, Nakamura and Takizawa 1961) Baumann, Baumann and Mandel 1971 (See also Vibrio alginolyticus)
Miyamoto, Y., K. Nakamura and K. Takizawa. 1961. Japanese Journal of Microbiology **5**:477-486; Baumann, P., L. Baumann and M. Mandel. 1971. Journal of Bacteriology **107**:268-294. **Type strain:** ATCC 17749
Description: Baumann, P., L. Baumann and M. Mandel. 1971. Ibid.

B. campbellii Baumann, Baumann and Mandel 1971
Baumann, P., L. Baumann and M. Mandel. 1971. Journal of Bacteriology **107**: 268-294. **Type strain:** ATCC 25920
Description: Baumann, P., L. Baumann and M. Mandel. 1971. Ibid.

B. harveyi (Johnson and Shunk 1936) Reichelt and Baumann 1973 (See also Lucibacterium harveyi)
Johnson, F.H. and I.V. Shunk. 1936. Journal of Bacteriology **31**:585-593; Reichelt, J.L. and P. Baumann. 1973. Archiv fur Mikrobiologie **94**:283-330. **Type strain:** ATCC 14126
Description: Reichelt, J.L. and P. Baumann. 1973. Ibid.

B. natriegens (Payne, Eagon and Williams 1961) Baumann, Baumann and Mandel 1971
Payne, W.J., R.G. Eagon and A.K. Williams. 1961. Antonie van Leeuwenhoek Journal of Microbiology and Serology **27**: 121-128; Baumann, P., L. Baumann and M. Mandel. 1971. Journal of Bacteriology **107**:268-294. **Type strain:** ATCC 14048
Description: Baumann, P., L. Baumann and M. Mandel. 1971. Ibid.

B. nigrapulchrituda Baumann, Baumann, Mandel and Allen 1971
Baumann, P., L. Baumann, M. Mandel and R.D. Allen. 1971. Journal of Bacteriology **108**:1380-1383. **Type strain:** ATCC 27043
Description: Baumann, P., L. Baumann, M. Mandel and R.D. Allen. 1971. Ibid.

B. parahaemolytica (Fujino, Okuno, Nakada, Aoyama, Fukai, Mukai and Ueho 1951) Baumann, Baumann and Mandel 1971 (See also Vibrio parahaemolyticus)
Fujino, T., Y. Okuno, D. Nakada, A. Aoyama, K. Fukai, T. Mukai and T. Ueho. 1951. Journal of the Japanese Association of Infectious Diseases **25**:11; Baumann, P., L. Baumann and M. Mandel. 1971. Journal of Bacteriology **107**: 268-294. **Type strain:** ATCC 17802
Description: Baumann, P., L. Baumann and J.L. Reichelt. 1973. Journal of Bacteriology **113**:1144-1155.

B. pelagia Baumann, Baumann and Mandel 1971
Baumann, P., L. Baumann and M. Mandel. 1971. Journal of Bacteriology **107**: 268-294. **Type strain:** ATCC 25916
Description: Baumann, P., L. Baumann and M. Mandel. 1971. Ibid.

B. splendida (Beijerinck 1900) Reichelt, Baumann and Baumann 1979
Beijerinck, M.W. 1900. Proceedings of the Academy of Sciences (Amsterdam) **3**: 352-365; Reichelt, J.L., P. Baumann and L. Baumann. 1979 in International Journal of Systematic Bacteriology **29**:79-80; Effective publication Reichelt, J.L., P. Baumann and L. Baumann. 1976. Archives of Microbiology **110**:101-120. **Type strain:** NCMB 1
Description: Reichelt, J.L., P. Baumann and L. Baumann. 1976. Ibid.

B. vulnifica Reichelt, Baumann and Baumann

1979
Reichelt, J.L., P. Baumann and L. Baumann. 1979 in International Journal of Systematic Bacteriology **29**:79-80; Effective publication Reichelt, J.L., P. Baumann and L. Baumann. 1976. Archives of Microbiology **110**:101-120. **Type strain:** ATCC 27562
Description: Reichelt, J.P., P. Baumann and L. Baumann. 1976. Ibid.

Bifidobacterium Orla-Jensen 1924
Orla-Jensen, S. 1924. Lait **4**:468-474. **Type species:** *B. bifidum* (Tissier 1900) Orla-Jensen 1924
Description: Bergey 8.

B. adolescentis Reuter 1963
Reuter, G. 1963. Zentralblatt fur Bakteriologie, Parasitenkunde, Infektionskrankheiten und Hygiene. Abteilung I. **191**: 486-507. **Type strain:** ATCC 15703 biotype a
Description: Reuter, G. 1971. International Journal of Systematic Bacteriology **21**: 273-275.

B. angulatum Scardovi and Crociani 1974
Scardovi, V. and F. Crociani. 1974. International Journal of Systematic Bacteriology **24**:6-20. **Type strain:** ATCC 27535
Description: Scardovi, V. and F. Crociani. 1974. Ibid.

B. animalis (Mitsuoka 1969) Scardovi and Trovatelli 1974
Mitsuoka, T. 1969. Zentralblatt fur Bakteriologie, Parasitenkunde, Infektionskrankheiten und Hygiene. Abteilung I. **210**:52-64; Scardovi, V. and L.D. Trovatelli. 1974. International Journal of Systematic Bacteriology **24**:21-28. **Type strain:** ATCC 25527
Description: Scardovi, V. and L.D. Trovatelli. 1974. Ibid.

B. asteroides Scardovi and Trovatelli 1969
Scardovi, V. and L.D. Trovatelli. 1969. Zentralblatt fur Bakteriologie, Parasitenkunde, Infektionskrankheiten und Hygiene. Abteilung II. **123**:64-88. **Type strain:** ATCC 25910
Description: Scardovi, V. and L.D. Trovatelli. 1969. Ibid.

B. bifidum (Tissier 1900) Orla-Jensen 1924
Tissier, H. 1900. Theses, Paris pp. 1-253; Orla-Jensen, S. 1924. Lait **4**:468-474. **Type strain:** Ti(Tissier) Anaerobe Collection, Institut Pasteur Lille
Description: Werner, H. and H.P.R. Seeliger. 1964. Pathologia et Microbiologia **27**:212-215.

B. boum Scardovi, Trovatelli, Biavati and Zani 1979
Scardovi, V., L.D. Trovarelli, B. Biavati and G. Zani. 1979. International Journal of Systematic Bacteriology **29**:291-311. **Type strain:** ATCC 27917
Description: Scardovi, V., L.D. Trovatelli, B. Biavati and G. Zani. 1979. Ibid.

B. breve Reuter 1963
Reuter, G. 1963. Zentralblatt fur Bakteriologie, Parasitenkunde, Infektionskrankheiten und Hygiene. Abteilung I. **191**: 486-507. **Type strain:** ATCC 15700
Description: Reuter, G. 1971. International Journal of Systematic Bacteriology **21**: 273-275.

B. catenulatum Scardovi and Crociani 1974
Scardovi, V. and F. Crociani. 1974. International Journal of Systematic Bacteriology **24**:6-20. **Type strain:** ATCC 27539
Description: Scardovi, V. and F. Crociani. 1974. Ibid.

B. choerinum Scardovi, Trovatelli, Biavati and Zani 1979
Scardovi, V., L.D. Trovatelli, B. Biavati and G. Zani. 1979. International Journal of Systematic Bacteriology **29**:291-311. **Type strain:** ATCC 27686
Description: Scardovi, V., L.D. Trovatelli, B. Biavati and G. Zani. 1979. Ibid.

B. cuniculi Scardovi, Trovatelli, Biavati and Zani 1979
Scardovi, V., L.D. Trovatelli, B. Biavati and G. Zani. 1979. International Journal of Systematic Bacteriology **29**:291-311. **Type strain:** ATCC 27916
Description: Scardovi, V., L.D. Trovatelli, B. Biavati and G. Zani. 1979. Ibid.

B. dentium Scardovi and Crociani 1974
Scardovi, V. and F. Crociani. 1974. International Journal of Systematic Bacteriology **24**:6-20. **Type strain:** ATCC 27534
Description: Scardovi, V. and F. Crociani. 1974. Ibid.

B. indicum Scardovi and Trovatelli 1969
Scardovi, V. and L.D. Trovatelli. 1969. Zentralblatt fur Bakteriologie, Parasiten-

kunde, Infektionskrankheiten und Hygiene. Abteilung II. **123**:64-88. **Type strain:** ATCC 25912
Description: Scardovi, V. and L.D. Trovatelli. 1969. Ibid.

B. infantis Reuter 1963
Reuter, G. 1963. Zentralblatt fur Bakteriologie, Parasitenkunde, Infektionskrankheiten und Hygiene. Abteilung I. **191**: 486-507. **Type strain:** ATCC 15697
Description: Reuter, G. 1971. International Journal of Systematic Bacteriology **21**: 273-275.

B. longum Reuter 1963
Reuter, G. 1963. Zentralblatt fur Bakteriologie, Parasitenkunde, Infektionskrankheiten und Hygiene. Abteilung I. **191**: 486-507. **Type strain:** ATCC 15707
Description: Reuter, G. 1971. International Journal of Systematic Bacteriology **21**: 273-275.

B. magnum Scardovi and Zani 1974
Scardovi, V. and G. Zani. 1974. International Journal of Systematic Bacteriology **24**:29-34. **Type strain:** ATCC 27540
Description: Scardovi, V. and G. Zani. 1974. Ibid.

B. pseudocatenulatum Scardovi, Trovatelli, Biavati and Zani 1979
Scardovi, V., L.D. Trovatelli, B. Biavati and G. Zani. 1979. International Journal of Systematic Bacteriology **29**:291-311. **Type strain:** ATCC 27919
Description: Scardovi, V., L.D. Trovatelli, B. Biavati and G. Zani. 1979. Ibid.

B. pseudolongum Mitsuoka 1969
Mitsuoka, T. 1969. Zentralblatt fur Bakteriologie, Parasitenkunde, Infektionskrankheiten und Hygiene. Abteilung I. **210**:52-64. **Type strain:** ATCC 25526
Description: Mitsuoka, T. 1969. Ibid.

B. pullorum Trovatelli, Crociani, Pedinotti and Scardovi 1974
Trovatelli, L.D., F. Crociani, M. Pedinotti and V. Scardovi. 1974. Archives of Microbiology **98**:187-198. **Type strain:** ATCC 27685
Description: Trovatelli, L.D., F. Crociani, M. Pedinotti and V. Scardovi. 1974. Ibid.

B. suis Matteuzzi, Crociani, Zani and Trovatelli 1971
Matteuzzi, D., F. Crociani, G. Zani and L.D. Trovatelli. 1971. Zeitschrift fur Allgemeine Mikrobiologie **11**:387-395. **Type strain:** ATCC 27533
Description: Matteuzzi, D., F. Crociani, G. Zani and L.D. Trovatelli. 1971. Ibid.

B. thermophilium Mitsuoka 1969
Mitsuoka, T. 1969. Zentralblatt fur Bakteriologie, Parasitenkunde, Infektionskrankheiten und Hygiene. Abteilung 1 **210**:52-64. **Type strain:** ATCC 25525
Description: Mitsuoka, T. 1969. Ibid.

Blastobacter Zavarzin 1961
Zavarzin, G.A. 1961. Mikrobiologiya **30**: 952-975 (English translation 774-791). **Type species:** *B. henricii* Zavarzin 1961
Description: Bergey 8.

B. henricii Zavarzin 1961
Zavarzin, G.A. 1961. Mikrobiologiya **30**: 952-975 (English translation 774-791). **Type strain:** no culture isolated.
Description: Bergey 8.

Blastococcus Ahrens and Moll 1970
Ahrens, R. and G. Moll. 1970. Archiv fur Mikrobiologie **70**:243-265. **Type species:** *B. aggregatus* Ahrens and Moll 1970
Description: Ahrens, R. and G. Moll. 1970. Ibid.

B. aggregatus Ahrens and Moll 1970
Ahrens, R. and G. Moll. 1970. Archiv fur Mikrobiologie **70**:243-265. **Type strain:** ATCC 25902
Description: Ahrens, R. and G. Moll. 1970. Ibid.

Blattabacterium Hollande and Favre 1931
Hollande, A.C. and R. Favre. 1931. Comptes Rendus des Seances de la Societe de Biologie (Paris) **107**:752-754. **Type species:** *B. cuenoti* (Mercier 1906) Hollande and Favre 1931
Description: Bergey 8.

B. cuenoti (Mercier 1906) Hollande and Favre 1931
Mercier, L. 1906. Comptes Rendus des Seances de la Societe de Biologie (Paris) **61**:682-684; Hollande, A.C. and R. Favre. 1931. Comptes Rendus des Seances de la Societe de Biologie (Paris) **107**:752-754. **Type strain:** no culture isolated.
Description: Brooks, M.A. 1970. Journal of Invertebrate Pathology **16**:249-258.

Bordetella Moreno-Lopez 1952
Moreno-Lopez, M. 1952. Microbiologia Espanola **5**:177-181. **Type species:** *B. pertussis* (Bergey, Harrison, Breed, Hammer and Huntoon 1923) Moreno-Lopez 1952
Description: Bergey 8.

B. bronchiseptica (Ferry 1912) Moreno-Lopez 1952
Ferry, N.S. 1912. Veterinary Journal **68**: 376-391; Moreno-Lopez, M. 1952. Microbiologia Espanola **5**:177-181. **Type strain:** ATCC 19395
Description: Bergey 8.

B. parapertussis (Eldering and Kendrick 1938) Moreno-Lopez 1952
Eldering, G. and P. Kendrick. 1938. Journal of Bacteriology **35**:561-572; Moreno-Lopez, M. 1952. Microbiologia Espanola **5**:177-181. **Type strain:** NCTC 5952
Description: Bergey 8.

B. pertussis (Bergey, Harrison, Breed, Hammer and Huntoon 1923) Moreno-Lopez 1952
Bergey, D.H., F.C. Harrison, R.S. Breed, B.W. Hammer and F.M. Huntoon. 1923. Bergey's Manual of Determinative Bacteriology, 1st ed. The Williams and Wilkins Co., Baltimore. pp. 1-442, Moreno-Lopez, M. 1952. Microbiologia Espanola **5**: 177-181. **Type strain:** ATCC 9797
Description: Bergey 8.

Borrelia Swellengrebel 1907
Swellengrebel, N.H. 1907. Annales de l'Institut Pasteur (Paris) **21**:562-586. **Type species:** *B. anserina* (Sakharoff 1891) Bergey, Harrison, Breed, Hammer and Huntoon 1925
Description: Bergey 8.

B. anserina (Sakharoff 1891) Bergey, Harrison, Breed, Hammer and Huntoon 1925
Sakharoff, M.N. 1891. Annales de l'Institut Pasteur (Paris) **5**:564-566; Bergey, D.H., F.C. Harrison, R.S. Breed, B.W. Hammer and F.M. Huntoon (eds). 1925. Bergey's Manual of Determinative Bacteriology, 2nd ed. The Williams and Wilkins Co., Baltimore. pp. 1-462. **Type strain:** no culture available.
Description: Bergey 8.

B. brasiliensis Davis 1952
Davis, G.E. 1952. Journal of Parasitology **38**:473-476. **Type strain:** no culture available.
Description: Bergey 8.

B. caucasica (Kandelaki 1945) Davis 1957
Kandelaki according to Maruashvili, G.M. 1945. Meditsinskaya Parazitologiya i Parazitarnye Bolezni **14**:24-27; Davis, G.E. 1957 in Breed, R.S., E.G.D. Murray and N.R. Smith (eds). Bergey's Manual of Determinative Bacteriology, 7th ed. The Williams and Wilkins Co., Baltimore. pp. 892-907. **Type strain:** no culture available.
Description: Bergey 8.

B. crocidurae (Leger 1917) Davis 1957
Leger, A. 1917. Bulletin de la Societe de Pathologie Exotique **10**:280-281; Davis, G.E. 1957 in Breed, R.S., E.G.D. Murray and N.R. Smith (eds). Bergey's Manual of Determinative Bacteriology, 7th ed. The Williams and Wilkins Co., Baltimore. pp. 892-907. **Type strain:** no culture available.
Description: Bergey 8.

B. dugesii (Mazzotti 1949) Davis 1957
Mazzotti, L. 1949. Revista del Instituto de Salubridad y Enfermedades Tropicales Mexico **10**:277-281; Davis, G.E. 1957 in Breed, R.S., E.G.D. Murray and N.R. Smith (eds). Bergey's Manual of Determinative Bacteriology, 7th ed. The Williams and Wilkins Co., Baltimore. pp. 892-907. **Type strain:** no culture available.
Description: Bergey 8.

B. duttonii (Novy and Knapp 1906) Bergey, Harrison, Breed, Hammer and Huntoon 1925
Novy, F.G. and R.E. Knapp. 1906. Journal of Infectious Diseases **3**:291-393; Bergey, D.H., F.C. Harrison, R.S. Breed, B.W. Hammer and F.M. Huntoon (eds). 1925. Bergey's Manual of Determinative Bacteriology, 2nd ed. The Williams and Wilkins Co., Baltimore. pp. 1-462. **Type strain:** no culture available.
Description: Bergey 8.

B. graingeri (Heisch 1953) Davis 1957
Heisch, R.B. 1953. Parasitology **43**:133-135; Davis, G.E. 1957 in Breed, R.S., E.G.D. Murray and N.R. Smith (eds). Bergey's Manual of Determinative Bacteriology, 7th ed. The Williams and Wilkins Co., Baltimore. pp. 892-907. **Type strain:** no culture available.
Description: Bergey 8.

B. harveyi (Garnham 1947) Davis 1948
Garnham, P.C.C. 1947. East African

Medical Journal **24**:47-51; Davis, G.E. 1948. Annual Review of Microbiology **2**: 305-334. **Type strain:** no culture available.
Description: Bergey 8.

B. hermsii (Davis 1942) Steinhaus 1946
Davis, G.E. 1942. Publications of the American Association for the Advancement of Science No. 18, pp. 41-47; Steinhaus, E.A. 1946. Insect Microbiology. Comstock Publishing Co., Ithaca, New York. **Type strain:** no culture available.
Description: Bergey 8.

B. hispanica (de Buen 1926) Steinhaus 1946
de Buen, S. 1926. Annales de Parasitologie Humaine et Comparee **4**:185-192; Steinhaus, E.A. 1946. Insect Microbiology. Comstock Publishing Co., Ithaca, New York. **Type strain:** no culture available.
Description: Bergey 8.

B. latyschewii (Sofiev 1941) Davis 1948
Sofiev, M.S. 1941. Medical Parasitology (Moscow) **10**:337-373; Davis, G.E. 1948. Annual Review of Microbiology **2**:305-334. **Type strain:** no culture available.
Description: Bergey 8.

B. mazzottii Davis 1956
Davis, G.E. 1956. American Journal of Hygiene **63**:13-17. **Type strain:** no culture available.
Description: Bergey 8.

B. parkeri (Davis 1942) Steinhaus 1946
Davis, G.E. 1942. Publications of the American Association for the Advancement of Science No. 18, pp. 41-47; Steinhaus, E.A. 1946. Insect Microbiology. Comstock Publishing Co., Ithaca, New York. **Type strain:** no culture available.
Description: Bergey 8.

B. persica (Dschunkowsky 1913) Steinhaus 1946
Dschunkowsky, E. 1913. Deutsche Medizinische Wochenschrift **39**:419-420; Steinhaus, E.A. 1946. Insect Microbiology. Comstock Publishing Co., Ithaca, New York. **Type strain:** no culture available.
Description: Bergey 8.

B. recurrentis (Lebert 1874) Bergey, Harrison, Breed, Hammer and Huntoon 1925
Lebert, H. 1874 in Ziemssen's Handbuch der Speciellen Pathologie und Therapie, Ed. 2. F.C.W. Vogel, Leipzig. pp. 267-304; Bergey, D.H., F.C. Harrison, R.S. Breed, B.W. Hammer and F.M. Huntoon (eds). 1925. Bergey's Manual of Determinative Bacteriology, 2nd ed. The Williams and Wilkins Co., Baltimore. pp. 1-462. **Type strain:** no culture available.
Description: Bergey 8.

B. theileri (Laveran 1903) Bergey, Harrison, Breed, Hammer and Huntoon 1925
Laveran, A. 1903. Compte Rendu de l'Academie des Sciences (Paris) **136**:939-941; Bergey, D.H., F.C. Harrison, R.S. Breed, B.W. Hammer and F.M. Huntoon (eds). 1925. Bergey's Manual of Determinative Bacteriology, 2nd ed. The Williams and Wilkins Co., Baltimore. pp. 1-462. **Type strain:** no culture available.
Description: Bergey 8.

B. tillae Zumpt and Organ 1961
Zumpt, F. and D. Organ. 1961. South African Journal of Laboratory and Clinical Medicine **7**:31-35. **Type strain:** no culture available.
Description: Bergey 8.

B. turicatae (Brumpt 1933) Steinhaus 1946
Brumpt, E. 1933. Comptes Rendus des Seances de la Societe de Biologie (Paris) **113**:1369-1372; Steinhaus, E.A. 1946. Insect Microbiology. Comstock Publishing Co., Ithaca, New York. **Type strain:** no culture available.
Description: Bergey 8.

B. venezuelensis (Brumpt 1921) Brumpt 1922
Brumpt, E. 1921 in Lavier, These, Paris. p. 207; Brumpt, E. 1922 in Roger, G.H., F. Widal and P.J. Teissier (eds). Nouveau Traite de Medecine. Masson, Paris **4**:491-531. **Type strain:** no culture available.
Description: Bergey 8.

Branhamella Catlin 1970
Catlin, B.W. 1970 International Journal of Systematic Bacteriology **20**:155-159. **Type species:** *B. catarrhalis* (Frosch and Kolle 1896) Catlin 1970
Description: Bergey 8.

B. catarrhalis (Frosch and Kolle 1896) Catlin 1970
Frosch, P. and W. Kolle. 1896 in Flugge, C. (ed). Die Mikroorganismen. F.C.W. Vogel, Leipzig. pp. 154-155; Catlin, B.W. 1970. International Journal of Systematic Bacteriology **20**:155-159. **Type strain:** ATCC 25238
Description: Bergey 8.

Brevibacterium Breed 1953
Breed, R.S. 1953. Riassunti delle Communicazione VI Congresso Internazionale di Microbiologia, Roma **1**:13-14. **Type species:** *B. linens* (Wolff 1910) Breed 1953
Description: Breed, R.S. 1953. Ibid.

B. acetylicum (Levine and Soppeland 1926) Breed 1957
Levine, M. and L. Soppeland. 1926. Bulletin of the Iowa State Agricultural College **77**:1-72; Breed, R.S. 1957 in Breed, R.S., E.G.D. Murray and N.R. Smith (eds). Bergey's Manual of Determinative Bacteriology, 7th ed. The Williams and Wilkins Co., Baltimore. pp. 490-503. **Type strain:** ATCC 953
Description: Keddie, R.M. and G.L. Cure. 1978 in Bousfield, I.J. and A.G. Callely (eds). Special Publications of the Society for General Microbiology I. Coryneform Bacteria. Academic Press, London. pp. 47-84.

B. albidum Komagata and Iizuka 1964
Komagata, K. and H. Iizuka. 1964. Journal of the Agricultural Chemical Society of Japan **38**:496-502. **Type strain:** ATCC 15831
Description: Komagata, K. and H. Iizuka. 1964. Ibid.

B. ammoniagenes (Cooke and Keith 1927) Breed 1953
Cooke, J.V. and H.R. Keith. 1927. Journal of Bacteriology **13**:315-319; Breed, R.S. 1953. Riassunti delle Communicazione VI Congresso Internazionale di Microbiologia, Roma **1**:13-14. **Type strain:** ATCC 6871
Description: Keddie, R.M. and G.L. Cure. 1978 in Bousfield, I.J. and A.G. Callely (eds). Special Publications of the Society for General Microbiology I. Coryneform Bacteria. Academic Press, London. pp. 47-84.

B. citreum Komagata and Iizuka 1964
Komagata, K. and H. Iizuka. 1964. Journal of the Agricultural Chemical Society of Japan **38**:496-502. **Type strain:** ATCC 15828
Description: Komagata, K. and H. Iizuka. 1964. Ibid.

B. divaricatum Su and Yamada 1960
Su, Y. and K. Yamada. 1960. Bulletin of the Agricultural Chemical Society of Japan **24**:69-74. **Type strain:** ATCC 14020
Description: Abe, S., K. Takayama and S. Kinoshita. 1967. Journal of General and Applied Microbiology **13**:279-301.

B. fermentans Chatelain and Second 1966
Chatelain, R. and L. Second. 1966. Annales de l'Institut Pasteur (Paris) **111**:630-644. **Type strain:** CIP 6611
Description: Chatelain, R. and L. Second. 1966. Ibid.

B. frigoritolerans Delaporte and Sasson 1967
Delaporte, B. and A. Sasson. 1967. Compte Rendu de l'Academie des Sciences. Paris, Series D, **264**:2257-2260. **Type strain:** ATCC 25097
Description: Delaporte, B. and A. Sasson. 1967. Ibid.

B. halotolerans Delaporte and Sasson 1967
Delaporte, B. and A. Sasson. 1967. Compte Rendu de l'Academie des Sciences. Paris, Series D, **264**:2257-2260. **Type strain:** ATCC 25096
Description: Delaporte, B. and A. Sasson. 1967. Ibid.

B. imperiale (Steinhaus 1941) Breed 1953
Steinhaus, E.A. 1941. Journal of Bacteriology **42**:757-790; Breed, R.S. 1953. Riassunti delle Communicazione VI Congresso Internazionale di Microbiologia, Roma **1**: 13-14. **Type strain:** ATCC 8365
Description: Steinhaus, E.A. 1941. Ibid.

B. incertum (Steinhaus 1941) Breed 1953
Steinhaus, E.A. 1941. Journal of Bacteriology **42**:757-790; Breed, R.S. 1953. Riassunti delle Communicazione VI Congresso Internazionale di Microbiologia, Roma **1**:13-14. **Type strain:** ATCC 8363
Description: Steinhaus, E.A. 1941. Ibid.

B. linens (Wolff 1910) Breed 1953
Wolff, M. 1910. Mitteilungen des Kaiser-Wilhelm-Instituts fur Landwirtschaft in Bromberg **3**:69-92; Breed, R.S. 1953. Riassunti delle Communicazione VI Congresso Internazionale di Microbiologia, Roma **1**:13-14. **Type strain:** ATCC 9172
Description: Crombach, W.H.J. 1974. Antonie van Leeuwenhoek Journal of Microbiology and Serology **40**:361-376.

B. liquefaciens Okabayashi and Masuo 1960
Okabayashi, T. and E. Masuo. 1960. Chemical and Pharmaceutical Bulletin (Tokyo) **8**:1084-1088. **Type strain:** ATCC 14929

Description: Okabayashi, T., M. Ide, T. Hattori and A. Yoshimoto. 1962. Annual Report of Shionogi Research Laboratory **12**:191-197.

B. luteum Komagata and Iizuka 1964
Komagata, K. and H. Iizuka. 1964. Journal of the Agricultural Chemical Society of Japan **38**:496-502. **Type strain:** ATCC 15830
Description: Komagata, K. and H. Iizuka. 1964. Ibid.

B. lyticum Takayama, Udagawa and Abe 1960
Takayama, K., K. Udagawa and S. Abe. 1960. Journal of the Agricultural Chemical Society of Japan **34**:652-656. **Type strain:** ATCC 15921
Description: Takayama, K., K. Udagawa and S. Abe. 1960. Ibid.

B. oxydans Chatelain and Second 1966
Chatelain, R. and L. Second. 1966. Annales de l'Institut Pasteur (Paris) **111**:630-644. **Type strain:** CIP 6612
Description: Chatelain, R. and L. Second. 1966. Ibid.

B. protophormiae Lysenko 1959
Lysenko, O. 1959. Journal of Insect Pathology **1**:34-42. **Type strain:** ATCC 19271
Description: Lysenko, O. 1959. Ibid.

B. pusillum Iizuka and Komagata 1965
Iizuka, H. and K. Komagata. 1965. Journal of General and Applied Microbiology **11**:1-14. **Type strain:** ATCC 19096
Description: Iizuka, H. and K. Komagata. 1965. Ibid.

B. saperdae Lysenko 1959
Lysenko, O. 1959. Journal of Insect Pathology **1**:34-42. **Type strain:** ATCC 19272
Description: Lysenko, O. 1959. Ibid.

B. stationis (ZoBell and Upham 1944) Breed 1953
ZoBell, C.E. and H.C. Upham. 1944. Bulletin Scripps Institution of Oceanography University of California (Technical Series) **5**:239-292; Breed, R.S. 1953. Riassunti delle Cmmunicazione Vi Congresso Internazionale di Microbiologia, Roma **1**: 13-14. **Type strain:** ATCC 14403
Description: ZoBell, C.E. and H.C. Upham. 1944. Ibid.

B. testaceum Komagata and Iizuka 1964
Komagata, K. and H. Iizuka. 1964. Journal of the Agricultural Chemical Society of Japan **38**:496-502. **Type strain:** ATCC 15829
Description: Komagata, K. and H. Iizuka. 1964. Ibid.

B. vitarumen (Bechdel, Honeywell, Dutscher and Knutsen 1928) Breed 1957
Bechdel, S.I., H.E. Honeywell, R.A. Dutscher and M.H. Knutsen. 1928. Journal of Biological Chemistry **80**:231-238; Breed, R.S. 1957 in Breed, R.S., E.G.D. Murray and N.R. Smith (eds). Bergey's Manual of Determinative Bacteriology, 7th ed. The Williams and Wilkins Co., Baltimore. pp. 490-503. **Type strain:** ATCC 10234
Description: Breed, R.S. 1957. Ibid.

Brochothrix Sneath and Jones 1976
Sneath, P.H.A. and D. Jones. 1976. International Journal of Systematic Bacteriology **26**:102-104. **Type species:** *B. thermosphacta* (McLean and Sulzbacher 1953) Sneath and Jones 1976
Description: Sneath, P.H.A. and D. Jones. 1976. Ibid.

B. thermosphacta (McLean and Sulzbacher 1953) Sneath and Jones 1976
McLean, R.A. and W.L. Sulzbacher. 1953. Journal of Bacteriology **65**:428-433; Sneath, P.H.A. and D. Jones. 1976. International Journal of Systematic Bacteriology **26**:102-104. **Type strain:** ATCC 11509
Description: Sneath, P.H.A. and D. Jones. 1976. Ibid.

Brucella Meyer and Shaw 1920
Meyer, K.F. and E.B. Shaw. 1920. Journal of Infectious Diseases **27**:173-184. **Type species:** *B. melitensis* (Hughes 1893) Meyer and Shaw 1920
Description: Bergey 8.

B. abortus (Schmidt 1901) Meyer and Shaw 1920
Schmidt, J. 1901 in Schmidt, J. and F. Weis. Bakterienne. Naturhistorisk Grundlag for det Bakteriologiske Studium, Morten Porsild, Kobenhavn 1899-1901. pp. 248-296; Meyer, K.F. and E.B. Shaw. 1920. Journal of Infectious Diseases **27**:173-184. **Type strain:** ATCC 23448

Description: Bergey 8.

B. canis Carmichael and Bruner 1968
Carmichael, L.E. and D.W. Bruner. 1968. Cornell Veterinarian **58**:579-592. **Type strain:** ATCC 23365
Description: Bergey 8.

B. melitensis (Hughes 1893) Meyer and Shaw 1920
Hughes, M.L. 1893. Mediterranean Nature **2**:325-327; Meyer, K.F. and E.B. Shaw. 1920. Journal of Infectious Diseases **27**:173-184. **Type strain:** ATCC 23456
Description: Bergey 8.

B. neotomae Stoenner and Lackman 1957
Stoenner, H.G. and D.B. Lackman. 1957. American Journal of Veterinary Research **18**:947-951. **Type strain:** ATCC 23459
Description: Bergey 8.

B. ovis Buddle 1956
Buddle, M.B. 1956. Journal of Hygiene **54**:351-364. **Type strain:** ATCC 25840
Description: Bergey 8.

B. suis Huddleson 1929
Huddleson, I.F. 1929. Michigan State College Agricultural Experimental Station Technical Bulletin **100**:1-16. **Type strain:** ATCC 23444
Description: Bergey 8.

Butyrivibrio Bryant and Small 1956
Bryant, M.P. and N. Small. 1956. Journal of Bacteriology **72**:16-21. **Type species:** *B. fibrisolvens* Bryant and Small 1956
Description: Bergey 8.

B. crossotus Moore, Johnson and Holdeman 1976
Moore, W.E.C., J.L. Johnson and L.V. Holdeman. 1976. International Journal of Systematic Bacteriology **26**:238-252. **Type strain:** VPI T9-40A
Description: Moore, W.E.C., J.L. Johnson and L.V. Holdeman. 1976. Ibid.

B. fibrisolvens Bryant and Small 1956
Bryant, M.P. and N. Small. 1956. Journal of Bacteriology **72**:16-21. **Type strain:** ATCC 19171
Description: Holdeman, L.V., E.P. Cato and W.E.C. Moore (eds). 1977. Anaerobe Laboratory Manual, 4th ed. Virginia Polytechnic Institute and State University, Blacksburg, Virginia. pp. 1-156.

Calymmatobacterium Aragao and Vianna 1913
Aragao, H. and G. Vianna. 1913. Memorias do Instituto Oswaldo Cruz Rio de Janeiro **5**:211-238. **Type species:** *C. granulomatis* Aragao and Vianna 1913
Description: Bergey 8.

C. granulomatis Aragao and Vianna 1913
Aragao, H. and G. Vianna. 1913. Memorias do Instituto Oswaldo Cruz Rio de Janeiro **5**:211-238. **Type strain:** no strain extant.
Description: Bergey 8.

Campylobacter Sebald and Veron 1963
Sebald, M. and M. Veron. 1963. Annales de l'Institut Pasteur (Paris) **105**:897-910. **Type species:** *C. fetus* (Smith and Taylor 1919) Sebald and Veron 1963
Description: Bergey 8.

C. coli (Doyle 1948) Veron and Chatelain 1973
Doyle, L.P. 1948. American Journal of Veterinary Research **9**:50-51; Veron, M. and R. Chatelain. 1973. International Journal of Systematic Bacteriology **23**:122-134. **Type strain:** CIP 7080
Description: Veron, M. and R. Chatelain. 1973. Ibid.

C. fetus (Smith and Taylor 1919) Sebald and Veron 1963
Smith, T. and M.S. Taylor. 1919. Journal of Experimental Medicine **30**:299-311; Sebald, M. and M. veron. 1963. Annales de l'Institut Pasteur (Paris) **105**:897-910. **Type strain:** CIP 5396
Description: Veron, M. and R. Chatelain. 1973. International Journal of Systematic Bacteriology **23**:133-134.

C. fetus subsp. *fetus* (Smith and Taylor 1919) Veron and Chatelain 1973
Smith, T. and M.S. Taylor. 1919. Journal of Experimental Medicine **30**:299-311; Veron, M. and R. Chatelain. 1973. International Journal of Systematic Bacteriology **23**:122-134. **Type strain:** CIP 5396
Description: Veron, M. and R. Chatelain. 1973. Ibid.

C. fetus subsp. *venerealis* (Florent 1959) Veron and Chatelain 1973
Florent, A. 1959. Proceedings 10th International Veterinary Congress Madrid **2**:953-957; Veron, M. and R. Chatelain. 1973. International Journal of Systematic Bacteriology **23**:122-134. **Type strain:**

ATCC 19483
Description: Veron, M. and R. Chatelain. 1973. Ibid.

C. jejuni (Jones, Orcutt and Little 1931) Veron and Chatelain 1973
Jones, F.S., M. Orcutt and R.B. Little. 1931. Journal of Experimental Medicine **53**:853-864; Veron, M. and R. Chatelain. 1973. International Journal of Systematic Bacteriology **23**:122-134. **Type strain:** CIP 702
Description: Veron, M. and R. Chatelain. 1973. Ibid.

C. sputorum (Prevot 1940) Veron and Chatelain 1973
Prevot, A.R. 1940. Annales de l'Institut Pasteur (Paris) **64**:117-125; Veron, M. and R. Chatelain. 1973. International Journal of Systematic Bacteriology **23**: 122-134. **Type strain:** strain Forsyth ER33 (Location not yet determined)
Description: Loesche, W.J., R.J. Gibbons and S.S. Socransky. 1965. Journal of Bacteriology **89**:1109-1116.

C. sputorum subsp. *bubulus* (Florent 1953) Veron and Chatelain 1973
Florent, A. 1953. Comptes Rendus des Seances de la Societe de Biologie **147**: 2066-2069; Veron, M. and R. Chatelain. 1973. International Journal of Systematic Bacteriology **23**:122-134. **Type strain:** CIP 53103
Description: Veron, M. and R. Chatelain. 1973. Ibid.

C. sputorum subsp. *sputorum* (Prevot 1940) Veron and Chatelain 1973
Prevot, A.R. 1940. Annales de l'Institut Pasteur (Paris) **64**:117-125; Veron, M. and R. Chatelain. 1973. International Journal of Systematic Bacteriology **23**: 122-134. **Type strain:** strain Forsyth ER33
Description: Loesche, W.J., R.J. Gibbons and S.S. Socransky. 1965. Journal of Bacteriology **89**:1109-1116.

Capsularis Prevot 1938
Prevot, A.R. 1938. Annales de l'Institut Pasteur (Paris) **60**:285-307. **Type species:** *C. zoogleiformans* (Weinberg, Nativelle and Prevot 1937) Prevot 1938
Description: Prevot, A.R., A. Turpin and P. Kaiser. 1967. Les bacteries anaerobies. Dunod, Paris. p. 272.

C. zoogleiformans (Weinberg, Nativelle and Prevot 1937) Prevot 1938
Weinberg, M., R. Nativelle and A.R. Prevot. 1937. Les microbes anaerobies. Masson et Cie; Prevot, A.R. 1938. Annales de l'Institut Pasteur (Paris) **60**:285-307. **Type strain:** VPI D28K-1
Description: Prevot, A.R., A. Turpin and P. Kaiser. 1967. Les bacteries anaerobies. Dunod, Paris. p. 272.

Cardiobacterium Slotnick and Dougherty 1964
Slotnick, I.J. and M. Dougherty. 1964. Antonie van Leeuwenhoek Journal of Microbiology and Serology **30**:261-272. **Type species:** *C. hominis* Slotnick and Dougherty 1964
Description: Bergey 8.

C. hominis Slotnick and Dougherty 1964
Slotnick, I.J. and M. Dougherty. 1964. Antonie van Leeuwenhoek Journal of Microbiology and Serology **30**:261-272. **Type strain:** ATCC 15826
Description: Bergey 8.

Caryophanon Peshkoff 1939
Peshkoff, M.A. 1939. Doklady Akademii Nauk SSSR **25**:239-242. **Type species:** *C. latum* Peshkoff 1939
Description: Bergey 8.

C. latum Peshkoff 1939
Peshkoff, M.A. 1939. Doklady Akademii Nauk SSSR **25**:239-242. **Type strain:** NCIB 9533
Description: Bergey 8.

Caseobacter Crombach 1978
Crombach, W.H.J. 1978. International Journal of Systematic Bacteriology **28**: 354-366. **Type species:** *C. polymorphus* Crombach 1978
Description: Crombach, W.H.J. 1978. Ibid.

C. polymorphus Crombach 1978
Crombach, W.H.J. 1978. International Journal of Systematic Bacteriology **28**: 354-366. **Type strain:** LMD AC 256
Description: Crombach, W.H.J. 1978. Ibid.

Caulobacter Henrici and Johnson 1935
Henrici, A.T. and D. Johnson. 1935. Journal of Bacteriology **30**:61-92. **Type**

species: *C. vibrioides* Henrici and Johnson 1935
Description: Bergey 8.

C. bacteroides Poindexter 1964
Poindexter, J.S. 1964. Bacteriological Reviews **28**:231-295. **Type strain:** ATCC 15254
Description: Poindexter, J.S. 1964. Ibid.

C. crescentus Poindexter 1964
Poindexter, J.S. 1964. Bacteriological Reviews **28**:231-295. **Type strain:** ATCC 15252
Description: Poindexter, J.S. 1964. Ibid.

C. fusiformis Poindexter 1964
Poindexter, J.S. 1964. Bacteriological Reviews **28**:231-295. **Type strain:** ATCC 15257
Description: Poindexter, J.S. 1964. Ibid.

C. halobacteroides Poindexter 1964
Poindexter, J.S. 1964. Bacteriological Reviews **28**:231-295. **Type strain:** ATCC 15269
Description: Poindexter, J.S. 1964. Ibid.

C. henricii Poindexter 1964
Poindexter, J.S. 1964. Bacteriological Reviews **28**:231-295. **Type strain:** ATCC 15253
Description: Poindexter, J.S. 1964. Ibid.

C. intermedius Poindexter 1964
Poindexter, J.S. 1964. Bacteriological Reviews **28**:231-295. **Type strain:** ATCC 15262
Description: Poindexter, J.S. 1964. Ibid.

C. leidyia Poindexter 1964
Poindexter, J.S. 1964. Bacteriological Reviews **28**:231-295. **Type strain:** ATCC 15260
Description: Poindexter, J.S. 1964. Ibid.

C. maris Poindexter 1964
Poindexter, J.S. 1964. Bacteriological Reviews **28**:231-295. **Type strain:** ATCC 15268
Description: Poindexter, J.S. 1964. Ibid.

C. subvibrioides Poindexter 1964
Poindexter, J.S. 1964. Bacteriological Reviews **28**:231-295. **Type strain:** ATCC 15264
Description: Poindexter, J.S. 1964. Ibid.

C. vibrioides Henrici and Johnson 1935
Henrici, A.T. and D. Johnson. 1935. Journal of Bacteriology **30**:61-92. **Type strain:** Stove CB51 (not extant)
Description: Poindexter, J.S. 1964. Bacteriological Reviews **28**:231-295.

Cellulomonas Bergey, Harrison, Breed, Hammer and Huntoon 1923
Bergey, D.H., F.C. Harrison, R.S. Breed, B.W. Hammer and F.M. Huntoon 1923. Bergey's Manual of Determinative Bacteriology, 1st ed. The Williams and Wilkins Co., Baltimore. pp. 1-442. **Type species:** *C. flavigena* (Kellerman and McBeth 1912) Bergey, Harrison, Breed, Hammer and Huntoon 1923
Description: Bergey 8.

C. biazotea (Kellerman, McBeth, Scales and Smith 1913) Bergey, Harrison, Breed, Hammer and Huntoon 1923
Kellerman, K.F., I.G. McBeth, F.M. Scales and N.R. Smith. 1913. Zentralblatt fur Bakteriologie, Parasitenkunde, Infektionskrankheiten und Hygiene. Abteilung II. **39**:502-522; Bergey, D.H., F.C. Harrison, R.S. Breed, B.W. Hammer and F.M. Huntoon. 1923. Bergey's Manual of Determinative Bacteriology, 1st ed. The Williams and Wilkins Co., Baltimore. pp. 1-442. **Type strain:** ATCC 486
Description: Bergey, D.H., F.C. Harrison, R.S. Breed, B.W. Hammer and F.M. Huntoon. 1923. Ibid.

C. cellasea (Kellerman, McBeth, Scales and Smith 1913) Bergey, Harrison, Breed, Hammer and Huntoon 1923
Kellerman, K.F., I.G. McBeth, F.M. Scales and N.R. Smith. 1913. Zentralblatt fur Bakteriologie, Parasitenkunde, Infektionskrankheiten und Hygiene. Abteilung II. **39**:502-522; Bergey, D.H., F.C. Harrison, R.S. Breed, B.W. Hammer and F.M. Huntoon. 1923. Bergey's Manual of Determinative Bacteriology, 1st ed. The Williams and Wilkins Co., Baltimore. pp. 1-442. **Type strain:** ATCC 487
Description: Kellerman, K.F., I.G. McBeth, F.M. Scales and N.R. Smith. 1913. Ibid.

C. fimi (McBeth and Scales 1913) Bergey, Harrison, Breed, Hammer and Huntoon 1923
McBeth, I.G. and F.M. Scales. 1913. United States Department of Agriculture Bureau of Plant Industry Bulletin **266**:1-52; Bergey, D.H., F.C. Harrison, R.S. Breed, B.W. Hammer and F.M. Huntoon. 1923. Bergey's Manual of Determi-

native Bacteriology, 1st ed. The Williams and Wilkins Co., Baltimore. pp. 1-442. **Type strain:** ATCC 484
Description: McBeth, I.G. and F.M. Scales. 1913. Ibid.

C. flavigena (Kellerman and McBeth 1912) Bergey, Harrison, Breed, Hammer and Huntoon 1923
Kellerman, K.F. and I.G. McBeth. 1912. Zentralblatt fur Bakteriologie, Parasitenkunde, Infektionskrankheiten und Hygiene. Abteilung II. **34**:485-494; Bergey, D.H., F.C. Harrison, R.S. Breed, B.W. Hammer and F.M. Huntoon 1923. Bergey's Manual of Determinative Bacteriology, 1st ed. The Williams and Wilkins Co., Baltimore. pp. 1-442. **Type strain:** ATCC 482
Description: Bergey 8.

C. gelida (Kellerman, McBeth, Scales and Smith 1913) Bergey, Harrison, Breed, Hammer and Huntoon 1923
Kellerman, K.F., I.G. McBeth, F.M. Scales and N.R. Smith. 1913. Zentralblatt fur Bakteriologie, Parasitenkunde, Infektionskrankheiten und Hygiene. Abteilung II. **39**:502-522; Bergey, D.H., F.C. Harrison, R.S. Breed, B.W. Hammer and F.M. Huntoon. 1923. Bergey's Manual of Determinative Bacteriology, 1st ed. The Williams and Wilkins Co., Baltimore. pp. 1-442. **Type strain:** ATCC 488
Description: Kellerman, K.F., I.G. McBeth, F.M. Scales and N.R. Smith. 1913. Ibid.

C. uda (Kellerman, McBeth, Scales and Smith 1913) Bergey, Harrison, Breed, Hammer and Huntoon 1923
Kellerman, K.F., I.G. McBeth, F.M. Scales and N.R. Smith. 1913. Zentralblatt fur Bakteriologie, Parasitenkunde, Infektionskrankheiten und Hygiene. Abteilung II. **39**:502-522; Bergey, D.H., F.C. Harrison, R.S. Breed, B.W. Hammer and F.M. Huntoon. 1923. Bergey's Manual of Determinative Bacteriology, 1st ed. The Williams and Wilkins Co., Baltimore. pp. 1-442. **Type strain:** ATCC 491
Description: Kellerman, K.F., I.G. McBeth, F.M. Scales and N.R. Smith. 1913. Ibid.

Chainia Thirumalachar 1955
Thirumalachar, M.J. 1955. Nature (London) **176**:934-935. **Type species:** *C. antibiotica* Thirumalachar 1955
Description: Thirulmalachar, M.J. 1955. Ibid.

C. antibiotica Thirumalachar 1955
Thirumalachar, M.J. 1955. Nature (London) **176**:934-935. **Type strain:** ATCC 15721
Description: Shirling, E.B. and D. Gottlieb. 1972. International Journal of Systematic Bacteriology **22**:265-394.

C. flava Thirumalachar and Sukapure 1964
Thirumalachar, M.J. and R.S. Sukapure. 1964. Hindustan Antibiotics Bulletin **6**: 157-166. **Type strain:** ATCC 19347
Description: Thirulmalachar, M.J. and R.S. Sukapure. 1964. Ibid.

C. fumigata Thirumalachar, Sukapure, Rahalkar and Gopalkrishnan 1966
Thirumalachar, M.J., R.S. Sukapure, P.W. Rahalkar and K.S. Gopalkrishnan. 1966. Hindustan Antibiotics Bulletin **9**:10-14. **Type strain:** ATCC 19345
Description: Thirulmalachar, M.J., R.S. Sukapure, P.W. Rahalkar and K.S. Gopalkrishnan. 1966. Ibid.

C. minutisclerotica Thirumalachar, Rahalkar, Deshmukh and Sukapure 1965
Thirumalachar, M.J., P.W. Rahalkar, P.V. Deshmukh and R.S. Sukapure. 1965. Hindustan Antibiotics Bulletin **8**:6-9. **Type strain:** ATCC 17757
Description: Shirling, E.B. and D. Gottlieb. 1972. International Journal of Systematic Bacteriology **22**:265-394.

C. nigra Thirumalachar 1955
Thirumalachar, M.J. 1955. Nature (London) **176**:934-935. **Type strain:** ATCC 17756
Description: Shirling, E.B. and D. Gottlieb. 1972. International Journal of Systematic Bacteriology **22**:265-394.

C. ochracea Kuznetsov 1962
Kuznetsov, V.D. 1962. Mikrobiologiya **31**:534-539. **Type strain:** ATCC 15814
Description: Shirling, E.B. and D. Gottlieb. 1972. International Journal of Systematic Bacteriology **22**:265-394.

C. olivacea Thirumalachar and Sukapure 1964
Thirumalachar, M.J. and R.S. Sukapure. 1964. Hindustan Antibiotics Bulletin **6**: 157-166. **Type strain:** ATCC 15722

Description: Shirling, E.B. and D. Gottlieb. 1972. International Journal of Systematic Bacteriology **22**:265-394.

C. poonensis Thirumalachar 1960
Thirumalachar, M.J. 1960 in Kalakoutskii, L.V. and N.A. Krassilnikov in Rautenshtein, Ya. I. Transactions of the Institute of Microbiology, Academy of Sciences USSR **8**:1-344. **Type strain:** ATCC 15723
Description: Shirling, E.B. and D. Gottlieb. 1972. International Journal of Systematic Bacteriology **22**:265-394.

C. purpurogena Thirumalachar and Sukapure 1964
Thirumalachar, M.J. and R.S. Sukapure. 1964. Hindustan Antibiotics Bulletin **6**: 157-166. **Type strain:** ATCC 19348
Description: Shirling, E.B. and D. Gottlieb. 1972. International Journal of Systematic Bacteriology **22**:265-394.

C. rosea Thirumalachar 1966
Thirumalachar, M.J. 1966 in Thirumalachar, M.J., R.S. Sukapure, P.W. Rahalkar and K.S. Gopalkrishnan. Hindustan Antibiotics Bulletin **9**:10-14. **Type strain:** ATCC 17755
Description: Shirling, E.B. and D. Gottlieb. 1972. International Journal of Systematic Bacteriology **22**:265-394.

C. rubra Thirumalachar 1955
Thirumalachar, M.J. 1955. Nature (London) **176**:934-935. **Type strain:** ATCC 17754
Description: Shirling, E.B. and D. Gottlieb. 1972. International Journal of Systematic Bacteriology 22-265-394.

C. violens Kalakoutskii and Krassilnikov 1960
Kalakoutskii, L.V. and N.A. Krassilnikov. 1960. Transactions of the Institute of Microbiology, Academy of Science USSR **8**:45-55. **Type strain:** ATCC 15898
Description: Shirling, E.B. and D. Gottlieb. 1972. International Journal of Systematic Bacteriology **22**:265-394.

Chlamydia Jones, Rake and Stearns 1945
Jones, H., G. Rake and B. Stearns. 1945. Journal of Infectious Diseases **76**:55-69. **Type species:** *C. trachomatis* (Busacca 1935) Rake 1957
Description: Bergey 8.

C. psittaci (Lillie 1930) Page 1968
Lillie, R.D. 1930. Public Health Report **45**:773-778; Page, L.A. 1968. International Journal of Systematic Bacteriology **18**: 51-66. **Type strain:** ATCC VR 125
Description: Bergey 8.

C. trachomatis (Busacca 1935) Rake 1957
Busacca, A. 1935. Archives d'Ophthalmologie **52**:567-572; Rake, G. 1957 in Breed, R.S., E.G.D. Murray and N.R. Smith (eds). Bergey's Manual of Determinative Bacteriology, 7th ed. The Williams and Wilkins Co., Baltimore. pp. 957-968. **Type strain:** ATCC VR 571
Description: Bergey 8.

Chlorobium Nadson 1906
Nadson, G.A. 1906. Bulletin du Jardin Botanique, St. Petersbourg **6**:190. **Type species:** *C. limicola* Nadson 1906
Description: Bergey 8.

C. chlorovibrioides Gorlenko, Chebotarev and Kachalkin 1974
Gorlenko, V.M., E.N. Chebotarev and V.I. Kachalkin. 1974. Mikrobiologiya **43**: 908-914. **Type strain:** DSM 1377
Description: Gorlenko, V.M., E.N. Chebotarev and V.I. Kachalkin. 1974. Ibid.

C. limicola Nadson 1906
Nadson, G.A. 1906. Bulletin du Jardin Botanique, St. Petersbourg **6**:190. **Type strain:** DSM 245
Description: Bergey 8.

C. phaeobacteroides Pfennig 1968
Pfennig, N. 1968. Archiv fur Mikrobiologie **63**:224-226. **Type strain:** DSM 266
Description: Bergey 8.

C. phaeovibrioides Pfennig 1968
Pfennig, N. 1968. Archiv fur Mikrobiologie **63**:224-226. **Type strain:** DSM 269
Description: Bergey 8.

C. vibrioforme Pelsh 1936
Pelsh, A.D. 1936. Trudy Solyanoi Laboratorii, Vsesoyuznogo Instituta Metallurgi NISNKT. Izdatelstvo Akademya Nauk SSSR, Moscow
eningrad **5**:49-126. **Type strain:** DSM 260
Description: Bergey 8.

Chloroflexus Pierson and Castenholz 1974
Pierson, B.K. and R.W. Castenholz. 1974. Archives of Microbiology **100**:5-24. **Type species:** *C. aurantiacus* Pierson and Castenholz 1974
Description: Pierson, B.K. and R.W. Cas-

tenholz. 1974. Ibid.

C. aurantiacus Pierson and Castenholz 1974
Pierson, B.K. and R.W. Castenholz. 1974. Archives of Microbiology **100**:5-24. **Type strain:** ATCC 29366
Description: Pierson, B.K. and R.W. Castenholz. 1974. Ibid.

Chloronema Dubinina and Gorlenko 1975
Dubinina, G.A. and V.M. Gorlenko. 1975. Mikrobiologiya **44**:511-517. **Type species:** *C. giganteum* Dubinina and Gorlenko 1975
Description: Dubinina, G.A. and V.M. Gorlenko. 1975. Ibid.

C. giganteum Dubinina and Gorlenko 1975
Dubinina, G.A. and V.M. Gorlenko. 1975. Mikrobiologiya **44**:511-517. **Type strain:** no pure culture.
Description: Dubinina, G.A. and V.M. Gorlenko. 1975. Ibid.

Chondromyces Berkeley and Curtis 1874
Berkeley, M.J. and M.A. Curtis. 1874 in Berkeley, M.J. Grevillea **3**:49-64. **Type species:** *C. crocatus* Berkeley and Curtis 1874
Description: Bergey 8.

C. apiculatus Thaxter 1897
Thaxter, R. 1897. Botanical Gazette **23**: 395-411. **Type strain:** TC 4481
Description: Bergey 8.

C. catenulatus Thaxter 1904
Thaxter, R. 1904. Botanical Gazette **37**: 405-416. **Type strain:** TC 4517
Description: Bergey 8.

C. crocatus Berkeley and Curtis 1874
Berkeley, M.J. and M.A. Curtis. 1874 in Berkeley, M.J. Grevillea **3**:49-64. **Type strain:** TC 601
Description: Bergey 8.

C. lanuginosus Kofler 1913
Kofler, L. 1913. Sitzungsberichte der Akademie der Wissenschaften in Wien. Mathematisch-naturwissenschaftliche Klasse Abteilung I. **122**:845-876. **Type strain:** TC 4494
Description: Kofler, L. 1913. Ibid.

C. pediculatus Thaxter 1904
Thaxter, R. 1904. Botanical Gazette **37**: 405-416. **Type strain:** TC 4524
Description: Bergey 8.

Chromatium Perty 1852
Perty, M. 1852. Zur Kenntnis kleinster Lebensformen. Jent and Reinert, Bern I-VIII, pp. 1-228. **Type species:** *C. okenii* (Ehrenberg 1838) Perty 1852
Description: Bergey 8.

C. buderi Truper and Jannasch 1968
Truper, H.G. and H.W. Jannasch. 1968. Archiv fur Mikrobiologie **61**:363-372. **Type strain:** ATCC 25588
Description: Bergey 8.

C. gracile Strzeszewski 1913
Strzeszewski, B. 1913. Bulletin de l'Academie des Sciences de Cracovie, Serie B. pp. 309-334. **Type strain:** DSM 203
Description: Bergey 8.

C. minus Winogradsky 1888
Winogradsky, S. 1888. Beitrage zur Morphologie und Physiologie der Bacterien. Heft I. Zur Morphologie und Physiologie der Schwefelbacterien. Arthur Felix, Leipzig. pp. 1-120. **Type strain:** DSM 178
Description: Bergey 8.

C. minutissimum Winogradsky 1888
Winogradsky, S. 1888. Beitrage zur Morphologie und Physiologie der Bacterien. Heft I. Zur Morphologie und Physiologie der Schwefelbacterien. Arthur Felix, Leipzig. pp. 1-120. **Type strain:** DSM 1376
Description: Bergey 8.

C. okenii (Ehrenberg 1838) Perty 1852
Ehrenberg, C.G. 1838. Die Infusionsthierchen als vollkommene Organismen. L. Voss, Leipzig; Perty, M. 1852. Zur Kenntnis kleinster Lebensformen. Jent and Reinert, Bern I-VIII, pp. 1-228. **Type strain:** DSM 169
Description: Bergey 8.

C. vinosum (Ehrenberg 1838) Winogradsky 1888
Ehrenberg, C.G. 1838. Die Infusionsthierchen als vollkommene Organismen. L. Voss, Leipzig; Winogradsky, S. 1888. Beitrage zur Morphologie und Physiologie der Bacterien. Heft I. Zur Morphologie und Physiologie der Schwefelbacterien. Arthur Felix, Leipzig. pp. 1-120. **Type strain:** ATCC 17899
Description: Bergey 8.

C. violascens Perty 1852
Perty, M. 1852. Zur Kenntnis kleinster Lebensformen. Jent and Reinert, Bern I-

VIII. pp. 1-228. **Type strain:** ATCC 17096
Description: Bergey 8.

C. warmingii (Cohn 1875) Migula 1900
Cohn, F. 1875. Beitrage zur Biologie der Pflanzen **1**:141-207; Migula, W. 1900. System der Bakterien, Vol. 2. Gustav Fischer, Jena. **Type strain:** ATCC 14959
Description: Bergey 8.

C. weissei Perty 1852
Perty, M. 1852. Zur Kenntnis kleinster Lebensformen. Jent and Reinert, Bern I-VIII. pp. 1-228. **Type strain:** DSM 171
Description: Bergey 8.

Chromobacterium Bergonzini 1881
Bergonzini, C. 1881. Annuar Soc. Nat. Modena, Series 2, **14**:149-158. **Type species:** *C. violaceum* Bergonzini 1881
Description: De Ley, J., P. Segers and M. Gillis. 1978. International Journal of Systematic Bacteriology **28**:154-168.

C. violaceum Bergonzini 1881
Bergonzini, C. 1881. Annuar. Soc. Nat. Modena, Series 2, **14**:149-158. **Type strain:** ATCC 12472
Description: De Ley, J., P. Segers and M. Gillis. 1978. International Journal of Systematic Bacteriology **28**:154-168.

Citrobacter Werkman and Gillen 1932
Werkman, C.H. and G.F. Gillen. 1932. Journal of Bacteriology **23**:167-182. **Type species:** *C. freundii* (Braak 1928) Werkmann and Gillen 1932
Description: Bergey 8.

C. diversus (Burkey 1928) Werkman and Gillen 1932
Burkey, L.A. 1928. Iowa State College Journal of Science **3**:57-100; Werkman, C.H. and G.F. Gillen. 1932. Journal of Bacteriology **23**:167-182. **Type strain:** ATCC 27156
Description: Ewing, W.H. and B.R. Davis. 1972. International Journal of Systematic Bacteriology **22**:12-18.

C. freundii (Braak 1928) Werkman and Gillen 1932
Braak, H.R. 1928. Onderzoekingen over vergisting van glycerine. Thesis. Meinema W.D. Uitgever- Delft. pp. 1-233; Werkman, C.H. and G.F. Gillen. 1932. Journal of Bacteriology **23**:167-182. **Type strain:** ATCC 8090
Description: Bergey 8.

C. koseri Frederiksen 1970
Frederiksen, W. 1970. Publication of the Faculty of Sciences University J.E. Purkyne, Brno. **47**:89-94. **Type strain:** ATCC 27028
Description: Frederiksen, W. 1970. Ibid.

Clostridium Prazmowski 1880
Prazmowski, A. 1880. Inaugural Dissertation Hugo Voigt, Leipzig. pp. 1-58. **Type species:** *C. butyricum* Prazmowski 1880
Description: Bergey 8.

C. absonum Nakamura, Shimamura, Hayase and Nishida 1973
Nakamura, S., T. Shimamura, M. Hayase and S. Nishida. 1973. International Journal of Systematic Bacteriology **23**:419-429. **Type strain:** ATCC 27555
Description: Nakamura, S., T. Shimamura, M. Hayase and S. Nishida. 1973. Ibid.

C. acetobutylicum McCoy, Fred, Peterson and Hastings 1926
McCoy, E., E.B. Fred, W.H. Peterson and E.G. Hastings. 1926. Journal of Infectious Diseases **39**:457-483. **Type strain:** ATCC 824
Description: Holdeman, L.V., E.P. Cato and W.E.C. Moore (eds). 1977. Anaerobe Laboratory Manual, 4th ed. Virginia Polytechnic Institute and State University, Blacksburg, Virginia. pp. 1-156.

C. acidiurici (Liebert 1909) Barker 1938
Liebert, F. 1909. Verslagen van de gewone vergadering der wis- en natuurkundige afdeeling. K. Akademie van wetenschappen te Amsterdam. **17**:990-1001; Barker, H.A. 1938. Journal of Bacteriology **36**:322-323. **Type strain:** ATCC 7906
Description: Bergey 8.

C. aminovalericum Hardman and Stadtman 1960
Hardman, J.K. and T.C. Stadtman. 1960. Journal of Bacteriology **79**:549-552. **Type strain:** ATCC 13725
Description: Holdeman, L.V., E.P. Cato and W.E.C. Moore (eds). 1977. Anaerobe Laboratory Manual, 4th ed. Virginia Polytechnic Institute and State University, Blacksburg, Virginia. pp. 1-156.

C. aurantibutyricum Hellinger 1944
Hellinger, E. 1944. Commemorative Volume to Dr. Weizmann's 70th Birthday

Private print November 1944. pp. 37-46. **Type strain:** ATCC 17777
Description: Holdeman, L.V., E.P. Cato and W.E.C. Moore (eds). 1977. Anaerobe Laboratory Manual, 4th ed. Virginia Polytechnic Institute and State University, Blacksburg, Virginia. pp. 1-156.

C. barati (Prevot 1938) Holdeman and Moore 1970
Prevot, A.R. 1938. Annales de l'Institut Pasteur (Paris) **61**:72-91; Holdeman, L.V. and W.E.C. Moore. 1970 in Cato, E.P., C.S. Cummins, L.V. Holdeman, J.L. Johnson, W.E.C. Moore, R.M. Smibert and L.DS. Smith (eds). Outline of clinical methods in anaerobic bacteriology, 2nd revision. Virginia Polytechnic Institute Anaerobe Laboratory, Blacksburg, Virginia. pp. 57-66. **Type strain:** ATCC 27638
Description: Holdeman, L.V., E.P. Cato and W.E.C. Moore (eds). 1977. Anaerobe Laboratory Manual, 4th ed. Virginia Polytechnic Institute and State University, Blacksburg, Virginia. pp. 1-156.

C. barkeri Stadtman, Stadtman, Pastan and Smith 1972
Stadtman, E.R., T.C. Stadtman, I. Pastan and L.DS. Smith. 1972. Journal of Bacteriology **110**:758-760. **Type strain:** ATCC 25849
Description: Holdeman, L.V., E.P. Cato and W.E.C. Moore (eds). 1977. Anaerobe Laboratory Manual, 4th ed. Virginia Polytechnic Institute and State University, Blacksburg, Virginia. pp. 1-156.

C. beijerinckii Donker 1926
Donker, H.J.L. 1926. Dissertations Delft. W.D. Meinema, Delft. **Type strain:** ATCC 25752
Description: Holdeman, L.V., E.P. Cato and W.E.C. Moore (eds). 1977. Anaerobe Laboratory Manual, 4th ed. Virginia Polytechnic Institute and State University, Blacksburg, Virginia. pp. 1-156.

C. bifermentans (Weinberg and Seguin 1918) Bergey, Harrison, Breed, Hammer and Huntoon 1923
Weinberg, M. and P. Seguin. 1918. La gangrene gazeuse-Bacteriologie, Reproduction experimentale, Sereotherapie. Masson et Cie, Paris, pp. 1-444; Bergey, D.H., F.C. Harrison, R.S. Breed, B.W. Hammer and F.M. Huntoon. 1923. Bergey's Manual of Determinative Bacteriology, 1st ed. The Williams and Wilkins Co., Baltimore. pp. 1-442. **Type strain:** ATCC 638
Description: Holdeman, L.V., E.P. Cato and W.E.C. Moore (eds). 1977. Anaerobe Laboratory Manual, 4th ed. Virginia Polytechnic Institute and State University, Blacksburg, Virginia. pp. 1-156.

C. botulinum (van Ermengem 1896) Bergey, Harrison, Breed, Hammer and Huntoon 1923
van Ermengem, E. 1896. Zentralblatt fur Bakteriologie, Parasitenkunde, Infektionskrankheiten und Hygiene. Abteilung I. **19**:442-444; Bergey, D.H., F.C. Harrison, R.S. Breed, B.W. Hammer and F.M. Huntoon. 1923. Bergey's Manual of Determinative Bacteriology, 1st ed. The Williams and Wilkins Co., Baltimore. pp. 1-442. **Type strain:** ATCC 25763
Description: Holdeman, L.V., E.P. Cato and W.E.C. Moore (eds). 1977. Anaerobe Laboratory Manual, 4th ed. Virginia Polytechnic Institute and State University, Blacksburg, Virginia. pp. 1-156.

C. butyricum Prazmowski 1880
Prazmowski, A. 1880. Inaugural Dissertation Hugo Voigt, Leipzig, pp. 1-58. **Type strain:** ATCC 19398
Description: Holdeman, L.V., E.P. Cato and W.E.C. Moore (eds). 1977. Anaerobe Laboratory Manual, 4th ed. Virginia Polytechnic Institute and State University, Blacksburg, Virginia. pp. 1-156.

C. cadaveris (Klein 1899) McClung and McCoy 1957
Klein, E. 1899. Zentralblatt fur Bakteriologie, Parasitenkunde, Infektionskrankheiten und Hygiene. Abteilung I. **25**:278-284; McClung, L.S. and E. McCoy. 1957 in Breed, R.S., E.G.D. Murray and N.R. Smith (eds). Bergey's Manual of Determinative Bacteriology, 7th ed. The Williams and Wilkins Co., Baltimore. pp. 634-693. **Type strain:** ATCC 25783
Description: Holdeman, L.V., E.P. Cato and W.E.C. Moore (eds). 1977. Anaerobe Laboratory Manual, 4th ed. Virginia Polytechnic Institute and State University, Blacksburg, Virginia. pp. 1-156.

C. carnis (Klein 1904) Spray 1939

Klein, E. 1904. Zentralblatt fur Bakteriologie, Parasitenkunde, Infektionskrankheiten und Hygiene. Abteilung I. **35**:459-461; Spray, R.S. 1939 in Bergey, D.H., R.S. Breed, E.G.D. Murray and A.P. Hitchens (eds). Bergey's Manual of Determinative Bacteriology, 5th ed. The Williams and Wilkins Co., Baltimore. pp. 743-790. **Type strain:** ATCC 25777

Description: Holdeman, L.V., E.P. Cato and W.E.C. Moore (eds). 1977. Anaerobe Laboratory Manual 4th ed. Virginia Polytechnic Institute and State University, Blacksburg, Virginia. pp. 1-156.

C. celatum Hauschild and Holdeman 1974

Hauschild, A.H.W. and L.V. Holdeman. 1974. International Journal of Systematic Bacteriology **24**:478-481. **Type strain:** ATCC 27791

Description: Hauschild, A.H.W. and L.V. Holdeman 1974. Ibid.

C. cellobioparum Hungate 1944

Hungate, R.E. 1944. Journal of Bacteriology **48**:499-513. **Type strain:** ATCC 15832

Description: Holdeman, L.V., E.P. Cato and W.E.C. Moore (eds). 1977. Anaerobe Laboratory Manual, 4th ed. Virginia Polytechnic Institute and State University, Blackburg, Virginia. pp. 1-156.

C. chauvoei (Arloing, Cornevin and Thomas 1887) Scott 1928

Arloing, S., Cornevin and Thomas. 1887. Le charbon symptomatique du boeuf. 2nd ed. Asselin and Houzeau, Paris. pp. 1-281; Scott, J.P. 1928. Cornell Veterinarian **18**:259-271. **Type strain:** ATCC 10092

Description: Holdeman, L.V., E.P. Cato and W.E.C. Moore (eds). 1977. Anaerobe Laboratory Manual, 4th ed. Virginia Polytechnic Institute and State University, Blacksburg, Virginia. pp. 1-156.

C. clostridiiforme (Burri and Ankersmit 1906) Kaneuchi, Watanabe, Terada, Benno and Mitsuoka 1976

Burri, R. and P. Ankersmit. 1906 in Ankersmit, P. Zentralblatt fur Bakteriologie, Parasitenkunde, Infektionskrankheiten und Hygiene. Abteilung I. **40**:100-118; Kaneuchi, C., K. Watanabe, A. Terada, Y. Benno and T. Mitsuoka. 1976. International Journal of Systematic Bacteriology **26**:195-204, 341. **Type strain:** ATCC 25537

Description: Cato, E.P. and C.W. Salmon. 1976. International Journal of Systematic Bacteriology **26**:205-211.

C. coccoides Kaneuchi, Benno and Mitsuoka 1976

Kaneuchi, C., Y. Benno and T. Mitsuoka. 1976. International Journal of Systematic Bacteriology **26**:482-486. **Type strain:** ATCC 29236

Description: Kaneuchi, C., Y. Benno and T. Mitsuoka. 1976. Ibid.

C. cochlearium (Douglas, Fleming and Colebrook 1919) Bergey, Harrison, Breed, Hammer and Huntoon 1923

Douglas, S.R., A. Fleming and L. Colebrook. 1919. Medical Research Council (Great Britain) Special Report Series **57**:1-159; Bergey, D.H., F.C. Harrison, R.S. Breed, B.W. Hammer and F.M. Huntoon. 1923. Bergey's Manual of Determinative Bacteriology, 1st ed. The Williams and Wilkins Co., Baltimore. pp. 1-442. **Type strain:** ATCC 17787

Description: Holdeman, L.V., E.P. Cato and W.E.C. Moore (eds). 1977. Anaerobe Laboratory Manual, 4th ed. Virginia Polytechnic Institute and State University, Blacksburg, Virginia. pp. 1-156.

C. cocleatum Kaneuchi, Miyazato, Shinjo and Mitsuoka 1979

Kaneuchi, C., T. Miyazato, T. Shinjo and T. Mitsuoka. 1979. International Journal of Systematic Bacteriology **29**:1-12. **Type strain:** ATCC 29902

Description: Kaneuchi, C., T. Miyazato, T. Shinjo and T. Mitsuoka. 1979. Ibid.

C. difficile (Hall and O'Toole 1935) Prevot 1938

Hall, I.C. and E. O'Toole. 1935. American Journal of Diseases of Children **49**: 390-402; Prevot, A.R. 1938. Annales de l'Institut Pasteur (Paris) **61**:72-91. **Type strain:** ATCC 9689

Description: Holdeman, L.V., E.P. Cato and W.E.C. Moore (eds). 1977. Anaerobe Laboratory Manual, 4th ed. Virginia Polytechnic Institute and State University, Blacksburg, Virginia. pp. 1-156.

C. durum Smith and Cato 1974

Smith, L.DS. and E.P. Cato. 1974. Canadian Journal of Microbiology **20**:1393-1397. **Type strain:** ATCC 27763

Description: Smith, L.DS. and E.P. Cato.

1974. Ibid.

C. fallax (Weinberg and Seguin 1915) Bergey, Harrison, Breed, Hammer and Huntoon 1923
Weinberg, M. and P. Seguin. 1915. Comptes Rendus des Seances de la Societe de Biologie, Paris **78**:686-689; Bergey, D.H., F.C. Harrison, R.S. Breed, B.W. Hammer and F.M. Huntoon. 1923. Bergey's Manual of Determinative Bacteriology, 1st ed. The Williams and Wilkins Co., Baltimore. pp. 1-442. **Type strain:** ATCC 19400
Description: Holdeman, L.V., E.P. Cato and W.E.C. Moore (eds). 1977. Anaerobe Laboratory Manual, 4th ed. Virginia Polytechnic Institute and State University, Blacksburg, Virginia. pp. 1-156.

C. felsineum (Carbone and Tombolato 1917) Bergey, Breed, Murray and Hitchens 1939
Carbone, D. and A. Tombolato. 1917. Stazioni Sperimentali Agrarie Italiane **50**: 563-575; Bergey, D.H., R.S. Breed, E.G.D. Murray and A.P. Hitchens (eds). 1939. Bergey's Manual of Determinative Bacteriology, 5th ed. The Williams and Wilkins Co., Baltimore. pp. 1-1032. **Type strain:** ATCC 17788
Description: Holdeman, L.V., E.P. Cato and W.E.C. Moore (eds). 1977. Anaerobe Laboratory Manual, 4th ed. Virginia Polytechnic Institute and State University, Blacksburg, Virginia. pp. 1-156.

C. formicoaceticum Andreesen, Gottschalk and Schlegel 1970
Andreesen, J.R., G. Gottschalk and H.G. Schlegel. 1970. Archiv fur Mikrobiologie **72**:154-174. **Type strain:** ATCC 27076
Description: Andreeson, J.R., G. Gottschalk and H.G. Schlegel. 1970. Ibid.

C. ghoni Prevot 1938
Prevot, A.R. 1938. Annales de l'Institut Pasteur (Paris) **61**:72-91. **Type strain:** ATCC 25757
Description: Holdeman, L.V., E.P. Cato and W.E.C. Moore (eds). 1977. Anaerobe Laboratory Manual, 4th ed. Virginia Polytechnic Institute and State University, Blacksburg, Virginia. pp. 1-156.

C. glycolicum Gaston and Stadtman 1963
Gaston, L.W. and E.R. Stadtman. 1963. Journal of Bacteriology **85**:356-362. **Type strain:** ATCC 14880
Description: Holdeman, L.V., E.P. Cato and W.E.C. Moore (eds). 1977. Anaerobe Laboratory Manual, 4th ed. Virginia Polytechnic Institute and State University, Blacksburg, Virginia. pp. 1-156.

C. haemolyticum (Hall 1929) Scott, Turner and Vawter 1935
Hall, I.C. 1929. Journal of Infectious Diseases **45**:156-162; Scott, J., A.W. Turner and L.R. Vawter. 1935. Twelfth International Veterinary Congress **2**:168-182. **Type strain:** ATCC 9650
Description: Holdeman, L.V., E.P. Cato and W.E.C. Moore (eds). 1977. Anaerobe Laboratory Manual, 4th ed. Virginia Polytechnic Institute and State University, Blacksburg, Virginia. pp. 1-156.

C. hastiforme MacLennan 1939
MacLennan, J.D. 1939. Journal of Pathology and Bacteriology **49**:535-548. **Type strain:** VPI 12193
Description: Holdeman, L.V., E.P. Cato and W.E.C. Moore (eds). 1977. Anaerobe Laboratory Manual, 4th ed. Virginia Polytechnic Institute and State University, Blacksburg, Virginia. pp. 1-156.

C. histolyticum (Weinberg and Seguin 1916) Bergey, Harrison, Breed, Hammer and Huntoon 1923
Weinberg, M. and P. Seguin. 1916. Comptes Rendus de l'Academie des Seances Paris **163**:449-451; Bergey, D.H., F.C. Harrison, R.S. Breed, B.W. Hammer and F.M. Huntoon. 1923. Bergey's Manual of Determinative Bacteriology, 1st ed. The Williams and Wilkins Co., Baltimore. pp. 1-442. **Type strain:** ATCC 19401
Description: Holdeman, L.V., E.P. Cato and W.E.C. Moore (eds). 1977. Anaerobe Laboratory Manual, 4th ed. Virginia Polytechnic Institute and State University, Blacksburg, Virginia. pp. 1-156.

C. indolis McClung and McCoy 1957
McClung, L.S. and E. McCoy. 1957 in Breed, R.S., E.G.D. Murray and N.R. Smith (eds). Bergey's Manual of Determinative Bacteriology, 7th ed. The Williams and Wilkins Co., Baltimore. pp. 634-693. **Type strain:** ATCC 25771
Description: Holdeman, L.V., E.P. Cato and W.E.C. Moore (eds). 1977. Anaerobe Laboratory Manual, 4th ed. Virginia Po-

lytechnic Institute and State University, Blacksburg, Virginia. pp. 1-156.

C. innocuum Smith and King 1962
Smith, L.DS. and E. King. 1962. Journal of Bacteriology **83**:938-939. **Type strain:** ATCC 14501
Description: Holdeman, L.V., E.P. Cato and W.E.C. Moore (eds). 1977. Anaerobe Laboratory Manual, 4th ed. Virginia Polytechnic Institute and State University, Blacksburg, Virginia. pp. 1-156.

C. irregularis (Choukevitch 1911) Prevot 1938
Choukevitch, J. 1911. Annales de l'Institut Pasteur (Paris) **25**:345-368; Prevot, A.R. 1938. Annales de l'Institut Pasteur (Paris) **61**:72-91. **Type strain:** ATCC 25756
Description: Holdeman, L.V., E.P. Cato and W.E.C. Moore (eds). 1977. Anaerobe Laboratory Manual, 4th ed. Virginia Polytechnic Institute and State University, Blacksburg, Virginia. pp. 1-156.

C. kluyveri Barker and Taha 1942
Barker, H.A. and S.M. Taha. 1942. Journal of Bacteriology **43**:347-363. **Type strain:** NCIB 10680
Description: Bergey 8.

C. lentoputrescens Hartsell and Rettger 1934
Hartsell, S.E. and L.F. Rettger. 1934. Journal of Bacteriology **27**:497-514. **Type strain:** ATCC 17794
Description: Bergey 8.

C. leptum Moore, Johnson and Holdeman 1976
Moore, W.E.C., J.L. Johnson and L.V. Holdeman. 1976. International Journal of Systematic Bacteriology **26**:238-252. **Type strain:** ATCC 29065
Description: Moore, W.E.C., J.L. Johnson and L.V. Holdeman. 1976. Ibid.

C. limosum Andre 1948
Andre, 1948 in Prevot, A.R. Annales de l'Institut Pasteur (Paris) **74**:157-170. **Type strain:** ATCC 25620
Description: Holdeman, L.V., E.P. Cato and W.E.C. Moore (eds). 1977. Anaerobe Laboratory Manual, 4th ed. Virginia Polytechnic Institute and State University, Blacksburg, Virginia. pp. 1-156.

C. lituseburense (Laplanche and Saissac 1948) McClung and McCoy 1957
Laplanche, J. and R. Saissac. 1948 in Prevot, A.R. Annales de l'Institut Pasteur (Paris) **74**:157-170; McClung, L.S. and E. McCoy. 1957 in Breed, R.S., E.G.D. Murray and N.R. Smith (eds). Bergey's Manual of Determinative Bacteriology, 7th ed. The Williams and Wilkins Co., Baltimore. pp. 634-693. **Type strain:** ATCC 25759
Description: Holdeman, L.V., E.P. Cato and W.E.C. Moore (eds). 1977. Anaerobe Laboratory Manual, 4th ed. Virginia Polytechnic Institute and State University, Blacksburg, Virginia. pp. 1-156.

C. malenominatum (Weinberg, Nativelle and Prevot 1937) Spray 1948
Weinberg, M., R. Nativelle and A.R. Prevot. 1937. Les Microbes Anaerobies. Masson et Cie, Paris. pp. 1-1186; Spray, R.S. 1948 in Breed, R.S., E.G.D. Murray and A.P. Hitchens (eds). Bergey's Manual of Determinative Bacteriology, 6th ed. The Williams and Wilkins Co., Baltimore. pp. 763-827. **Type strain:** ATCC 25776
Description: Holdeman, L.V., E.P. Cato and W.E.C. Moore (eds). 1977. Anaerobe Laboratory Manual, 4th ed. Virginia Polytechnic Institute and State University, Blacksburg, Virginia. pp. 1-156.

C. mangenotii (Prevot and Zimmes-Chaverou 1947) McClung and McCoy 1957
Prevot, A.R. and Zimmes-Chaverou. 1947. Annales de l'Institut Pasteur (Paris) **73**:602-604; McClung, L.S. and E. McCoy. 1957 in Breed, R.S., E.G.D. Murray and N.R. Smith (eds). Bergey's Manual of Determinative Bacteriology. 7th ed. The Williams and Wilkins Co., Baltimore. pp. 634-693. **Type strain:** ATCC 25761
Description: Holdeman, L.V., E.P. Cato and W.E.C. Moore (eds). 1977. Anaerobe Laboratory Manual, 4th ed. Virginia Polytechnic Institute and State University, Blacksburg, Virginia. pp. 1-156.

C. nexile Holdeman and Moore 1974
Holdeman, L.V. and W.E.C. Moore. 1974. International Journal of Systematic Bacteriology **24**:260-277. **Type strain:** ATCC 27757
Description: Holdeman, L.V. and W.E.C. Moore. 1974. Ibid.

C. novyi (Migula 1894) Bergey, Harrison, Breed, Hammer and Huntoon 1923
Migula, W. 1894. Zeitschrift fur Hygiene **17**:209; Bergey, D.H., F.C. Harrison,

R.S. Breed, B.W. Hammer and F.M. Huntoon. 1923. Bergey's Manual of Determinative Bacteriology, 1st ed. The Williams and Wilkins Co., Baltimore. pp. 1-442. **Type strain:** ATCC 17861
Description: Holdeman, L.V., E.P. Cato and W.E.C. Moore (eds). 1977. Anaerobe Laboratory Manual, 4th ed. Virginia Polytechnic Institute and State University, Blacksburg, Virginia. pp. 1-156.

C. oceanicum Smith 1970
Smith, L.DS. 1970. Journal of Bacteriology **103**:811-813. **Type strain:** ATCC 25647
Description: Holdeman, L.V., E.P. Cato and W.E.C. Moore (eds). 1977. Anaerobe Laboratory Manual, 4th ed. Virginia Polytechnic Institute and State University, Blacksburg, Virginia. pp. 1-156.

C. oroticum (Wachsman and Barker 1954) Cato, Moore and Holdeman 1968
Wachsman, J.T. and H.A. Barker. 1954. Journal of Bacteriology **68**:400-404; Cato, E., W.E.C. Moore and L.V. Holdeman. 1968. International Journal of Systematic Bacteriology **18**:9-13. **Type strain:** ATCC 13619
Description: Holdeman, L.V., E.P. Cato and W.E.C. Moore (eds). 1977. Anaerobe Laboratory Manual, 4th ed. Virginia Polytechnic Institute and State University, Blacksburg, Virginia. pp. 1-156.

C. paraperfringens Nakamura, Tamai and Nishida 1970
Nakamura, S., K. Tamai and S. Nishida. 1970. Medical Biology **80**:137-140. **Type strain:** ATCC 27639
Description: Nakamura, S., T. Shimamura, M. Hayase and S. Nishida. 1973. Ibid.

C. paraputrificum (Bienstock 1906) Snyder 1936
Bienstock, B. 1906. Annales de l'Institut Pasteur (Paris) **20**:407-415; Snyder, M.L. 1936. Journal of Bacteriology **32**:401-410. **Type strain:** ATCC 25780
Description: Holdeman, L.V., E.P. Cato and W.E.C. Moore (eds). 1977. Anaerobe Laboratory Manual, 4th ed. Virginia Polytechnic Institute and State University, Blacksburg, Virginia. pp. 1-156.

C. pasteurianum Winogradsky 1895
Winogradsky, S. 1894-95. Archives of Science and Biology, St. Petersbourg **3**:297-352. **Type strain:** ATCC 6013
Description: Holdeman, L.V., E.P. Cato and W.E.C. Moore (eds). 1977. Anaerobe Laboratory Manual, 4th ed. Virginia Polytechnic Institute and State University, Blacksburg, Virginia. pp. 1-156.

C. perenne (Prevot 1940) McClung and McCoy 1957
Prevot, A.R. 1940. Comptes Rendus des Seances de la Societe de Biologie (Paris) **133**:574-577; McClung, L.S. and E. McCoy. 1957 in Breed, R.S., E.G.D. Murray and N.R. Smith (eds). Bergey's Manual of Determinative Bacteriology, 7th ed. The Williams and Wilkins Co., Baltimore. pp. 634-693. **Type strain:** ATCC 25782
Description: Holdeman, L.V., E.P. Cato and W.E.C. Moore (eds). 1977. Anaerobe Laboratory Manual, 4th ed. Virginia Polytechnic Institute and State University, Blacksburg, Virginia. pp. 1-156.

C. perfringens (Veillon and Zuber 1898) Hauduroy, Ehringer, Urbain, Guillot and Magrou 1937
Veillon, A. and A. Zuber. 1898. Archives de Medicine Experimentale et d'Anatomie Pathologique **10**:517-545; Hauduroy, P., G. Ehringer, A. Urbain, G. Guillot and J. Magrou. 1937. Dictionnaire des bacteries pathogenes. Masson et Cie, Paris. pp. 1-597. **Type strain:** ATCC 13124
Description: Holdeman, L.V., E.P. Cato and W.E.C. Moore (eds). 1977. Anaerobe Laboratory Manual, 4th ed. Virginia Polytechnic Institute and State University, Blacksurg, Virginia. pp. 1-156.

C. propionicum Cardon and Barker 1946
Cardon, B.P. and H.A. Barker. 1946. Journal of Bacteriology **52**:629-634. **Type strain:** ATCC 25522
Description: Holdeman, L.V., E.P. Cato and W.E.C. Moore (eds). 1977. Anaerobe Laboratory Manual, 4th ed. Virginia Polytechnic Institute and State University, Blacksburg, Virginia. pp. 1-156.

C. putrefaciens (McBryde 1911) Sturges and Drake 1927
McBryde, C.N. 1911. United States Bureau of Animal Industry **132**:1-55; Sturges, W.S. and E.T. Drake. 1927. Journal of Bacteriology **14**:175-179. **Type strain:** ATCC 25786
Description: Holdeman, L.V., E.P. Cato and W.E.C. Moore (eds). 1977. Anaerobe

Laboratory Manual, 4th ed. Virginia Polytechnic Institute and State University, Blacksburg, Virginia. pp. 1-156.

C. putrificum (Trevisan 1889) Reddish and Rettger 1922
Trevisan, V. 1889. I Generi e le Specie delle Battieriacee. Zanaboni and Gabuzzi, Milano; Reddish, G. and L. Rettger. 1922. Abstracts of Bacteriology **6**:9. **Type strain:** ATCC 25784
Description: Holdeman, L.V., E.P. Cato and W.E.C. Moore (eds). 1977. Anaerobe Laboratory Manual, 4th ed. Virginia Polytechnic Institute and State University, Blacksburg, Virginia. pp. 1-156.

C. quercicolum Stankewich, Cosenza and Shigo 1971
Stankewich, J.P., B.J. Cosenza and A.L. Shigo. 1971. Antonie van Leeuwenhoek Journal of Microbiology and Serology **37**: 299-302. **Type strain:** ATCC 25974
Description: Holdeman, L.V., E.P. Cato and W.E.C. Moore (eds). 1977. Anaerobe Laboratory Manual, 4th ed. Virginia Polytechnic Institute and State University, Blacksburg, Virginia. pp. 1-156.

C. ramosum (Veillon and Zuber 1898) Holdeman, Cato and Moore 1971
Veillon, A. and A. Zuber. 1898. Archives de Medecine Experimentale et d'Anatomie Pathologique **10**:517-545; Holdeman, L.V., E.P. Cato and W.E.C. Moore. 1971. International Journal of Systematic Bacteriology **21**:35-39. **Type strain:** ATCC 25582
Description: Holdeman, L.V., E.P. Cato and W.E.C. Moore (eds). 1977. Anaerobe Laboratory Manual, 4th ed. Virginia Polytechnic Institute and State University, Blacksburg, Virginia. pp. 1-156.

C. rectum (Heller 1922) Holdeman and Moore 1972
Heller, H.H. 1922. Journal of Bacteriology **7**:1-38; Holdeman, L.V. and W.E.C. Moore. 1972. Anaerobe Laboratory Manual Virginia Polytechnic Institute Anaerobe Laboratory, Blacksburg, Virginia. **Type strain:** ATCC 25751
Description: Holdeman, L.V., E.P. Cato and W.E.C. Moore (eds). 1977. Anaerobe Laboratory Manual, 4th ed. Virginia Polytechnic Institute and State University, Blacksburg, Virginia. pp. 1-156.

C. sardiniensis Prevot 1938
Prevot, A.R. 1938. Annales de l'Institut Pasteur (Paris) **61**:72-91. **Type strain:** VPI 2971
Description: Holdeman, L.V., E.P. Cato and W.E.C. Moore (eds). 1977. Anaerobe Laboratory Manual, 4th ed. Virginia Polytechnic Institute and State University, Blacksburg, Virginia. pp. 1-156.

C. sartagoformum Partansky and Henry 1935
Partansky, A.M. and B.S. Henry. 1935. Journal of Bacteriology **30**:559-571. **Type strain:** ATCC 25778
Description: Holdeman, L.V., E.P. Cato and W.E.C. Moore (eds). 1977. Anaerobe Laboratory Manual, 4th ed. Virginia Polytechnic Institute and State University, Blacksburg, Virginia. pp. 1-156.

C. scatologenes (Weinberg and Ginsbourg 1927) Prevot 1948
Weinberg, M. and B. Ginsbourg. 1927. Donnees recentes dur les microbes anaerobies et leur role en pathologie. Masson et Cie, Paris. pp. 1-291; Prevot, A.R. 1948. Manuel de classification et de determination des bacteries anaerobies, 2nd ed. Masson et Cie, Paris. pp. 1-290. **Type strain:** ATCC 25775
Description: Holdeman, L.V., E.P. Cato and W.E.C. Moore (eds). 1977. Anaerobe Laboratory Manual, 4th ed. Virginia Polytechnic Institute and State University, Blacksburg, Virginia. pp. 1-156.

C. septicum (Mace 1889) Ford 1927
Mace, E. 1889. Traite Pratique de Bacteriologie, 1st ed. J.-B. Balliere and Sons, Paris I-III. pp. 1-711; Ford, W.W. 1927. Text-book of Bacteriology. Saunders, Philadelphia. pp. 1-1069. **Type strain:** ATCC 12464
Description: Holdeman, L.V., E.P. Cato and W.E.C. Moore (eds). 1977. Anaerobe Laboratory Manual, 4th ed. Virginia Polytechnic Institute and State University, Blacksburg, Virginia. pp. 1-156.

C. sordellii (Hall and Scott 1927) Prevot 1938
Hall, I.C. and J.P. Scott. 1927. Journal of Infectious Diseases **41**:329-335; Prevot, A.R. 1938. Annales de l'Institut Pasteur (Paris) **61**:72-91. **Type strain:** NCIB 10717
Description: Holdeman, L.V., E.P. Cato and W.E.C. Moore (eds). 1977. Anaerobe Laboratory Manual, 4th ed. Virginia Po-

lytechnic Institute and State University, Blacksburg, Virginia. pp. 1-156.

C. sphenoides (Douglas, Fleming and Colebrook 1919) Bergey, Harrison, Breed, Hammer and Huntoon 1923
Douglas, S.R., A. Fleming and L. Colebrook. 1919. Medical Research Council (Great Britain) Special Report Series **57**:1-159; Bergey, D.H., F.C. Harrison, R.S. Breed, B.W. Hammer and F.M. Huntoon. 1923. Bergey's Manual of Determinative Bacteriology, 1st ed. The Williams and Wilkins Co., Baltimore. pp. 1-442. **Type strain:** ATCC 19403
Description: Holdeman, L.V., E.P. Cato and W.E.C. Moore (eds). 1977. Anaerobe Laboratory Manual, 4th ed. Virginia Polytechnic Institute and State University, Blacksburg, Virginia. pp. 1-156.

C. spiroforme Kaneuchi, Miyazato, Shinjo and Mitsuoka 1979
Kaneuchi, C., T. Miyazato, T. Shinjo and T. Mitsuoka. 1979. International Journal of Systematic Bacteriology **29**:1-12. **Type strain:** ATCC 29900
Description: Kaneuchi, C., T. Miyazato, T. Shinjo and T. Mitsuoka. 1979. Ibid.

C. sporogenes (Metchnikoff 1908) Bergey, Harrison, Breed, Hammer and Huntoon 1923
Metchnikoff, E. 1908. Annales de l'Institut Pasteur (Paris) **22**:929-955; Bergey, D.H., F.C. Harrison, R.S. Breed, B.W. Hammer and F.M. Huntoon. 1923. Bergey's Manual of Determinative Bacteriology 1st ed. The Williams and Wilkins Co., Baltimore. pp. 1-442. **Type strain:** ATCC 3584
Description: Holdeman, L.V., E.P. Cato and W.E.C. Moore (eds). 1977. Anaerobe Laboratory Manual, 4th ed. Virginia Polytechnic Institute and State University, Blacksburg, Virginia. pp. 1-156.

C. sporosphaeroides Soriano and Soriano 1948
Soriano, S. and A. Soriano. 1948. Revista de la Asocion Argentina de Dietologia **6**: 36-41. **Type strain:** ATCC 25781
Description: Holdeman, L.V., E.P. Cato and W.E.C. Moore (eds). 1977. Anaerobe Laboratory Manual, 4th ed. Virginia Polytechnic Institute and State University, Blacksburg, Virginia. pp. 1-156.

C. sticklandii Stadtman and McClung 1957
Stadtman, T.C. and L.S. McClung. 1957. Journal of Bacteriology **73**:218-219. **Type strain:** ATCC 12662
Description: Holdeman, L.V., E.P. Cato and W.E.C. Moore (eds). 1977. Anaerobe Laboratory Manual, 4th ed. Virginia Polytechnic Institute and State University, Blacksburg, Virginia. pp. 1-156.

C. subterminale (Hall and Whitehead 1927) Spray 1948
Hall, I.C. and R.W. Whitehead. 1927. Journal of Infectious Diseases **41**:51-69; Spray, R.S. 1948 in Breed, R.S., E.G.D. Murray and A.P. Hitchens (eds). Bergey's Manual of Determinative Bacteriology, 6th ed. The Williams and Wilkins Co., Baltimore. pp. 763-827. **Type strain:** ATCC 25774
Description: Holdeman, L.V., E.P. Cato and W.E.C. Moore (eds). 1977. Anaerobe Laboratory Manual, 4th ed. Virginia Polytechnic Institute and State University, Blacksburg, Virginia. pp. 1-156.

C. symbiosum (Stevens 1956) Kaneuchi, Watanabe, Terada, Benno and Mitsuoka 1976
Stevens, W.C. 1956. Taxonomic studies on the genus Bacteroides and similar forms. Thesis, Vanderbilt University; Kaneuchi, C., K. Watanabe, A. Terada, Y. Benno and T. Mitsuoka. 1976. International Journal of Systematic Bacteriology **26**:195-204. **Type strain:** ATCC 14940
Description: Kaneuchi, C., K. Watanabe, A. Terada, Y. Benno and T. Mitsuoka. 1976. Ibid.

C. tertium (Henry 1917) Bergey, Harrison, Breed, Hammer and Huntoon 1923
Henry, H. 1917. Journal of Pathology and Bacteriology **21**:344-385; Bergey, D.H., F.C. Harrison, R.S. Breed, B.W. Hammer and F.M. Huntoon. 1923. Bergey's Manual of Determinative Bacteriology, 1st ed. The Williams and Wilkins Co., Baltimore. pp. 1-442. **Type strain:** ATCC 14573
Description: Holdeman, L.V., E.P. Cato and W.E.C. Moore (eds). 1977. Anaerobe Laboratory Manual, 4th ed. Virginia Polytechnic Institute and State University, Blacksburg, Virginia. pp. 1-156.

C. tetani (Flugge 1886) Bergey, Harrison, Breed, Hammer and Huntoon 1923
Flugge, C. 1886. Die Mikroorganismen. F.C.W. Vogel, Leipzig. pp. 1-692; Ber-

gey, D.H., F.C. Harrison, R.S. Breed, B.W. Hammer and F.M. Huntoon. 1923. Bergey's Manual of Determinative Bacteriology, 1st ed. The Williams and Wilkins Co., Baltimore. pp. 1-442. **Type strain:** ATCC 19406
Description: Holdeman, L.V., E.P. Cato and W.E.C. Moore (eds). 1977. Anaerobe Laboratory Manual, 4th ed. Virginia Polytechnic Institute and State University, Blacksburg, Virginia. pp. 1-156.

C. thermoaceticum Fontaine, Peterson, McCoy, Johnson and Ritter 1942
Fontaine, F.E., W.H. Peterson, E. McCoy, M.J. Johnson and G.J. Ritter. 1942. Journal of Bacteriology **43**:701-715. **Type strain:** DSM 521
Description: Fontaine, F.E., W.H. Peterson, E. McCoy, M.J. Johnson and G.J. Ritter. 1942. Ibid.

C. thermocellum Viljoen, Fred and Peterson 1926
Viljoen, J.A., E.B. Fred and W.H. Peterson. 1926. Journal of Agricultural Science Cambridge **16**:1-17. **Type strain:** ATCC 27405
Description: Bergey 8.

C. thermohydrosulfuricum Klaushofer and Parkkinen 1965
Klaushofer, H. and E. Parkkinen. 1965. Zeitschrift fur Zuckerindustrien Boehmen **15**:445-449. **Type strain:** DSM 567
Description: Matteuzzi, D., F. Hollaus and B. Bravati. 1978. International Journal of Systematic Bacteriology **28**:528-531.

C. thermosaccharolyticum McClung 1935
McClung, L.S. 1935. Journal of Bacteriology **29**:189-203. **Type strain:** ATCC 7956
Description: Holdeman, L.V., E.P. Cato and W.E.C. Moore (eds). 1977. Anaerobe Laboratory Manual, 4th ed. Virginia Polytechnic Institute and State University, Blacksburg, Virginia. pp. 1-156.

C. tyrobutyricum van Beynum and Pette 1935
van Beynum, J. and J.W. Pette. 1935. Zentralblatt fur Bakteriologie, Parasitenkunde, Infektionskrankheiten und Hygiene. Abteilung II. **93**:198-212. **Type strain:** ATCC 25755
Description: Holdeman, L.V., E.P. Cato and W.E.C. Moore (eds). 1977. Anaerobe Laboratory Manual, 4th ed. Virginia Polytechnic Institute and State University, Blacksburg, Virginia. pp. 1-156.

C. villosum Love, Jones and Bailey 1979
Love, D.N., R.F. Jones and M. Bailey. 1979. International Journal of Systematic Bacteriology **29**:241-244. **Type strain:** NCTC 11220
Description: Love, D.N., R.F. Jones and M. Bailey. 1979. Ibid.

Coprococcus Holdeman and Moore 1974
Holdeman, L.V. and W.E.C. Moore. 1974. International Journal of Systematic Bacteriology **24**:260-277. **Type species:** *C. eutactus* Holdeman and Moore 1974
Description: Holdeman, L.V. and W.E.C. Moore. 1974. Ibid.

C. catus Holdeman and Moore 1974
Holdeman, L.V. and W.E.C. Moore. 1974. International Journal of Systematic Bacteriology **24**:260-277. **Type strain:** VPI C6-61
Description: Holdeman, L.V. and W.E.C. Moore. 1974. Ibid.

C. comes Holdeman and Moore 1974
Holdeman, L.V. and W.E.C. Moore. 1974. International Journal of Systematic Bacteriology **24**:260-277. **Type strain:** VPI C1-38
Description: Holdeman, L.V. and W.E.C. Moore. 1974. Ibid.

C. eutactus Holdeman and Moore 1974
Holdeman, L.V. and W.E.C. Moore. 1974. International Journal of Systematic Bacteriology **24**:260-277. **Type strain:** ATCC 27759
Description: Holdeman, L.V. and W.E.C. Moore. 1974. Ibid.

Corynebacterium Lehmann and Neumann 1896
Lehmann, K.B. and R. Neumann. 1896. Atlas und Grundriss der Bakteriologie und Lehrbuch der speciellen bakteriologischen Diagnostik. 1st ed. J.F. Lehmann, Munchen. **Type species:** *C. diphtheriae* (Kruse 1886) Lehmann and Neumann 1896
Description: Bergey 8.

C. betae Keyworth, Howell and Dowson 1956
Keyworth, W.G., J.T.S. Howell and W.J. Dowson. 1956. Plant Pathology **5**:88-90. **Type strain:** NCPPB 374
Description: Dye, D.W. and W.J. Kemp.

1977. New Zealand Journal of Agricultural Research **20**:563-582.

C. beticola Abdou 1969
Abdou, M. A-F. 1969. Phytopathologische Zeitschrift **66**:147-167. **Type strain:** NCPPB 2256
Description: Bergey 8.

C. bovis Bergey, Harrison, Breed, Hammer and Huntoon 1923
Bergey, D.H., F.C. Harrison, R.S. Breed, B.W. Hammer and F.M. Huntoon. 1923. Bergey's Manual of Determinative Bacteriology, 1st ed. The Williams and Wilkins Co., Baltimore. pp. 1-442. **Type strain:** ATCC 7715
Description: Bergey 8.

C. callunae (Lee and Good 1963) Yamada and Komagata 1972
Lee, W.H. and R.C. Good. 1963. Official Gazette of the United States Patent Office **789**:1349; Yamada, K. and K. Komagata. 1972. Journal of General and Applied Microbiology **18**:417-431. **Type strain:** ATCC 15991
Description: Yamada, K. and K. Komagata. 1972. Journal of General and Applied Microbiology **18**:399-416.

C. cystitidis Yanagawa and Honda 1978
Yanagawa, R. and E. Honda. 1978. International Journal of Systematic Bacteriology **28**:209-216. **Type strain:** ATCC 29593
Description: Yanagawa, R. and E. Honda. 1978. Ibid.

C. diphtheriae (Kruse 1886) Lehmann and Neumann 1896
Kruse, W. 1886 in Flugge, C. Die Mikroorganismen. F.C.W. Vogel, Leipzig; Lehmann, K.B. and R. Neumann. 1896. Atlas und Grundriss der Bakteriologie und Lehrbuch der speziellen bakteriologischen Diagnostik. 1st ed. J.F. Lehmann, Munchen. **Type strain:** ATCC 27010
Description: Barksdale, L. 1970. Bacteriological Reviews **34**:414.

C. equi Magnusson 1923
Magnusson, M. 1923. Archiv fur wissenschaftliche u. praktische Tierheilkunde **50**: 22-38. **Type strain:** ATCC 6939
Description: Bergey 8.

C. fascians (Tilford 1936) Dowson 1942
Tilford, P.E. 1936. Journal of Agricultural Research **53**:383-394; Dowson, W.J. 1942. Transactions of the British Mycological Society **25**:311-314. **Type strain:** ATCC 12974
Description: Bergey 8.

C. flaccumfaciens (Hedges 1922) Dowson 1942
Hedges, F. 1922. Science (Washington) **55**:433-434; Dowson, W.J. 1942. Transactions of the British Mycological Society **25**:311-314. **Type strain:** NCPPB 1446
Description: Bergey 8.

C. flavescens Barksdale, Laneelle, Pollice, Asselineau, Welby and Norgard 1979
Barksdale, L., M.-A. Leneelle, M.C. Pollice, J. Asselineau, M. Welby and M.V. Norgard. 1979. International Journal of Systematic Bacteriology **29**:222-233. **Type strain:** ATCC 10340
Description: Barksdale, L., M.-A. Laneelle, M.C. Pollice, J. Asselineau, M. Welby and M.V. Norgard. 1979. Ibid.

C. glutamicum (Kinoshita, Nakayama and Akita 1958) Abe, Takayama and Kinoshita 1967
Kinoshita, S., S. Nakayama and S. Akita. 1958. Bulletin of the Agricultural Chemical Society of Japan **22**:176-185; Abe, S., K. Takayama and S. Kinoshita. 1967. Journal of General and Applied Microbiology **13**:279-301. **Type strain:** ATCC 13032
Description: Yamada, K. and K. Komagata. 1972. Journal of General and Applied Microbiology **18**:399-416.

C. hoagii (Morse 1912) Eberson 1918
Morse, M.E. 1912. Journal of Infectious Diseases **11**:253-285; Eberson, F. 1918. Journal of Infectious Diseases **23**:1-42. **Type strain:** ATCC 7005
Description: Yamada, K. and K. Komagata. 1972. Journal of General and Applied Microbiology **18**:399-416.

C. ilicis Mandel, Guba and Litsky 1961
Mandel, M., E.F. Guba and W. Litsky. 1961. Bacteriological Proceedings **1961**:61. **Type strain:** ATCC 14264
Description: Dye, D.W. and W.J. Kemp. 1977. New Zealand Journal of Agricultural Research **20**:563-582.

C. insidiosum (McCulloch 1925) Jensen 1934
McCulloch, L. 1925. Phytopathology **15**: 496-497; Jensen, H.J. 1934. Proceedings of the Linnean Society of New South Wales **59**:19-61. **Type strain:** NCPPB 1109

Description: Bergey 8.

C. kutscheri (Migula 1900) Bergey, Harrison, Breed, Hammer and Huntoon 1925
Migula, W. 1900. System der Bakterien, Vol. 2. Gustav Fischer, Jena; Bergey, D.H., F.C. Harrison, R.S. Breed, B.W. Hammer and F.M. Huntoon (eds). 1925. Bergey's Manual of Determinative Bacteriology, 2nd ed. The Williams and Wilkins Co., Baltimore. pp. 1-462. **Type strain:** ATCC 15677
Description: Bergey 8.

C. lilium Lee and Good 1963
Lee, W.H. and R.C. Good. 1963. Official Gazette of the United States Patent Office **789**:1349. **Type strain:** ATCC 15990
Description: Yamada, K. and K. Komagata. 1972. Journal of General and Applied Microbiology **18**:399-416.

C. michiganense (Smith 1910) Jensen 1934
Smith, E.F. 1910. Science (Washington) **31**:794-796; Jensen, H.L. 1934. Proceedings of the Linnean Society of New South Wales **59**:19-61. **Type strain:** NCPPB 2979
Description: Bergey 8.

C. nebraskense Vidaver and Mandel 1974
Vidaver, A.K. and M. Mandel. 1974. International Journal of Systematic Bacteriology **24**:482-485. **Type strain:** NCPPB 2581
Description: Vidaver, A.K. and M. Mandel. 1974. Ibid.

C. oortii Saaltink and Maas Geesteranus 1969
Saaltink, G.J. and H.P. Maas Geesteranus. 1969. Netherlands Journal of Plant Pathology **75**:123-128. **Type strain:** ATCC 25283
Description: Saaltink, G.J. and H.P. Maas Geesteranus. 1969. Ibid.

C. paurometabolum Steinhaus 1941
Steinhaus, E.A. 1941. Journal of Bacteriology **42**:757-790. **Type strain:** ATCC 8368
Description: Bergey 8.

C. pilosum Yanagawa and Honda 1978
Yanagawa, R. and E. Honda. 1978. International Journal of Systematic Bacteriology **28**:209-216. **Type strain:** ATCC 29592
Description: Yanagawa, R. and E. Honda. 1978. Ibid.

C. poinsettiae (Starr and Pirone 1942) Burkholder 1948
Starr, M.P. and P.P. Pirone. 1942. Phytopathology **32**:1076-1081; Burkholder, W.H. 1948 in Breed, R.S., E.G.D. Murray and A.P. Hitchens (eds). Bergey's Manual of Determinative Bacteriology, 6th ed. The Williams and Wilkins Co., Baltimore. pp. 381-408. **Type strain:** ATCC 9682
Description: Bergey 8.

C. pseudodiphtheriticum Lehmann and Neumann 1896
Lehmann, K.B. and R. Neumann. 1896. Atlas und Grundriss der Bakteriologie und Lehrbuch der speziellen bakteriologischen Diagnostik. 1st ed. J.F. Lehmann, Munchen. **Type strain:** ATCC 10700
Description: Bergey 8.

C. pseudotuberculosis (Buchanan 1911) Eberson 1918
Buchanan, R.E. 1911. Veterinary Bacteriology. W.B. Saunders Co., Philadelphia; Eberson, F. 1918. Journal of Infectious Diseases **23**:1-42. **Type strain:** ATCC 19410
Description: Bergey 8.

C. pyogenes (Glage 1903) Eberson 1918
Glage, F. 1903. Zeitschrift fur Fleisch und Milchhygiene **13**:166-175; Eberson, F. 1918. Journal of Infectious Diseases **23**:1-42. **Type strain:** ATCC 19411
Description: Bergey 8.

C. rathayi (Smith 1913) Dowson 1942
Smith, E.F. 1913. Science (Washington) **38**:926; Dowson, W.J. 1942. Transactions of the British Mycological Society **25**:311-314. **Type strain:** NCPPB 2980
Description: Dye, D.W. and W.J. Kemp. 1977. New Zealand Journal of Agricultural Research **20**:563-582.

C. renale (Migula 1900) Ernst 1906
Migula, W. 1900. System der Bakterien, Vol. 2. Gustav Fischer, Jena; Ernst, W. 1906. Zentralblatt fur Bakteriologie, Parasitenkunde, Infektionskrankheiten und Hygiene. Abteilung I. **40**:79-91. **Type strain:** ATCC 19412
Description: Bergey 8.

C. sepedonicum (Spieckermann and Kotthoff 1914) Skaptason and Burkholder 1942
Spieckermann, A. and P. Kotthoff. 1914. Landwirtschaftliche Jahrbucher Berlin **46**: 659-732; Skaptason, J.B. and W.H. Burkholder. 1942. Phytopathology **32**:439-

441. **Type strain:** ATCC 33113
Description: Bergey 8.

C. striatum (Chester 1901) Eberson 1918
Chester, F.D. 1901. A manual of determinative bacteriology. The Macmillan Co., New York. pp. 1-401; Eberson, F. 1918. Journal of Infectious Diseases **23**:1-42. **Type strain:** ATCC 6940
Description: Bergey 8.

C. xerosis Lehmann and Neumann 1899
Lehmann, K.B. and R. Neumann. 1899. Lehmann's Medizin, Handatlanten X. Atlas und Grundriss der Bakteriologie und Lehrbuch der speziellen bakteriologischen Diagnostik. 2. Auflage. **2**:1-497. **Type strain:** ATCC 373
Description: Bergey 8.

Cowdria Moshkovski 1947
Moshkovski, S.D. 1947. Science (Washington) **106**:62. **Type species:** *C. ruminantium* (Cowdry 1925) Moshkovski 1947
Description: Bergey 8.

C. ruminantium (Cowdry 1925) Moshkovski 1947
Cowdry, E.V. 1925. Journal of Experimental Medicine **42**:231-252; Moshkovski, S.D. 1947. Science (Washington) **106**:62. **Type strain:** no culture isolated.
Description: Bergey 8.

Coxiella (Philip 1943) Philip 1948
Philip, C.B. 1943. American Journal of Hygiene **37**:301-309; Philip, C.B. 1948. Public Health Report **63**:58. **Type species:** *C. burnetii* (Derrick 1939) Philip 1948
Description: Bergey 8.

C. burnetii (Derrick 1939) Philip 1948
Derrick, E.H. 1939. Medical Journal of Australia **1**:14; Philip, C.B. 1948. Public Health Report **63**:58. **Type strain:** ATCC VR 615
Description: Bergey 8.

Crenothrix Cohn 1870
Cohn, F. 1870. Beitrage zur Biologie der Pflanzen **1**:108-131. **Type species:** *C. polyspora* Cohn 1870
Description: Bergey 8.

C. polyspora Cohn 1870
Cohn, F. 1870. Beitrage zur Biologie der Pflanzen **1**:108-131. **Type strain:** no culture isolated.
Description: Bergey 8.

Cristispira Gross 1910
Gross, J. 1910. Mitteilungen aus der Zoologischen Station zu Neapel **20**:41-93. **Type species:** *C. pectinis* Gross 1910
Description: Bergey 8.

C. pectinis Gross 1910
Gross, J. 1910. Mitteilungen aus der Zoologischen Station zu Neapel **20**:41-93. **Type strain:** no culture isolated.
Description: Bergey 8.

Curtobacterium Yamada and Komagata 1972
Yamada, K. and K. Komagata. 1972. Journal of General and Applied Microbiology **18**:417-431. **Type species:** *C. citreum* (Komagata and Iizuka 1964) Yamada and Komagata 1972
Description: Yamada, K. and K. Komagata. 1972. Ibid.

C. albidum (Komagata and Iizuka 1964) Yamada and Komagata 1972
Komagata, K. and H. Iizuka. 1964. Journal of the Agricultural Chemical Society of Japan **38**:496-502; Yamada, K. and K. Komagata. 1972. Journal of General and Applied Microbiology **18**:417-431. **Type strain:** ATCC 15831
Description: Yamada, K. and K. Komagata. 1962. Ibid.

C. citreum (Komagata and Iizuka 1964) Yamada and Komagata 1972
Komagata, K. and H. Iizuka. 1964. Journal of the Agricultural Chemical Society of Japan **38**:496-502; Yamada, K. and K. Komagata. 1972. Journal of General and Applied Microbiology **18**:417-431. **Type strain:** ATCC 15828
Description: Yamada, K. and K. Komagata. 1972. Ibid.

C. luteum (Komagata and Iizuka 1964) Yamada and Komagata 1972
Komagata, K. and H. Iizuka. 1964. Journal of the Agricultural Chemical Society of Japan **38**:496-502; Yamada, K. and K. Komagata. 1972. Journal of General and Applied Microbiology **18**:417-431. **Type strain:** IAM 1623
Description: Yamada, K. and K. Komagata. 1972. Ibid.

C. pusillum (Iizuka and Komagata 1965) Yamada and Komagata 1972

Iizuka, H. and K. Komagata. 1965. Journal of General and Applied Microbiology **11**:1-14; Yamada, K. and K. Komagata. 1972. Journal of General and Applied Microbiology **18**:417-431. **Type strain:** ATCC 19096
Description: Yamada, K. amd K. Komagata. 1972. Ibid.

C. saperdae (Lysenko 1959) Yamada and Komagata 1972
Lysenko, O. 1959. Journal of Insect Pathology **1**:34-42; Yamada, K. and K. Komagata. 1972. Journal of General and Applied Microbiology **18**:417-431. **Type strain:** ATCC 19272
Description: Yamada, K. and K. Komagata. 1972. Ibid.

C. testaceum (Komagata and Iizuka 1964) Yamada and Komagata 1972
Komagata, K. and H. Iizuka. 1964. Journal of the Agricultural Chemical Society of Japan **38**:496-502; Yamada, K. and K. Komagata. 1972. Journal of General and Applied Microbiology **18**:417-431. **Type strain:** ATCC 15829
Description: Yamada, K. and K. Komagata. 1972. Ibid.

Cystobacter Schroeter 1886
Schroeter, J. 1885-1889 in Cohn, F. Kryptogamenflora von Schlesien. Band 3, Heft 3, Pilze J.U. Kern's Verlag, Breslau. pp. 1-814. **Type species:** *C. fuscus* Schroeter 1886
Description: Bergey 8.

C. ferrugineus (Krzemieniewska and Krzemieniewski 1927) McCurdy 1970
Krzemieniewska, H. and S. Krzemieniewski. 1927. Acta Societatis Botanicorum Poloniae **5**:79-98; McCurdy, H.D. 1970. International Journal of Systematic Bacteriology **20**:283-296. **Type strain:** Windsor M-203
Description: Bergey 8.

C. fuscus Schroeter 1886
Schroeter, J. 1885-1889 in Cohn, F. Kryptogamenflora von Schlesien. Band 3, Heft 3, Pilze J.U. Kern's Verlag, Breslau. pp. 1-814. **Type strain:** ATCC 25194
Description: Bergey 8.

C. minus (Krzemieniewska and Krzemieniewski 1926) McCurdy 1970
Krzemieniewska, H. and S. Krzemieniewski. 1926. Acta Societatis Botanicorum Poloniae **4**:1-54; McCurdy, H.D. 1970. International Journal of Systematic Bacteriology **20**:283-296. **Type strain:** Windsor M-307
Description: Bergey 8.

Cytophaga Winogradsky 1929
Winogradsky, S. 1929. Annales de l'Institut Pasteur (Paris) **43**:549-633. **Type species:** *C. hutchinsonii* Winogradsky 1929
Description: Bergey 8.

C. aquatilis Strohl and Tait 1978
Strohl, W.R. and L.R. Tait. 1978. International Journal of Systematic Bacteriology **28**:293-303. **Type strain:** ATCC 29551
Description: Strohl, W.R. and L.R. Tait. 1978. Ibid.

C. fermentans Bachmann 1955
Bachmann, B.J. 1955. Journal of General Microbiology **13**:541-551. **Type strain:** ATCC 19072
Description: Bergey 8.

C. flevensis van der Meulen, Harder and Veldkamp 1974
van der Meulen, H.J., W. Harder and H. Veldkamp. 1974. Antonie van Leeuwenhoek Journal of Microbiology and Serology **40**:329-346. **Type strain:** ATCC 27944
Description: van der Meulen, H.J., W. Harder and H. Veldkamp. 1974. Ibid.

C. hutchinsonii Winogradsky 1929
Winogradsky, S. 1929. Annales de l'Institut Pasteur (Paris) **43**:549-633. **Type strain:** NCIB 9469
Description: Bergey 8.

C. johnsonae Stanier 1947
Stanier, R.Y. 1947. Journal of Bacteriology **53**:297-315. **Type strain:** ATCC 17061
Description: Bergey 8.

C. latercula Lewin 1969
Lewin, R.A. 1969. Journal of General Microbiology **58**:189-206. **Type strain:** ATCC 23177
Description: Lewin, R.A. 1969. Ibid.

C. lytica Lewin 1969
Lewin, R.A. 1969. Journal of General Microbiology **58**:189-206. **Type strain:** ATCC 23178
Description: Lewin, R.A. 1969. Ibid.

C. salmonicolor Veldkamp 1961
Veldkamp, H. 1961. Journal of General

Microbiology **26**:331-342. **Type strain:** ATCC 19041
Description: Veldkamp, H. 1961. Ibid.

Dactylosporangium Thiemann, Pagani and Beretta 1967
Thiemann, J.E., H. Pagani and G. Beretta. 1967. Archiv fur Mikrobiologie **58**:42-52. **Type species:** *D. aurantiacum* Thiemann, Pagani and Beretta 1967
Description: Bergey 8.

D. aurantiacum Thiemann, Pagani and Beretta 1967
Thiemann, J.E., H. Pagani and G. Beretta. 1967. Archiv fur Mikrobiologie **58**:42-52. **Type strain:** ATCC 23491
Description: Bergey 8.

D. thailandense Thiemann, Pagani and Beretta 1967
Thiemann, J.E., H. Pagani and G. Beretta. 1967. Archiv fur Mikrobiologie **58**:42-52. **Type strain:** ATCC 23490
Description: Bergey 8.

Dermatophilus (Van Saceghem 1915) Gordon 1964
Van Saceghem, R. 1915. Bulletin de la Societe de Pathologie Exotique **8**:354-359; Gordon, M.A. 1964. Journal of Bacteriology **88**:509-522. **Type species:** *D. congolensis* (Van Saceghem 1915) Gordon 1964
Description: Bergey 8.

D. congolensis (Van Saceghem 1915) Gordon 1964
Van Saceghem, R. 1915. Bulletin de la Societe de Pathologie Exotique **8**:354-359; Gordon, M.A. 1964. Journal of Bacteriology **88**:509-522. **Type strain:** ATCC 14637
Description: Bergey 8.

Derxia Jensen, Petersen, De and Bhattacharya 1960
Jensen, H.L., E.J. Petersen, P.K. De and R. Bhattacharya. 1960. Archiv fur Mikrobiologie **36**:182-195. **Type species:** *D. gummosa* Jensen, Petersen, De and Bhattacharya 1960
Description: Bergey 8.

D. gummosa Jensen, Petersen, De and Bhattacharya 1960
Jensen, H.L., E.J. Petersen, P.K. De and R. Bhattacharya. 1960. Archiv fur Mikrobiologie **36**:182-195. **Type strain:** ATCC 15994
Description: Thompson, J.P. and V.B.D. Skerman. 1979. Azotobacteraceae - the taxonomy and ecology of the aerobic nitrogen-fixing bacteria. Academic Press, London. pp. 310-313.

Desulfomonas Moore, Johnson and Holdeman 1976
Moore, W.E.C., J.L. Johnson and L.V. Holdeman. 1976. International Journal of Systematic Bacteriology **26**:238-252. **Type species:** *D. pigra* Moore, Johnson and Holdeman 1976
Description: Moore, W.E.C., J.L. Johnson and L.V. Holdeman. 1976. Ibid.

D. pigra Moore, Johnson and Holdeman 1976
Moore, W.E.C., J.L. Johnson and L.V. Holdeman. 1976. International Journal of Systematic Bacteriology **26**:238-252. **Type strain:** ATCC 29098
Description: Moore, W.E.C., J.L. Johnson and L.V. Holdeman. 1976. Ibid.

Desulfotomaculum Campbell and Postgate 1965
Campbell, L.L. and J.R. Postgate. 1965. Bacteriological Reviews **29**:359-363. **Type species:** *D. nigrificans* (Werkman and Weaver 1927) Campbell and Postgate 1965
Description: Bergey 8.

D. acetoxidans Widdel and Pfennig 1977
Widdel, F. and N. Pfennig. 1977 in International Journal of Systematic Bacteriology **27**:306; Effective publication Widdel, F. and N. Pfennig. 1977. Archives of Microbiology **112**:119-122. **Type strain:** DSM 771
Description: Widdel, F. and N. Pfennig. 1977. Ibid.

D. nigrificans (Werkman and Weaver 1927) Campbell and Postgate 1965
Werkman, C.H. and H.J. Weaver. 1927. Iowa State Journal of Science **2**:57-67; Campbell, L.L. and J.R. Postgate. 1965. Bacteriological Reviews **29**:359-363. **Type strain:** ATCC 19998
Description: Bergey 8.

D. orientis Campbell and Postgate 1965
Campbell, L.L. and J.R. Postgate. 1965. Bacteriological Reviews **29**:359-363. **Type strain:** ATCC 19365
Description: Bergey 8.

D. ruminis Campbell and Postgate 1965

Campbell, L.L. and J.R. Postgate. 1965. Bacteriological Reviews **29**:359-363. **Type strain:** ATCC 23193
Description: Bergey 8.

Desulfovibrio Kluyver and van Niel 1936
Kluyver, A.J. and B.C. van Niel. 1936. Zentralblatt fur Bakteriologie, Parasitenkunde, Infektionskrankheiten und Hygiene. Abteilung II. **94**:369-403. **Type species:** *D. desulfuricans* (Beijerinck 1895) Kluyver and van Niel 1936
Description: Bergey 8.

D. africanus Campbell, Kasprzycki and Postgate 1966
Campbell, L.L., M.A. Kasprzycki and J.R. Postgate. 1966. Journal of Bacteriology **92**:1122-1127. **Type strain:** NCIB 8401
Description: Campbell, L.L., M.A. Kasprzycki and J.R. Postgate. 1966. Ibid.

D. desulfuricans (Beijerinck 1895) Kluyver and van Niel 1936
Beijerinck, M.W. 1895. Zentralblatt fur Bakteriologie, Parasitenkunde, Infektionskrankheiten und Hygiene. Abteilung I. **1**: 104-114; Kluyver, A.J. and C.B. van Niel. 1936. Zentralblatt fur Bakteriologie, Parasitenkunde, Infektionskrankheiten und Hygiene. Abteilung II. **94**:369-403. **Type strain:** NCIB 8307
Description: Bergey 8.

D. desulfuricans subsp. *aestuarii* Postgate and Campbell 1966
Postgate, J.R. and L.L. Campbell. 1966. Bacteriological Reviews **30**:732-738. **Type strain:** NCIB 9335
Description: Bergey 8.

D. desulfuricans subsp. *desulfuricans* (Beijerinck 1895) Kluyver and van Niel 1936
Beijerinck, M.W. 1895. Zentralblatt fur Bakteriologie, Parasitenkunde, Infektionskrankheiten und Hygiene. Abteilung I. **1**: 104-114; Kluyver, A.J. and C.B. van Niel. 1936. Zentralblatt fur Bakteriologie, Parasitenkunde, Infektionskrankheiten und Hygiene. Abteilung II. **94**:369-403. **Type strain:** NCIB 8307
Description: Bergey 8.

D. gigas Le Gall 1963
Le Gall, J. 1963. Journal of Bacteriology **86**:1120. **Type strain:** NCIB 9332
Description: Le Gall, J. 1963. Ibid.

D. salexigens Postgate and Campbell 1966
Postgate, J.R. and L.L. Campbell. 1966. Bacteriological Reviews **30**:732-738. **Type strain:** ATCC 14822
Description: Bergey 8.

D. thermophilus Rozanova and Khudyakova 1974
Rozanova, E.P. and A.I. Khudyakova. 1974. Microbiology **43**:908-912. **Type strain:** VKM V-1128
Description: Rozanova, E.P. and A.I. Khudyakova. 1974. Ibid.

D. vulgaris Postgate and Campbell 1966
Postgate, J.R. and L.L. Campbell. 1966. Bacteriological Reviews **30**:732-738. **Type strain:** DSM 644
Description: Bergey 8.

D. vulgaris subsp. *oxamicus* Postgate and Campbell 1966
Postgate, J.R. and L.L. Campbell. 1966. Bacteriological Reviews **30**:732-738. **Type strain:** NCIB 9442
Description: Bergey 8.

D. vulgaris subsp. *vulgaris* Postgate and Campbell 1966
Postgate, J.R. and L.L. Campbell. 1966. Bacteriological Reviews **30**:732-738. **Type strain:** DSM 644
Description: Bergey 8.

Desulfuromonas Pfennig and Biebl 1977
Pfennig, N. and H. Biebl. 1977 in International Journal of Systematic Bacteriology **27**:306; Effective publication Pfennig, N. and H. Biebl. 1976. Archives of Microbiology **110**:3-12. **Type species:** *D. acetoxidans* Pfennig and Biebl 1977
Description: Pfennig, N. and H. Biebl. 1976. Ibid.

D. acetoxidans Pfennig and Biebl 1977
Pfennig, N. and H. Biebl. 1977 in International Journal of Systematic Bacteriology **27**:306; Effective publication Pfennig, N. and H. Biebl. 1976. Archives of Microbiology **110**:3-12. **Type strain:** DSM 684
Description: Pfennig, N. and H. Biebl. 1976. Ibid.

Ectothiorhodospira Pelsh 1936
Pelsh, A.O. 1936. Trudy Solyanoi Laboratorii Vsesoyuznogo Instituta Metallurgii NISNKT. Izdatelstvo Akademya Nauk

SSSR, Moscow
eningrad **5**: 49-126. **Type species:** *E. mobilis* Pelsh 1936
Description: Bergey 8.

E. halochloris Imhoff and Truper 1979
Imhoff, J.F. and H.G. Truper. 1979 in International Journal of Systematic Bacteriology **29**:79-80; Effective publication Imhoff, J.F. and H.G. Truper. 1977. Archives of Microbiology **114**:114-121. **Type strain:** DSM 1059
Description: Imhoff, J.F. and H.G. Truper. 1977. Ibid.

E. halophila Raymond and Sistrom 1969
Raymond, J.C. and W.R. Sistrom. 1969. Archiv fur Mikrobiologie **69**:121-126. **Type strain:** DSM 244
Description: Bergey 8.

E. mobilis Pelsh 1936
Pelsh, A.D. 1936. Trudy Solyanoi Laboratorii. Vsesoyuznogo Instituta Metallurgii NISNKT. Izdatelstvo Akademya Nauk SSSR, Moscow
eningrad **5**:49-126. **Type strain:** DSM 237
Description: Bergey 8.

E. shaposhnikovii Cherni, Soloveva, Fedorova and Kondrateva 1969
Cherni, N.E., J.V. Soloveva, V.D. Fedorova and E.N. Kondrateva. 1969. Mikrobiologiya **38**:479-484. **Type strain:** DSM 243
Description: Bergey 8.

Edwardsiella Ewing and McWhorter 1965
Ewing, W.H. and A.C. McWhorter. 1965 in Ewing, W.H., A.C. McWhorter, M.R. Escobar and A.H. Lubin. International Bulletin of Bacteriological Nomenclature and Taxonomy **15**:33-38. **Type species:** *E. tarda* Ewing and McWhorter 1965
Description: Bergey 8.

E. anguillimortifera (Hoshina 1962) Sakazaki and Tamura 1975
Hoshina, T. 1962. Bulletin of the Society of Scientific Fisheries (Japan) **28**:162; Sakazaki, R. and K. Tamura. 1975. International Journal of Systematic Bacteriology **25**:219-220. **Type strain:** ATCC 15947
Description: Sakazaki, R. and K. Tamura. 1975. Ibid.

E. tarda Ewing and McWhorter 1965
Ewing, W.H. and A.C. McWhorter. 1965 in Ewing, W.H., A.C. McWhorter, M.R. Escobar and A.H. Lubin. International Bulletin of Bacteriological Nomenclature and Taxonomy **15**:33-38. **Type strain:** ATCC 15947
Description: Ewing, W.H., A.C. McWhorter, M.R. Escobar and A.H. Lubin. 1965. Ibid.

Ehrlichia Moshkovski 1945
Moshkovski, S.D. 1945. Advances in Modern Biology (Moscow) **19**:1-44. **Type species:** *E. canis* (Donatien and Lestoquard 1935) Moshkovski 1945
Description: Bergey 8.

E. canis (Donatien and Lestoquard 1935) Moshkovski 1945
Donatien, A. and F. Lestoquard. 1935. Bulletin de la Societe de Pathologie Exotique **28**:418-419; Moshkovski, S.D. 1945. Advances in Modern Biology (Moscow) **19**:1-44. **Type strain:** no culture isolated.
Description: Bergey 8.

E. phagocytophila (Foggie 1949) Philip 1962
Foggie, A. 1949. Journal of General Microbiology **3**:5-6; Philip, C.B. 1962. Report 2nd Meeting FAOIE Expert Panel on Tick-borne Diseases of livestock, Cairo. United Nations, Rome 1962. pp. 41-43. **Type strain:** no culture isolated.
Description: Bergey 8.

Eikenella Jackson and Goodman 1972
Jackson, F.L. and Y.E. Goodman. 1972. International Journal of Systematic Bacteriology **22**:73-77. **Type species:** *E. corrodens* (Eiken 1958) Jackson and Goodman 1972
Description: Jackson, F.L. and Y.E. Goodman. 1972. Ibid.

E. corrodens (Eiken 1958) Jackson and Goodman 1972
Eiken, M. 1958. Acta Pathologica et Microbiologica Scandinavica **43**:404-416; Jackson, F.L. and Y.E. Goodman. 1972. International Journal of Systematic Bacteriology **22**:73-77. **Type strain:** ATCC 23834
Description: Jackson, F.L. and Y.E. Goodman. 1972. Ibid. Jackson and Goodman 1972
Jackson, F.L. and Y.E. Goodman. 1972.

International Journal of Systematic Bacteriology **22**:73-77. **Type species:** *E. corrodens* (Eiken 1958) Jackson and Goodman 1972
Description: Bergey 8.

E. corrodens (Eiken 1958) Jackson and Goodman 1972
Eiken, M. 1958. Acta Pathologica et Microbiologica Scandinavica **43**:404-416; Jackson, F.L. and Y.E. Goodman. 1972. International Journal of Systematic Bacteriology **22**:73-77. **Type strain:** ATCC 23834
Description: Bergey 8.

Elytrosporangium Falcao de Morais, Batista and Massa 1966
Falcao de Morais, J.O., A.C. Batista and D.M.G. Massa. 1966. Mycopathologia et Mycologia Applicata **30**:161-171. **Type species:** *E. brasiliense* Falcao de Morais, Batista and Massa 1966
Description: Falcao de Morais, J.O., C. Batista and D.M.G. Massa. 1966. Ibid.

E. brasiliense Falcao de Morais, Batista and Massa 1966
Falcao de Morais, J.O., C. Batista and D.M.G. Massa. 1966. Mycopathologia et Mycologia Applicata **30**:161-171. **Type strain:** ATCC 23727
Description: Falcao de Morais, J.O., C. Batista and D.M.G. Massa. 1966. Ibid.

E. carpinense Falcao de Morais, Oliveira da Silva and Machado 1971
Falcao de Morais, J.O., J. Oliveira da Silva and C. Machado. 1971. Revista Microbiologia Brazil **2**:203-206. **Type strain:** 70-6-2 (Lab. Microbiol. Inst. Biociencias Univer. Federal Pernambucco, Recife, Brazil)
Description: Falcao de Morais, J.O., J. Oliveira da Silva and C. Machado. 1971. Ibid.

E. spirale Falcao de Morais 1970
Falcao de Morais, J.O. 1970. Revista Microbiologia Brazil **1**:79-84. **Type strain:** ATCC 25664
Description: Falcao de Morais, J.O. 1970. Ibid.

Enterobacter Hormaeche and Edwards 1960
Hormaeche, E. and P.R. Edwards. 1960. International Bulletin of Bacteriological Nomenclature and Taxonomy **10**:71-74. **Type species:** *E. cloacae* (Jordan 1890) Hormaeche and Edwards 1960
Description: Bergey 8.

E. aerogenes Hormaeche and Edwards 1960
Hormaeche, E. and P.R. Edwards. 1960. International Bulletin of Bacteriological Nomenclature and Taxonomy **10**:71-74. **Type strain:** ATCC 13048
Description: Hormaeche, E. and P.R. Edwards. 1960. Ibid.

E. agglomerans Ewing and Fife 1972
Ewing, W.H. and M.A. Fife. 1972. International Journal of Systematic Bacteriology **22**:4-11. **Type strain:** ATCC 27155
Description: Ewing, W.H. and M.A. Fife. 1972. Ibid.

E. cloacae (Jordon 1890) Hormaeche and Edwards 1960
Jordan, E.O. 1890 in Sedgewick, W.T. Report of the Massachusettes Board of Public Health, pp. 821-844; Hormaeche, E. and P.R. Edwards. 1960. International Bulletin of Bacteriological Nomenclature and Taxonomy **10**:71-74. **Type strain:** ATCC 13047
Description: Hormaeche, E. and P.R. Edwards. 1960. Ibid.

Eperythrozoon Schilling 1928
Schilling, V. 1928. Klinische Wochenschrift **7**:1853-1855. **Type species:** *E. coccoides* Schilling 1928
Description: Bergey 8.

E. coccoides Schilling 1928
Schilling, V. 1928. Klinische Wochenschrift **7**:1854-1855. **Type strain:** no culture isolated.
Description: Bergey 8.

E. ovis Neitz, Alexander and du Toit 1934
Neitz, W.O., R.A. Alexander and P.J. du Toit. 1934. Onderstepoort Journal of Veterinary Science **3**:263-274. **Type strain:** no culture isolated.
Description: Bergey 8.

E. parvum Splitter 1950
Splitter, E.J. 1950. Science (Washington) **111**:513-514. **Type strain:** no culture isolated.
Description: Bergey 8.

E. suis Splitter 1950
Splitter, E.J. 1950. Science (Washington) **111**:513-514. **Type strain:** no culture iso-

lated.
Description: Bergey 8.

E. wenyonii Adler and Ellenbogen 1934
Adler, S. and V. Ellenbogen. 1934. Journal of Comparative Pathology **47**: 220-221. **Type strain:** no culture isolated.
Description: Bergey 8.

Erwinia Winslow, Broadhurst, Buchanan, Krumwiede, Rogers and Smith 1920
Winslow, C.E.A., J. Broadhurst, R.E. Buchanan, C. Krumwiede Jr., L.A. Rogers and G.H. Smith. 1920. Journal of Bacteriology **5**:191-229. **Type species:** *E. amylovora* (Burrill 1882) Winslow, Broadhurst, Buchanan, Krumwiede, Rogers and Smith 1920
Description: Bergey 8.

E. amylovora (Burrill 1882) Winslow, Broadhurst, Buchanan, Krumwiede, Rogers and Smith 1920
Burrill, T.J. 1882. Illinois Industrial University 11th Report pp. 93-157; Winslow, C.E.A., J. Broadhurst, R.E. Buchanan, C. Krumwiede Jr., L.A. Rogers and G.H. Smith. 1920. Journal of Bacteriology **5**: 191-229. **Type strain:** ATCC 15580
Description: Bergey 8.

E. ananas Serrano 1928
Serrano, F.B. 1928. Philippine Journal of Science **36**:271-305. **Type strain:** NCPPB 1846
Description: Bergey 8.

E. cancerogena Urosevic 1966
Urosevic, B. 1966. Lesnicky casopsis **12**: 493-505. **Type strain:** NCPPB 2176
Description: Urosevic, B. 1966. Ibid.

E. carnegieana Standring 1942 (See also Pectobacterium carnegieana)
Standring, E.T. 1942 in Lightle, P.C., E.T. Standring and J.G. Brown. Phytopathology **32**:303-313. **Type strain:** NCPPB 439
Description: Bergey 8.

E. carotovora (Jones 1901) Bergey, Harrison, Breed, Hammer and Huntoon 1923 (See also Pectobacterium carotovorum)
Jones, L.R. 1901. Zentralblatt fur Bakteriologie, Parasitenkunde, Infektionskrankheiten und Hygiene. Abteilung II. **7**:12-21; Bergey, D.H., F.C. Harrison, R.S. Breed, B.W. Hammer and F.M. Huntoon. 1923. Bergey's Manual of Determinative Bacteriology, 1st ed. The Williams and Wilkins Co., Baltimore. pp. 1-442. **Type strain:** ATCC 15713
Description: Bergey 8.

E. carotovora subsp. *atroseptica* (van Hall 1902) Dye 1969
van Hall, C.J.J. 1902. Inaugural Dissertation Amsterdam; Dye, D.W. 1969. New Zealand Journal of Science **12**:81-97. **Type strain:** NCPPB 549
Description: Bergey 8.

E. carotovora subsp. *carotovora* (Jones 1901) Bergey, Harrison, Breed, Hammer and Huntoon 1923
Jones, L.R. 1901. Zentralblatt fur Bakteriologie, Parasitenkunde, Infektionskrankheiten und Hygiene. Abteilung II. **7**:12-21; Bergey, D.H., F.C. Harrison, R.S. Breed, B.W. Hammer and F.M. Huntoon. 1923. Bergey's Manual of Determinative Bacteriology, 1st ed. The Williams and Wilkins Co., Baltimore. pp. 1-442. **Type strain:** ATCC 15713
Description: Bergey 8.

E. chrysanthemi Burkholder, McFadden and Dimock 1953 (See also Pectobacterium chrysanthemi)
Burkholder, W.H., L.A. McFadden and A.V. Dimock. 1953. Phytopathology **43**: 522-526. **Type strain:** ATCC 11663
Description: Bergey 8.

E. cypripedii (Hori 1911) Bergey, Harrison, Breed, Hammer and Huntoon 1923 (See also Pectobacterium cypripedii)
Hori, S. 1911. Zentralblatt fur Bakteriologie, Parasitenkunde, Infektionskrankheiten und Hygiene. Abteilung II. **31**:85-92; Bergey, D.H., F.C. Harrison, R.S. Breed, B.W. Hammer and F.M. Huntoon. 1923. Bergey's Manual of Determinative Bacteriology, 1st ed. The Williams and Wilkins Co., Baltimore. pp. 1-442. **Type strain:** PDDCC 1591
Description: Bergey 8.

E. dissolvens (Rosen 1922) Burkholder 1948
Rosen, H.R. 1922. Phytopathology **12**: 497-499; Burkholder, W.H. 1948 in Breed, R.S., E.G.D. Murray and A.P. Hitchens (eds). Bergey's Manual of Determinative Bacteriology, 6th ed. The Williams and Wilkins Co., Baltimore. pp. 463-478. **Type strain:** ATCC 23373
Description: Dye, D.W. 1969. New Zea-

land Journal of Science **12**:833-839.

E. herbicola (Lohnis 1911) Dye 1964
Lohnis, F. 1911. Landwirtschaftlich-bakteriologisches Praktikum. Gebruder Borntraeger, Berlin. pp. 1-156; Dye, D.W. 1964. New Zealand Journal of Science **7**: 261-269. **Type strain:** NCPPB 2971
Description: Bergey 8.

E. mallotivora Goto 1976
Goto, M. 1976. International Journal of Systematic Bacteriology **26**:467-473. **Type strain:** ATCC 29573
Description: Goto, M. 1976. Ibid.

E. milletiae (Kawakami and Yoshida 1920) Magrou 1937
Kawakami, K. and S. Yoshida. 1920. Botanical Magazine Tokyo **34**:110-115; Magrou, J. 1937 in Hauduroy, P., G. Ehringer, A. Urbain, G. Guillot and J. Magrou (eds). Dictionnaire des bacteries pathogenes. Masson et Cie, Paris. p. 60. **Type strain:** NCPPB 2519
Description: Dye, D.W. 1969. New Zealand Journal of Science **12**:223-236.

E. nigrifluens Wilson, Starr and Berger 1957
Wilson, E.E., M.P. Starr and J.A. Berger. 1957. Phytopathology **47**:669-673. **Type strain:** ATCC 13028
Description: Wilson, E.E., M.P. Starr and J.A. Berger. 1957. Ibid.

E. nimipressuralis (Carter 1945) Dye 1969
Carter, J.C. 1945. Illinois Natural History Survey Bulletin **23**:407-448; Dye, D.W. 1969. New Zealand Journal of Science **12**: 833-839. **Type strain:** ATCC 9912
Description: Dye, D.W. 1969. Ibid.

E. paradisiaca Fernandez-Borrero and Lopez-Duque 1970
Fernandez-Borrero, O. and S. Lopez-Duque. 1970. Cenicafe **21**:3-44. **Type strain:** NCPPB 2511
Description: Fernandez-Borrero, O. and S. Lopez-Duque. 1970. Ibid.

E. quercina Hildebrand and Schroth 1967
Hildebrand, D.C. and M.N. Schroth. 1967. Phytopathology **57**:250-253. **Type strain:** ATCC 29281
Description: Hildebrand, D.C. and M.N. Schroth. 1967. Ibid.

E. rhapontici (Millard 1924) Burkholder 1948
(See also Pectobacterium rhapontici)
Millard, W.A. 1924. Bulletin of the University of Leeds and the Yorkshire Council for Agricultural Education **134**:1-28; Burkholder, W.H. 1948 in Breed, R.S., E.G.D. Murray and A.P. Hitchens (eds). Bergey's Manual of Determinative Bacteriology, 6th ed. The Williams and Wilkins Co., Baltimore. pp. 463-478. **Type strain:** ATCC 29283
Description: Bergey 8.

E. rubrifaciens Wilson, Zeitoun and Fredrickson 1967
Wilson, E.E., F.M. Zeitoun and D.L. Fredrickson. 1967. Phytopathology **57**: 618-621. **Type strain:** ATCC 29291
Description: Wilson, E.E., F.M. Zeitoun and D.L. Fredrickson. 1967. Ibid.

E. salicis (Day 1924) Chester 1939
Day, W.R. 1924. Oxford Forestry Memoirs **3**:1-30; Chester, F.D. 1939 in Bergey, R.S., E.G.D. Murray and A.P. Hitchens (eds). Bergey's Manual of Determinative Bacteriology, 5th ed. The Williams and Wilkins Co., Baltimore. pp. 404-420. **Type strain:** ATCC 15712
Description: Bergey 8.

E. stewartii (Smith 1898) Dye 1963
Smith, E.F. 1898. Proceedings of the American Association for the Advancement of Science **47**:422-426; Dye, D.W. 1963. New Zealand Journal of Science **6**: 495-506. **Type strain:** ATCC 8199
Description: Bergey 8.

E. tracheiphila (Smith 1895) Bergey, Harrison, Breed, Hammer and Huntoon 1923
Smith, E.F. 1895. Zentralblatt fur Bakteriologie, Parasitenkunde, Infektionskrankheiten und Hygiene. Abteilung II. **1**:364-373; Bergey, D.H., F.C. Harrison, R.S. Breed, B.W. Hammer and F.M. Huntoon. 1923. Bergey's Manual of Determinative Bacteriology, 1st ed. The Williams and Wilkins Co., Baltimore. pp. 1-442. **Type strain:** NCPPB 2452
Description: Bergey 8.

E. uredovora (Pon, Townsend, Wessman, Schmitt and Kingsolver 1954) Dye 1963
Pon, D.S., C.E. Townsend, G.E. Wessman, C.G. Schmitt and C.H. Kingsolver. 1954. Phytopathology **44**:707-710; Dye, D.W. 1963. New Zealand Journal of Science **6**:146-149. **Type strain:** ATCC 19321
Description: Bergey 8.

Erysipelothrix Rosenbach 1909

Rosenbach, F.J. 1909. Zeitschrift fur Hygiene und Infektionskrankheiten Medizinische Mikrobiologie, Immunologie und Virologie **63**:343-371. **Type species:** *E. rhusiopathiae* (Migula 1900) Buchanan 1918
Description: Bergey 8.

E. rhusiopathiae (Migula 1900) Buchanan 1918
Migula, W. 1900. System der Bakterien, Vol. 2. Gustav Fischer, Jena; Buchanan, R.E. 1918. Journal of Bacteriology **3**:27-61. **Type strain:** ATCC 19414
Description: Bergey 8.

Escherichia Castellani and Chalmers 1919
Castellani, A. and A.J. Chalmers. 1919. Manual of tropical medicine, 3rd ed. Williams, Wood and Co., New York. **Type species:** *E. coli* (Migula 1895) Castellani and Chalmers 1919
Description: Bergey 8.

E. adecarboxylata Leclerc 1962
Leclerc, H. 1962. Annales de l'Institut Pasteur (Paris) **102**:726-741. **Type strain:** ATCC 23216
Description: Leclerc, H. 1962. Ibid.

E. blattae Burgess, McDermott and Whiting 1973
Burgess, N.R.H., S.N. McDermott and J. Whiting. 1973. Journal of Hygiene (Cambridge) **71** l-7. **Type strain:** CDC 9005-74
Description: Burgess, N.R.H., S.N. McDermott and J. Whiting. 1973. Ibid.

E. coli (Migula 1895) Castellani and Chalmers 1919
Migula, W. 1895 in Engler, A. and K. Prantl (eds). Die Naturlichen Pfanzenfamilien, W. Engelmann, Leipzig, Teil I, Abteilung Ia. pp. 20-30; Castellani, A. and A.J. Chalmers. 1919. Manual of tropical medicine. 3rd ed. Williams, Wood and Co., New York. **Type strain:** ATCC 11775
Description: Bergey 8.

Eubacterium Prevot 1938
Prevot, A.R. 1938. Annales de l'Institut Pasteur (Paris) **60**:285-307. **Type species:** *E. foedans* (Klein 1908) Prevot 1938
Description: Bergey 8.

E. aerofaciens (Eggerth 1935) Prevot 1938
Eggerth, A.H. 1935. Journal of Bacteriology **30**:277-299; Prevot, A.R. 1938. Annales de l'Institut Pasteur (Paris) **60**:285-307. **Type strain:** ATCC 25986
Description: Holdeman, L.V., E.P. Cato and W.E.C. Moore (eds). 1977. Anaerobe Laboratory Manual, 4th ed. Virginia Polytechnic Institute and State University, Blacksburg, Virginia. pp. 1-156.

E. alactolyticum (Prevot and Taffanel 1942) Holdeman and Moore 1970
Prevot, A.R. and J. Taffanel. 1942. Annales de l'Institut Pasteur (Paris) **68**:259-262; Holdeman, L.V. and W.E.C. Moore. 1970. in Cato, E.P., C.S. Cummins, L.V. Holdeman, J.L. Johnson, W.E.C. Moore, R.M. Smibert and L.DS. Smith (eds). Outline of clinical methods in anaerobic bacteriology, 2nd revision. Virginia Polytechnic Institute Anaerobe Laboratory, Blacksburg, Virginia. pp. 23-30. **Type strain:** ATCC 23263
Description: Holdeman, L.V., E.P. Cato and W.E.C. Moore (eds). 1977. Anaerobe Laboratory Manual, 4th ed. Virginia Polytechnic Institute and State University, Blacksburg, Virginia. pp. 1-156.

E. biforme (Eggerth 1935) Prevot 1938
Eggerth, A.H. 1935. Journal of Bacteriology **30**:277-299; Prevot, A.R. 1938. Annales de l'Institut Pasteur (Paris) **60**:285-307. **Type strain:** ATCC 27806
Description: Holdeman, L.V. and W.E.C. Moore. 1974. International Journal of Systematic Bacteriology **24**:260-277.

E. budayi (Le Blaye and Guggenheim 1914) Holdeman and Moore 1970
Le Blaye, R. and H. Guggenheim. 1914. Manuel pratique de diagnostic bacteriologique et de technique appliquee a la determination des bacteries. Vigot Freres (eds). Paris. pp. 1-444; Holdeman, L.V. and W.E.C. Moore. 1970 in Cato, E.P., C.S. Cummins, L.V. Holdeman, J.L. Johnson, W.E.C. Moore, R.M. Smibert and L.DS. Smith (eds). Outline of clinical methods in anaerobic bacteriology, 2nd revision. Virginia Polytechnic Institute Anaerobe Laboratory, Blacksburg, Virginia. pp. 23-30. **Type strain:** ATCC 25541
Description: Holdeman, L.V., E.P. Cato and W.E.C. Moore (eds). 1977. Anaerobe Laboratory Manual, 4th ed. Virginia Polytechnic Institute and State University, Blacksburg, Virginia. pp. 1-156.

E. cellulosolvens (Bryant, Small, Bouma and Robinson 1958) Holdeman and Moore 1972
Bryant, M.P., N. Small, C. Bouma and I.M. Robinson. 1958. Journal of Bacteriology **76**:529-537; Holdeman, L.V. and W.E.C. Moore. 1972 in Holdeman, L.V. and W.E.C. Moore (eds). Anaerobe Laboratory Manual. Virginia Polytechnic Institute Anaerobe Laboratory, Blacksburg, Virginia. pp. 39-47. **Type strain:** Bryant B348
Description: Holdeman, L.V., E.P. Cato and W.E.C. Moore (eds). 1977. Anaerobe Laboratory Manual, 4th ed. Virginia Polytechnic Institute and State University, Blacksburg, Virginia. pp. 1-156.

E. combesii (Prevot and Laplanche 1947) Holdeman and Moore 1970
Prevot, A.R. and J. Laplanche. 1947. Annales de l'Institut Pasteur (Paris) **73**:687-688; Holdeman, L.V. and W.E.C. Moore. 1970 in Cato, E.P., C.S. Cummins, L.V. Holdeman, J.L. Johnson, W.E.C. Moore, R.M. Smibert and L.DS. Smith (eds). Outline of clinical methods in anaerobic bacteriology, 2nd revision. Virginia Polytechnic Institute Anaerobe Laboratory, Blacksburg, Virginia. pp. 23-30. **Type strain:** ATCC 25545
Description: Holdeman, L.V., E.P. Cato and W.E.C. Moore (eds). 1977. Anaerobe Laboratory Manual, 4th ed. Virginia Polytechnic Institute and State University, Blacksburg, Virginia. pp. 1-156.

E. contortum (Prevot 1947) Holdeman, Cato and Moore 1971
Prevot, A.R. 1947. Annales de l'Institut Pasteur (Paris) **73**:409-418; Holdeman, L.V., E.P. Cato and W.E.C. Moore. 1971. International Journal of Systematic Bacteriology **21**:304-306. **Type strain:** ATCC 25540
Description: Holdeman, L.V., E.P. Cato and W.E.C. Moore (eds). 1977. Anaerobe Laboratory Manual, 4th ed. Virginia Polytechnic Institute and State University, Blacksburg, Virginia. pp. 1-156.

E. cylindroides (Rocchi 1908) Holdeman and Moore 1970
Rocchi, G. 1908. Bulletino delle Scienze Mediche (Societa di Bologna) **8**:457-528; Holdeman, L.V. and W.E.C. Moore. 1970 in Cato, E.P., C.S. Cummins, L.V. Holdeman, J.L. Johnson, W.E.C. Moore, R.M. Smibert and L.DS. Smith (eds). Outline of clinical methods in anaerobic bacteriology, 2nd revision. Virginia Polytechnic Institute Anaerobe Laboratory, Blacksburg, Virginia. pp. 23-30. **Type strain:** ATCC 27803
Description: Holdeman, L.V., E.P. Cato and W.E.C. Moore (eds). 1977. Anaerobe Laboratory Manual, 4th ed. Virginia Polytechnic Institute and State University, Blacksburg, Virginia. pp. 1-156.

E. dolichum Moore, Johnson and Holdeman 1976
Moore, W.E.C., J.L. Johnson and L.V. Holdeman. 1976. International Journal of Systematic Bacteriology **26**:238-252. **Type strain:** ATCC 29143
Description: Moore, W.E.C., J.L. Johnson and L.V. Holdeman. 1976. Ibid.

E. eligens Holdeman and Moore 1974
Holdeman, L.V. and W.E.C. Moore. 1974. International Journal of Systematic Bacteriology **24**:260-277. **Type strain:** VPI C15-48
Description: Holdeman, L.V. and W.E.C. Moore. 1974. Ibid.

E. fissicatena Taylor 1972
Taylor, M.M. 1972. Journal of General Microbiology **71**:457-463. **Type strain:** NCIB 10446
Description: Taylor, M.M. 1972. Ibid.

E. formicigenerans Holdeman and Moore 1974
Holdeman, L.V. and W.E.C. Moore. 1974. International Journal of Systematic Bacteriology **24**:260-277. **Type strain:** ATCC 27755
Description: Holdeman, L.V. and W.E.C. Moore. 1974. Ibid.

E. hadrum Moore, Johnson and Holdeman 1976
Moore, W.E.C., J.L. Johnson and L.V. Holdeman. 1976. International Journal of Systematic Bacteriology **26**:238-252. **Type strain:** VPI B2-52
Description: Moore, W.E.C., J.L. Johnson and L.V. Holdeman. 1976. Ibid.

E. hallii Holdeman and Moore 1974
Holdeman, L.V. and W.E.C. Moore. 1974. International Journal of Systematic Bacteriology **24**:260-277. **Type strain:** VPI B4-27

Description: Holdeman, L.V. and W.E.C. Moore. 1974. Ibid.

E. lentum (Eggerth 1935) Prevot 1938
Eggerth, A.H. 1935. Journal of Bacteriology **30**:277-299; Prevot, A.R. 1938. Annales de l'Institut Pasteur (Paris) **60**:285-307. **Type strain:** ATCC 25559
Description: Holdeman, L.V., E.P. Cato and W.E.C. Moore (eds). 1977. Anaerobe Laboratory Manual, 4th ed. Virginia Polytechnic Institute and State University, Blacksburg, Virginia. pp. 1-156.

E. limosum (Eggerth 1935) Prevot 1938
Eggerth, A.H. 1935. Journal of Bacteriology **30**:277-299; Prevot, A.R. 1938. Annales de l'Institut Pasteur (Paris) **60**:285-307. **Type strain:** ATCC 8486
Description: Holdeman, L.V., E.P. Cato and W.E.C. Moore (eds). 1977. Anaerobe Laboratory Manual, 4th ed. Virginia Polytechnic Institute and State University, Blacksburg, Virginia. pp. 1-156.

E. moniliforme (Repaci 1910) Holdeman and Moore 1970
Repaci, G. 1910. Comptes Rendus des Seances de la Societe de Biologie (Paris) **68**:410-412; Holdeman, L.V. and W.E.C. Moore. 1970 in Cato, E.P., C.S. Cummins, L.V. Holdeman, J.L. Johnson, W.E.C. Moore, R.M. Smibert and L.DS. Smith (eds). Outline of clinical methods in anaerobic bacteriology, 2nd revision. Virginia Polytechnic Institute Anaerobe Laboratory, Blacksburg, Virginia. pp. 23-30. **Type strain:** ATCC 25546
Description: Holdeman, L.V., E.P. Cato and W.E.C. Moore (eds). 1977. Anaerobe Laboratory Manual, 4th ed. Virginia Polytechnic Institute and State University, Blacksburg, Virginia. pp. 1-156.

E. multiforme (Distaso 1911) Holdeman and Moore 1970
Distaso, A. 1911. Zentralblatt fur Bakteriologie, Parasitenkunde, Infektionskrankheiten und Hygiene. Abteilung I. **59**:97-103; Holdeman, L.V. and W.E.C. Moore. 1970. in Cato, E.P., C.S. Cummins, L.V. Holdeman, J.L. Johnson, W.E.C. Moore, R.M. Smibert and L.DS. Smith (eds). Outline of clinical methods in anaerobic bacteriology, 2nd revision. Virginia Polytechnic Institute Anaerobe Laboratory, Blacksburg, Virginia. pp. 23-30. **Type strain:** ATCC 25552
Description: Holdeman, L.V., E.P. Cato and W.E.C. Moore (eds). 1977. Anaerobe Laboratory Manual, 4th ed. Virginia Polytechnic Institute and State University, Blacksburg, Virginia. pp. 1-156.

E. nitritogenes Prevot 1940
Prevot, A.R. 1940. Comptes Rendus des Seances de la Societe de Biologie (Paris) **134**:353-355. **Type strain:** ATCC 25547
Description: Holdeman, L.V., E.P. Cato and W.E.C. Moore (eds). 1977. Anaerobe Laboratory Manual, 4th ed. Virginia Polytechnic Institute and State University, Blacksburg, Virginia. pp. 1-156.

E. plexicaudatum Wilkins, Fulghum and Wilkins 1974
Wilkins, T.D., R.S. Fulghum and J.H. Wilkins. 1974. International Journal of Systematic Bacteriology **24**:408-411. **Type strain:** VPI 7582
Description: Wilkins, T.D., R.S. Fulghum and J.H. Wilkins. 1974. Ibid.

E. ramulus Moore, Johnson and Holdeman 1976
Moore, W.E.C., J.L. Johnson and L.V. Holdeman. 1976. International Journal of Systematic Bacteriology **26**:238-252. **Type strain:** ATCC 29099
Description: Moore, W.E.C., J.L. Johnson and L.V. Holdeman. 1976. Ibid.

E. rectale (Hauduroy, Ehringer, Urbain, Guillot and Magrou 1937) Prevot 1938
Hauduroy, P., G. Ehringer, A. Urbain, G. Guillot and J. Magrou. 1937. Dictionnaire des bacteries pathogenes. Masson et Cie, Paris; Prevot, A.R. 1938. Annales de l'Institut Pasteur (Paris) **60**:285-307. **Type strain:** VPI 0989
Description: Holdeman, L.V., E.P. Cato and W.E.C. Moore (eds). 1977. Anaerobe Laboratory Manual, 4th ed. Virginia Polytechnic Institute and State University, Blacksburg, Virginia. pp. 1-156.

E. ruminantium Bryant 1959
Bryant, M.P. 1959. Bacteriological Reviews **23**:125-153. **Type strain:** ATCC 17233
Description: Holdeman, L.V., E.P. Cato and W.E.C. Moore (eds). 1977. Anaerobe Laboratory Manual, 4th ed. Virginia Polytechnic Institute and State University, Blacksburg, Virginia. pp. 1-156.

E. saburreum (Prevot 1966) Holdeman and Moore 1970
Prevot, A.R. 1966. Manual for the classification and determination of the anaerobic bacteria. 1st American ed. transl. by V. Fredette, Lea and Febiger, Philadelphia. pp. 1-402; Holdeman, L.V. and W.E.C. Moore. 1970. In Cato, E.P., C.S. Cummins, L.V. Holdeman, J.L. Johnson, W.E.C. Moore, R.M. Smibert and L.DS. Smith (eds). Outline of clinical methods in anaerobic bacteriology, 2nd revision. Virginia Polytechnic Institute Anaerobe Laboratory, Blacksburg, Virginia. pp. 23-30. **Type strain:** VPI 11763
Description: Holdeman, L.V., E.P. Cato and W.E.C. Moore (eds). 1977. Anaerobe Laboratory Manual, 4th ed. Virginia Polytechnic Institute and State University, Blacksburg, Virginia. pp. 1-156.

E. siraeum Moore, Johnson and Holdeman 1976
Moore, W.E.C., J.L. Johnson and L.V. Holdeman. 1976. International Journal of Systematic Bacteriology **26**:238-252. **Type strain:** ATCC 29066
Description: Moore, W.E.C., J.L. Johnson and L.V. Holdeman. 1976. Ibid.

E. tarantellus Udey, Young and Sallman 1977
Udey, L.R., E. Young and B. Sallman. 1977 in International Journal of Systematic Bacteriology **27**:306; Effective publication Udey, L.R., E. Young and B. Sallman. 1977. Journal of the Fisheries Research Board of Canada **34**:402-409. **Type strain:** ATCC 29255
Description: Udey, L.R., E. Young and B. Sallman. 1977. Ibid.

E. tenue (Bergey, Harrison, Breed, Hammer and Huntoon 1923) Holdeman and Moore 1970
Bergey, D.H., F.C. Harrison, R.S. Breed, B.W. Hammer and F.M. Huntoon. 1923. Bergey's Manual of Determinative Bacteriology, 1st ed. The Williams and Wilkins Co., Baltimore. pp. 1-442; Holdeman, L.V. and W.E.C. Moore. 1970 in Cato, E.P., C.S. Cummins, L.V. Holdeman, J.L. Johnson, W.E.C. Moore, R.M. Smibert and L.DS. Smith (eds). Outline of clinical methods in anaerobic bacteriology, 2nd revision. Virginia Polytechnic Institute Anaerobe Laboratory, Blacksburg, Virginia. pp. 23-30. **Type strain:** ATCC 25553
Description: Holdeman, L.V., E.P. Cato and W.E.C. Moore (eds). 1977. Anaerobe Laboratory Manual, 4th ed. Virginia Polytechnic Institute and State University, Blacksburg, Virginia. pp. 1-156.

E. tortuosum (Debono 1912) Prevot 1938
Debono, M. 1912. Zentralblatt fur Bakteriologie, Parasitenkunde, Infektionskrankheiten und Hygiene. Abteilung I. **62**:229-234; Prevot, A.R. 1938. Annales de l'Institut Pasteur (Paris) **60**:285-307. **Type strain:** ATCC 25548
Description: Holdeman, L.V., E.P. Cato and W.E.C. Moore (eds). 1977. Anaerobe Laboratory Manual, 4th ed. Virginia Polytechnic Institute and State University, Blacksburg, Virginia. pp. 1-156.

E. ventriosum (Tissier 1908) Prevot 1938
Tissier, H. 1908. Annales de l'Institut Pasteur (Paris) **22**:189-208; Prevot, A.R. 1938. Annales de l'Institut Pasteur (Paris) **60**:285-307. **Type strain:** ATCC 27560
Description: Holdeman, L.V., E.P. Cato and W.E.C. Moore (eds). 1977. Anaerobe Laboratory Manual, 4th ed. Virginia Polytechnic Institute and State University, Blacksburg, Virginia. pp. 1-156.

Excellospora Agre and Guzeva 1975
Agre, N.S. and L.N. Guzeva. 1975. Mikrobiologiya **44**:459-463. **Type species:** *E. viridilutea* Agre and Guzeva 1975
Description: Agre, N.S. and L.N. Guzeva. 1975. Ibid.

E. viridilutea Agre and Guzeva 1975
Agre, N.S. and L.N. Guzeva. 1975. Mikrobiologiya **44**:459-463. **Type strain:** INMI 187
Description: Agre, N.S. and L.N. Guzeva. 1975. Ibid.

Flavobacterium Bergey, Harrison, Breed, Hammer and Huntoon 1923
Bergey, D.H., F.C. Harrison, R.S. Breed, B.W. Hammer and F.M. Huntoon. 1923. Bergey's Manual of Determinative Bacteriology, 1st ed. The Williams and Wilkins Co., Baltimore. pp. 1-442. **Type species:** *F. aquatile* (Frankland and Frankland 1889) Bergey, Harrison, Breed, Hammer and Huntoon 1923

Description: Bergey 8.

F. acidificum Steinhaus 1941
Steinhaus, E.A. 1941. Journal of Bacteriology **42**:757-790. **Type strain:** NCIB 9891
Description: Steinhaus, E.A. 1941. Ibid.

F. acidurans Millar 1973
Millar, W.N. 1973. International Journal of Systematic Bacteriology **23**:142-150. **Type strain:** ATCC 27383
Description: Millar, W.N. 1973. Ibid.

F. aquatile (Frankland and Frankland 1889) Bergey, Harrison, Breed, Hammer and Huntoon 1923
Frankland, G.C. and P.F. Frankland. 1889. Zeitschrift fur Hygiene und Infektionskrankheiten Medizinische Mikrobiologie, Immunologie und Virologie **6**:373-400; Bergey, D.H., F.C. Harrison, R.S. Breed, B.W. Hammer and F.M. Huntoon. 1923. Bergey's Manual of Determinative Bacteriology, 1st ed. The Williams and Wilkins Co., Baltimore. pp. 1-442. **Type strain:** ATCC 11947
Description: Weeks, O.B. 1955. Journal of Bacteriology **69**:649-658.

F. balustinum Harrison 1929
Harrison, F.C. 1929. Canadian Journal of Research **1**:214-239. **Type strain:** NCTC 11212
Description: Harrison, F.C. 1929. Ibid.

F. capsulatum Leifson 1962
Leifson, E. 1962. International Bulletin of Bacteriological Nomenclature and Taxonomy **12**:161-170. **Type strain:** ATCC 14666
Description: Leifson, E. 1962. Ibid.

F. devorans (Zimmermann 1890) Bergey, Harrison, Breed, Hammer and Huntoon 1923
Zimmermann, O.E.R. 1890. 1. Reihe, Elfter Bericht der Naturwissenschaftlichen Gesellschaft Chemnitz, pp. 54-154; Bergey, D.H., F.C. Harrison, R.S. Breed, B.W. Hammer and F.M. Huntoon. 1923. Bergey's Manual of Determinative Bacteriology, 1st ed. The Williams and Wilkins Co., Baltimore. pp. 1-442. **Type strain:** ATCC 10829
Description: Bergey 8.

F. esteraromaticum (Omelianski 1923) Bergey, Harrison, Breed, Hammer and Huntoon 1930
Omelianski, V.L. 1923. Journal of Bacteriology **8**:393-419; Bergey, D.H., F.C. Harrison, R.S. Breed, B.W. Hammer and F.M. Huntoon (eds). 1930. Bergey's Manual of Determinative Bacteriology, 3rd ed. The Williams and Wilkins Co., Baltimore. pp. 1-589. **Type strain:** ATCC 8091
Description: Bergey, D.H., F.C. Harrison, R.S. Breed, B.W. Hammer and F.M. Huntoon. 1930. Ibid.

F. ferrugineum Sickles and Shaw 1934
Sickles, G.M. and M. Shaw. 1934. Journal of Bacteriology **28**:415-431. **Type strain:** ATCC 13524
Description: Sickles, G.M. and M. Shaw. 1934. Ibid.

F. halmephilium Elazari-Volcani 1940
Elazari-Volcani, B. 1940. Studies on the microflora of the Dead Sea. In Hebrew, English summary. Thesis, Hebrew University Jerusalem. **Type strain:** ATCC 19717
Description: Elazari-Volcani, B. 1940. Ibid.

F. heparinum Payza and Korn 1956
Payza, A.N. and E.D. Korn. 1956. Journal of Biological Chemistry **223**:853-858. **Type strain:** ATCC 13125
Description: Payza, A.N. and E.D. Korn. 1956. Ibid.

F. indoltheticum Campbell and Williams 1951
Campbell, L.L. and O.B. Williams. 1951. Journal of General Microbiology **5**:894-905. **Type strain:** ATCC 27950
Description: Campbell, L.L. and O.B. Williams. 1951. Ibid.

F. marinotypicum ZoBell and Upham 1944
ZoBell, C.E. and H.C. Upham. 1944. Bulletin Scripps Institution of Oceanography University of California (Technical Series) **5**:239-292. **Type strain:** ATCC 19260
Description: Zobell, C.E. and H.C. Upham. 1944. Ibid.

F. meningosepticum King 1959
King, E.O. 1959. American Journal of Clinical Pathology **31**:241-247. **Type strain:** ATCC 13253
Description: King, E.O. 1959. Ibid.

F. oceanosedimentum Carty and Litchfield 1978
Carty, C.E. and C.D. Litchfield. 1978. International Journal of Systematic Bacteriology **28**:561-566. **Type strain:** ATCC 31317
Description: Carty, C.E. and C.D. Litch-

field. 1978. Ibid.

F. odoratum Stutzer 1929
Stutzer, M. 1929. Zentralblatt fur Bakteriologie, Parasitenkunde, Infektionskrankheiten und Hygiene. Abteilung I. **113**:219-225. **Type strain:** ATCC 4651
Description: Stutzer, M. 1929. Ibid.

F. okeanokoites ZoBell and Upham 1944
ZoBell, C.E. and H.C. Upham. 1944. Bulletin Scripps Institution of Oceanography University of California (Technical Series) **5**:239-292. **Type strain:** CCM 320
Description: Zobell, C.E. and H.C. Upham. 1944. Ibid.

F. resinovorum Delaporte and Daste 1956
Delaporte, B. and P. Daste. 1956. Compte Rendu de l'Academie des Sciences (Paris) **242**:831-834. **Type strain:** NCIB 8767
Description: Delaporte, B. and P. Daste. 1956. Ibid.

F. uliginosum ZoBell and Upham 1944
ZoBell, C.E. and H.C. Upham. 1944. Bulletin Scripps Institution of Oceanography University of California (Technical Series) **5**:239-292. **Type strain:** ATCC 14397
Description: Zobell, C.E. and H.C. Upham. 1944. Ibid.

Flectobacillus Larkin, Williams and Taylor 1977
Larkin, J.M., P.M. Williams and R. Taylor. 1977. International Journal of Systematic Bacteriology **27**:147-156. **Type species:** *F. major* Larkin, Williams and Taylor 1977
Description: Larkin, J.M., P.M. Williams and R. Taylor. 1977. Ibid.

F. major (Gromov 1963) Larkin, Williams and Taylor 1977
Gromov, B.V. 1963. Doklady Akademii Nauk SSSR **152**:733-734; Larkin, J.M., P.M. Williams and R. Taylor. 1977. International Journal of Systematic Bacteriology **27**:147-156. **Type strain:** DSM 103
Description: Larkin, J.M., P.M. Williams and R. Taylor. 1977. Ibid.

F. marinus (Raj 1976) Borrall and Larkin 1978
Raj, H.D. 1976. International Journal of Systematic Bacteriology **26**:528-544; Borrall, R. and J.M. Larkin. 1978. International Journal of Systematic Bacteriology **28**:341-343. **Type strain:** ATCC 25205
Description: Raj, H.D. 1976. Ibid.

Flexibacter Soriano 1945
Soriano, S. 1945. Ciencia e Investigacion (Buenos Aires) **1**:92-93. **Type species:** *F. flexilis* Soriano 1945
Description: Bergey 8.

F. aggregans (Lewin 1969) Leadbetter 1974
Lewin, R.A. 1969. Journal of General Microbiology **58**:189-206; Leadbetter, E.R. 1974 Bergey 8. **Type strain:** ATCC 23162
Description: Lewin, R.A. 1969. Ibid.

F. aurantiacus Lewin 1969
Lewin, R.A. 1969. Journal of General Microbiology **58**:189-206. **Type strain:** ATCC 23107
Description: Lewin, R.A. 1969. Ibid.

F. flexilis Soriano 1945
Soriano, S. 1945. Ciencia e Investigacion (Buenos Aires) **1**:92-93. **Type strain:** ATCC 23079
Description: Soriano, S. 1945. Ibid.

F. litoralis Lewin 1969
Lewin, R.A. 1969. Journal of General Microbiology **58**:189-206. **Type strain:** ATCC 23117
Description: Lewin, R.A. 1969. Ibid.

F. polymorphus Lewin 1974
Lewin, R.A. 1974. Journal of General Microbiology **82**:393-403. **Type strain:** ATCC 27820
Description: Lewin, R.A. 1974. Ibid.

F. roseolus Lewin 1969
Lewin, R.A. 1969. Journal of General Microbiology **58**:189-206. **Type strain:** ATCC 23088
Description: Lewin, R.A. 1969. Ibid.

F. ruber Lewin 1969
Lewin, R.A. 1969. Journal of General Microbiology **58**:189-206. **Type strain:** ATCC 23103
Description: Lewin, R.A. 1969. Ibid.

F. sancti Lewin 1969
Lewin, R.A. 1969. Journal of General Microbiology **58**:189-206. **Type strain:** ATCC 23092
Description: Lewin, R.A. 1969. Ibid.

F. tractuosus (Lewin 1969) Leadbetter 1974
Lewin, R.A. 1969. Journal of General Microbiology **58**:189-206; Leadbetter,

E.R. 1974 Bergey 8. **Type strain:** ATCC 23168
Description: Lewin, R.A. 1969. Ibid.

Flexithrix Lewin 1970
Lewin, R.A. 1970. Canadian Journal of Microbiology **16**:511-515. **Type species:** *F. dorotheae* Lewin 1970
Description: Bergey 8.

F. dorotheae Lewin 1970
Lewin, R.A. 1970. Canadian Journal of Microbiology **16**:511-515. **Type strain:** ATCC 23163
Description: Lewin, R.A. 1970. Ibid.

Francisella Dorofeev 1947
Dorofeev, K.A. 1947. Symposium Research Works Institute Epidemiology and Microbiology Chita **1**:170-180. **Type species:** *F. tularensis* (McCoy and Chapin 1912) Dorofeev 1947
Description: Bergey 8.

F. novicida (Larson, Wicht and Jellison 1955) Olsufiev, Emelyanova and Dunaeva 1959
Larson, C.L., W. Wicht and W.L. Jellison. 1955. Public Health Report **70**:253-258; Olsufiev, N.G., O.S. Emelyanova and T.N. Dunaeva. 1959. Journal of Hygiene, Epidemiology, Microbiology and Immunology (Prague) **3**:138-149. **Type strain:** ATCC 15482
Description: Bergey 8.

F. tularensis (McCoy and Chapin 1912) Dorofeev 1947
McCoy, G.W. and C.W. Chapin. 1912. Journal of Infectious Diseases **10**:61-72; Dorofeev, K.A. 1947. Symposium Research Works Institute Epidemiology and Microbiology Chita **1**:170-180. **Type strain:** ATCC 6223
Description: Bergey 8.

Frankia Brunchorst 1886
Brunchorst, J. 1886. Botanische Institut Tubingen **2**:151-177. **Type species:** *F. alni* (Woronin 1866) Von Tubeuf 1895
Description: Bergey 8.

F. alni (Woronin 1866) Von Tubeuf 1895
Woronin, M. 1866. Memoires de l'Academie des Sciences St Petersburg **10**: 1-10; Von Tubeuf, K. 1895. Pflanzenkrankheiten durch kryptogame Parasiten verursacht. Verlag J Springer, Berlin. pp. 1-599. **Type strain:** no culture isolated.
Description: Bergey 8.

Fusobacterium Knorr 1922
Knorr, M. 1922. Zentralblatt fur Bakteriologie, Parasitenkunde, Infektionskrankheiten und Hygiene. Abteilung I. **89**:4-22. **Type species:** *F. nucleatum* Knorr 1922
Description: Bergey 8.

F. gonidiaformans (Tunnicliff and Jackson 1925) Moore and Holdeman 1970
Tunnicliff, R. and L. Jackson. 1925. Journal of Infectious Diseases **36**:430-438; Moore, W.E.C. and L.V. Holdeman. 1970 in Cato, E.P., C.S. Cummins, L.V. Holdeman, J.L. Johnson, W.E.C. Moore, R.M. Smibert and L.DS. Smith (eds). Outline of clinical methods in anaerobic bacteriology 2nd revision. Virginia Polytechnic Institute Anaerobe Laboratory, Blacksburg, Virginia. **Type strain:** ATCC 25563
Description: Holdeman, L.V., E.P. Cato and W.E.C. Moore (eds). 1977. Anaerobe Laboratory Manual, 4th ed. Virginia Polytechnic Institute and State University, Blacksburg, Virginia. pp. 1-156.

F. mortiferum (Harris 1901) Moore and Holdeman 1970
Harris, N.M. 1901. Journal of Experimental Medicine **6**:519-547; Moore, W.E.C. and L.V. Holdeman 1970 in Cato, E.P., C.S. Cummins, L.V. Holdeman, J.L. Johnson, W.E.C. Moore, R.M. Smibert and L.DS. Smith (eds). Outline of clinical methods in anaerobic bacteriology 2nd revision. Virginia Polytechnic Institute Anaerobe Laboratory, Blacksburg, Virginia. **Type strain:** ATCC 25557
Description: Holdeman, L.V., E.P. Cato and W.E.C. Moore (eds). 1977. Anaerobe Laboratory Manual 4th ed. Virginia Polytechnic Institute and State University, Blacksubrg, Virginia. pp. 1-156.

F. naviforme (Jungano 1909) Moore and Holdeman 1970
Jungano, M. 1909. Comptes Rendus des Seances de la Societe de Biologie (Paris) **66**:122-124; Moore, W.E.C. and L.V. Holdeman. 1970 in Cato, E.P., C.S. Cummins, L.V. Holdeman, J.L. Johnson, W.E.C. Moore, R.M. Smibert and L.DS. Smith (eds). Outline of clinical in anaero-

bic bacteriology, 2nd revision. Virginia Polytechnic Institute Anaerobe Laboratory, Blacksburg, Virginia. **Type strain:** ATCC 25832

Description: Holdeman, L.V., E.P. Cato and W.E.C. Moore (eds). 1977. Anaerobe Laboratory Manual 4th ed. Virginia Polytechnic Institute and State University, Blacksburg, Virginia. pp. 1-156.

F. necrogenes (Weinberg, Nativelle and Prevot 1937) Moore and Holdeman 1970

Weinberg, M., R. Nativelle and A.R. Prevot. 1937. Masson et Cie, Paris. pp. 1-1186; Moore, W.E.C. and L.V. Holdeman. 1970 in Cato, E.P., C.S. Cummins, L.V. Holdeman, J.L. Johnson, W.E.C. Moore, R.M. Smibert and L.DS. Smith (eds). Outline of clinical methods in anaerobic bacteriology, 2nd revision. Virginia Polytechnic Institute Anaerobe Laboratory, Blacksburg, Virginia. **Type strain:** ATCC 25556

Description: Holdeman, L.V., E.P. Cato and W.E.C. Moore (eds). 1977. Anaerobe Laboratory Manual 4th ed. Virginia Polytechnic Institute and State University, Blacksburg, Virginia. pp. 1-156.

F. necrophorum (Flugge 1886) Moore and Holdeman 1969

Flugge, C. 1886. Die Mikroorganismen F.C.W. Vogel, Leipzig. pp. 1-692; Moore, W.E.C. and L.V. Holdeman. 1969 in Cato, E.P., C.S. Cummins, L.V. Holdeman, J.L. Johnson, W.E.C. Moore, R.M. Smibert and L.DS. Smith (eds). Outline of clinical methods in anaerobic bacteriology, lst revision. Virginia Polytechnic Institute Anaerobe Laboratory, Blacksburg, Virginia. pp. 1-107. **Type strain:** ATCC 25286

Description: Holdeman, L.V., E.P. Cato and W.E.C. Moore (eds). 1977. Anaerobe Laboratory Manual, 4th ed. Virginia Polytechnic Institute and State University, Blacksburg, Virginia. pp. 1-156.

F. nucleatum Knorr 1922

Knorr, M. 1922. Zentralblatt fur Bakteriologie, Parasitenkunde, Infektionskrankheiten und Hygiene. Abteilung I. **89**:4-22. **Type strain:** ATCC 25586

Description: Holdeman, L.V., E.P. Cato and W.E.C. Moore (eds). 1977. Anaerobe Laboratory Manual, 4th ed. Virginia Polytechnic Institute and State University, Blacksburg, Virginia. pp. 1-156.

F. perfoetens (Tissier 1905) Moore and Holdeman 1973

Tissier, H. 1905. Annales de l'Institut Pasteur (Paris) **19**:109-123; Moore, W.E.C. and L.V. Holdeman. 1973. International Journal of Systematic Bacteriology **23**:69-74. **Type strain:** ATCC 29250

Description: Van Assche, P.F. and A.T. Wilssens. 1977. International Journal of Systematic Bacteriology **27**:1-5.

F. plauti Seguin 1928

Seguin, P. 1928. Comptes Rendus des Seances de la Societe de Biologie (Paris) **99**:439-442. **Type strain:** ATCC 29863

Description: Holdeman, L.V., E.P. Cato and W.E.C. Moore (eds). 1977. Anaerobe Laboratory Manual, 4th ed. Virginia Polytechnic Institute and State University, Blacksburg, Virginia. pp. 1-156.

F. prausnitzii (Hauduroy, Ehringer, Urbain, Guillot and Magrou 1937) Moore and Holdeman 1970

Hauduroy, P., G. Ehringer, A. Urbain, G. Guillot, and J. Magrou. 1937. Dictionnaire des bacteries pathogenes. Masson et Cie, Paris. pp. 1-597; Moore, W.E.C. and L.V. Holdeman. 1970 in Cato, E.P., C.S. Cummins, L.V. Holdeman, J.L. Johnson, W.E.C. Moore, R.M. Smibert and L.DS. Smith (eds). Outline of clinical methods in anaerobic bacteriology, 2nd revision. Virginia Polytechnic Institute Anaerobe Laboratory, Blacksburg, Virginia. **Type strain:** ATCC 27768

Description: Holdeman, L.V., E.P. Cato and W.E.C. Moore (eds). 1977. Anaerobe Laboratory Manual, 4th ed. Virginia Polytechnic Institute and State University, Blacksburg, Virginia. pp. 1-156.

F. russii (Hauduroy, Ehringer, Urbain, Guillot and Magrou 1937) Moore and Holdeman 1970

Hauduroy, P., G. Ehringer, A. Urbain, G. Guillot, and J. Magrou. 1937. Dictionnaire des bacteries pathogenes. Masson et Cie, Paris. pp. 1-597; Moore, W.E.C. and L.V. Holdeman. 1970 in Cato, E.P., C.S. Cummins, L.V. Holdeman, J.L. Johnson, W.E.C. Moore, R.M. Smibert and L.DS. Smith (eds). Outline of clinical methods in anaerobic bacteriology, 2nd revision. Vir-

ginia Polytechnic Institute Anaerobe Laboratory, Blacksburg, Virginia. **Type strain:** ATCC 25533
Description: Holdeman, L.V., E.P. Cato and W.E.C. Moore (eds). 1977. Anaerobe Laboratory Manual, 4th ed. Virginia Polytechnic Institute and State University, Blacksburg, Virginia. pp. 1-156.

F. varium (Eggerth and Gagnon 1933) Moore and Holdeman 1969
Eggerth, A.H. and B.H. Gagnon. 1933. Journal of Bacteriology **25**:389-413; Moore, W.E.C. and L.V. Holdeman. 1969 in Cato, E.P., C.S. Cummins, L.V. Holdeman, J.L. Johnson, W.E.C. Moore, R.M. Smibert and L.DS Smith (eds). Outline of clinical methods in anaerobic bacteriology, 1st revision. Virginia Polytech Institute Anaerobe Laboratory, Blacksburg, Virginia. pp. 1-107. **Type strain:** ATCC 8501
Description: Holdeman, L.V., E.P. Cato and W.E.C. Moore (eds). 1977 . Anaerobe Laboratory Manual, 4th ed. Virginia Polytechnic Institute and State University, Blacksburg, Virginia. pp. 1-156.

Gallionella Ehrenberg 1838
Ehrenberg, C.G. 1838. Die Infusionsthierchen als vollkommene Organismen. L. Voss, Leipzig. **Type species:** *G. ferruginea* Ehrenberg 1838
Description: Bergey 8.

G. ferruginea Ehrenberg 1838
Ehrenberg, C.G. 1838. Die Infusionsthierchen als vollkommene Organismen. L. Voss, Leipzig. **Type strain:** no culture isolated.
Description: Bergey 8.

Gemella Berger 1960
Berger, U. 1960. Zeitschrift fur Hygiene und Infektionskrankheiten Medizinische Mikrobiologie, Immunologie und Virologie **146**:253-259. **Type species:** *G. haemolysans* (Thjotta and Boe 1938) Berger 1960
Description: Bergey 8.

G. haemolysans (Thjotta and Boe 1938) Berger 1960
Thjotta, T. and J. Boe. 1938. Acta Pathologica et Bacteriologica Scandinavica Supplement **37**:527-531; Berger, U. 1960. Zeitschrift fur Hygiene und Infektionskrankheiten Medizinische Mikrobiologie, Immunologie und Virologie **146**:253-259. **Type strain:** ATCC 10379
Description: Bergey 8.

Gemmiger Gossling and Moore 1975
Gossling, J. and W.E.C. Moore. 1975. International Journal of Systematic Bacteriology **25**:202-207. **Type species:** *G. formicilis* Gossling and Moore 1975
Description: Gossling, J. and W.E.C. Moore. 1975. Ibid.

G. formicilis Gossling and Moore 1975
Gossling, J. and W.E.C. Moore. 1975. International Journal of Systematic Bacteriology **25**:202-207. **Type strain:** ATCC 27749
Description: Gossling, J. and W.E.C. Moore. 1975. Ibid.

Geodermatophilus Luedemann 1968
Luedemann, G.M. 1968. Journal of Bacteriology **96**:1848-1858. **Type species:** *G. obscurus* Luedemann 1968
Description: Bergey 8.

G. obscurus Luedemann 1968
Luedemann, G.M. 1968. Journal of Bacteriology **96**:1848-1858. **Type strain:** ATCC 25078
Description: Luedemann, G.M. 1968. Ibid.

Gluconobacter Asai 1935
Asai, T. 1935. Journal of the Agricultural Chemical Society of Japan **11**:674-708. **Type species:** *G. oxydans* (Henneberg 1897) De Ley 1961
Description: Bergey 8.

G. oxydans (Henneberg 1897) De Ley 1961
Henneberg, W. 1897. Zentralblatt fur Bakteriologie, Parasitenkunde, Infektionskrankheiten und Hygiene. Abteilung II. **3**:223-231; De Ley, J. 1961. Journal of General Microbiology **24**:31-50. **Type strain:** ATCC 19357
Description: Bergey 8.

G. oxydans subsp. *industrius* (Henneberg 1898) De Ley and Frateur 1974
Henneberg, W. 1898. Deutsche Essigindustrie **2**:145-148; De Ley, J. and J. Frateur. 1974 in Bergey 8. **Type strain:** LMD 24.1
Description: Bergey 8.

G. oxydans subsp. *melanogenes* (Beijerinck 1911) De Ley and Frateur 1974
Beijerinck, M. 1911. Zentralblatt fur Bakteriologie, Parasitenkunde, Infektionskrankheiten und Hygiene. Abteilung II. **29**:169-176; De Ley, J. and J. Frateur. 1974 in Bergey 8. **Type strain:** LMD 29.2
Description: Bergey 8.

G. oxydans subsp. *oxydans* (Henneberg 1897) De Ley 1961
Henneberg, W. 1897. Zentralblatt fur Bakteriologie, Parasitenkunde, Infektionskrankheiten und Hygiene. Abteilung II. **3**:223-231; De Ley, J. 1961. Journal of General Microbiology **24**:31-50. **Type strain:** ATCC 19357
Description: Bergey 8.

G. oxydans subsp. *sphaericus* Ameyama 1975
Ameyama, M. 1975. International Journal of Systematic Bacteriology **25**:365-370. **Type strain:** IFO 12467
Description: Ameyama, M. 1975. Ibid.

G. oxydans subsp. *suboxydans* (Kluyver and de Leeuw 1924) De Ley and Frateur 1974
Kluyver, A.J. and F.J.G. de Leeuw. 1924. Tijdschrift voor vergeeijkende geneeskunde, gezondheidsleer en parasitaire en infectieziekten **10**:170-281; De Ley, J. and J. Frateur. 1974 in Bergey 8. **Type strain:** LMD 23.2
Description: Bergey 8.

Haemobartonella Tyzzer and Weinman 1939
Tyzzer, E.E. and D. Weinman. 1939. American Journal of Hygiene **30**:141-157. **Type species:** *H. muris* (Mayer 1921) Tyzzer and Weinman 1939
Description: Bergey 8.

H. muris (Mayer 1921) Tyzzer and Weinman 1939
Mayer, M. 1921. Archiv fur Schiffs und Tropenhygiene **25**:150-152; Tyzzer, E.E. and D. Weinman. 1939. American Journal of Hygiene **30**:141-157. **Type strain:** no culture isolated.
Description: Bergey 8.

Haemophilus Winslow, Broadhurst, Buchanan, Krumwiede, Rogers and Smith 1917
Winslow, C.E.A., J. Broadhurst, R.E. Buchanan, C. Krumwiede Jr., L.A. Rogers and G.H. Smith. 1917. Journal of Bacteriology **2**:505-566. **Type species:** *H. influenzae* (Lehmann and Neumann 1896) Winslow, Broadhurst, Buchanan, Krumwiede, Rogers and Smith 1917
Description: Bergey 8.

H. aegyptius (Trevisan 1889) Pittman and Davis 1950
Trevisan, V. 1889. I. Generi e le Specie delle Battieriacee. Zanaboni and Gabuzzi, Milano; Pittman, M. and D.J. Davis. 1950. Journal of Bacteriology **59**:413-426. **Type strain:** ATCC 11116
Description: Pittman, M. and D.J. Davis. 1950. Ibid.

H. aphrophilus Khairat 1940
Khairat, O. 1940. Journal of Pathology and Bacteriology **50**:497-505. **Type strain:** NCTC 5906
Description: Kilian, M. 1976. Journal of General Microbiology **93**:9-62.

H. avium Hinz and Kunjara 1977
Hinz, K.H. and C. Kunjara. 1977. International Journal of Systematic Bacteriology **27**:324-329. **Type strain:** ATCC 29546
Description: Hinz, K.H. and C. Kunjara. 1977. Ibid.

H. ducreyi (Neveu-Lemaire 1921) Bergey, Harrison, Breed, Hammer and Huntoon 1923
Neveu-Lemaire, M. 1921. Precis de la parasitologie humaine, 5th ed. J. Lemaire Paris; Bergey, D.H., F.C. Harrison, R.S. Breed, B.W. Hammer and F.M. Huntoon. 1923. Bergey's Manual of Determinative Bacteriology, 1st ed. The Williams and Wilkins Co., Baltimore. p. 271. **Type strain:** CIP 542
Description: Kilian, M. 1976. Journal of General Microbiology **93**:9-62.

H. haemoglobinophilus (Lehmann and Neumann 1907) Murray 1939
Lehmann, K.B. and R. Neumann. 1907. Lehmann's Medizin, Handatlanten. X Atlas und Grundriss der Bakteriologie und Lehrbuch der speziellen bakteriologischen Diagnostik. Vol. 2, 4th ed. p. 270; Murray, R.G.E. 1938 in Bergey, D.H., R.S. Breed, E.G.D. Murray and A.P. Hitchens (eds). Bergey's Manual of Determinative Bacteriology, 5th ed. The Williams and Wilkins Co., Baltimore. p. 309. **Type strain:** NCTC 1659
Description: Kilian, M. 1976. Journal of General Microbiology **93**:9-62.

H. haemolyticus Bergey, Harrison, Breed,

Hammer and Huntoon 1923
Bergey, D.H., F.C. Harrison, R.S. Breed, B.W. Hammer and F.M. Huntoon. 1923. Bergey's Manual of Determinative Bacteriology, 1st ed. The Williams and Wilkins Co., Baltimore. p. 271. **Type strain:** NCTC 10659
Description: Kilian, M. 1976. Journal of General Microbiology **93**:9-62.

H. influenzae (Lehmann and Neumann 1896) Winslow, Broadhurst, Buchanan, Krumwiede, Rogers and Smith 1917
Lehmann, K.B. and R. Neumann. 1896. Atlas und Grundriss der Bakteriologie und Lehrbuch der speziellen bakteriologischen Diagnostik. 1st ed. J.F. Lehmann, Munchen; Winslow, C.E.A., J. Broadhurst, R.E. Buchanan, C. Krumwiede Jr., L.A. Rogers and G.H. Smith. 1917. Journal of Bacteriology **2**:505-566. **Type strain:** NCTC 8143
Description: Kilian, M. 1976. Journal of General Microbiology **93**:9-63.

H. paragallinarum Biberstein and White 1969
Biberstein, E.L. and D.C. White. 1969. Journal of Medical Microbiology **2**:75-78. **Type strain:** ATCC 29545
Description: Hinz, K.H. and C. Kunjara. 1977. International Journal of Systematic Bacteriology **27**:324-329.

H. parahaemolyticus Pittman 1953
Pittman, M. 1953. Journal of Bacteriology **65**:750-751. **Type strain:** ATCC 10014
Description: Kilian, M. 1976. Journal of General Microbiology **93**:9-62.

H. parainfluenzae Rivers 1922
Rivers, T.M. 1922. Johns Hopkins Hospital Bulletin **33**:429-431. **Type strain:** NCTC 7857
Description: Kilian, M. 1976. Journal of General Microbiology **93**:9-62.

H. paraphrohaemolyticus Zinnemann, Rogers, Frazer and Devaraj 1971
Zinnemann, K., K.B. Rogers, J. Frazer and S.K. Devaraj. 1971. Journal of Medical Microbiology **4**:139-143. **Type strain:** NCTC 10670
Description: Zinnemann, K., K.B. Rogers, J. Frazer and S.K. Devaraj. 1971. Ibid.

H. paraphrophilus Zinnemann, Rogers, Frazer and Boyce 1968
Zinnemann, K., K.B. Rogers, J. Frazer and J.M.H. Boyce. 1968. Journal of Pathology and Bacteriology **96**:413-419. **Type strain:** NCTC 10557
Description: Kilian, M. 1976. Journal of General Microbiology **93**:9-62.

H. parasuis Biberstein and White 1969
Biberstein E.L. and D.C. White. 1969. Journal of Medical Microbiology **2**:75-78. **Type strain:** ATCC 19417
Description: Kilian, M. 1976. Journal of General Microbiology **93**:9-62.

H. piscium Sniezko, Griffin and Friddle 1950
Sniezko, S.F., P.J. Griffin and S.B. Friddle. 1950. Journal of Bacteriology **59**:699-710. **Type strain:** ATCC 10801
Description: Kilian, M. 1976. Journal of General Microbiology **93**:9-62.

H. pleuropneumoniae (Matthews and Pattison 1961) Shope 1964
Matthews, P.R.J. and I.H. Pattison. 1961. Journal of Comparative Pathology **71**:44-52; Shope, R.E. 1964. Journal of Experimental Medicine **119**:357-368. **Type strain:** ATCC 27088
Description: Kilian, M., J. Nicolet and E.L. Biberstein. 1978. International Journal of Systematic Bacteriology **28**:20-26.

H. segnis Kilian 1977
Kilian, M. 1977 in International Journal of Systematic Bacteriology **27**:306; Effective publication Kilian, M. 1976. Journal of General Microbiology **93**:9-62. **Type strain:** NCTC 10977
Description: Kilian, M. and J. Theilade. 1978. International Journal of Systematic Bacterioloy **28**:411-415.

H. vaginalis Gardner and Dukes 1955
Gardner, H.L. and C.D. Dukes. 1955. American Journal of Obstetrics and Gynecology **69**:962-976. **Type strain:** ATCC 14018
Description: Bergey 8.

Hafnia Moller 1954
Moller, V. 1954. Acta Pathologica et Microbiologica Scandinavica **35**:259-277. **Type species:** *H. alvei* Moller 1954
Description: Bergey 8.

H. alvei Moller 1954
Moller, V. 1954. Acta Pathologica et Microbiologica Scandinavica **35**:259-277. **Type strain:** ATCC 13337
Description: Ewing, W.H. and M.A. Fife.

1968. International Journal of Systematic Bacteriology **18**:263-271.

Haliscomenobacter van Veen, van der Kooij, Geuze and van der Vlies 1973
van Veen, W.L., D. van der Kooij, E.C.W.A. Geuze and A.W. van der Vlies. 1973. Antonie van Leeuwenhoek Journal of Microbiology and Serology **39**:207-216. **Type species:** *H. hydrossis* van Veen, van der Kooij, Geuze and van der Vlies 1973
Description: van Veen, W.L., D. van der Kooij, E.C.W.A. Geuze and A.W. van der Vlies. 1973. Ibid.

H. hydrossis van Veen, van der Kooij, Geuze and van der Vlies 1973
van Veen, W.L., D. van der Kooij, E.C.W.A. Geuze and A.W. van der Vlies. 1973. Antonie van Leeuwenhoek Journal of Microbiology and Serology **39**:207-216. **Type strain:** ATCC 27775
Description: van Veen, W.L., D. van der Kooij, E.C.W.A. Geuze and A.W. van der Vlies. 1970. Ibid.

Halobacterium Elazari-Volcani 1957
Elazari-Volcani, B. 1957 in Breed, R.S., E.G.D. Murray and N.R. Smith (eds). Bergey's Manual of Determinative Bacteriology, 7th ed. The Williams and Wilkins Co., Baltimore. pp. 207-212. **Type species:** *H. salinarium* (Harrison and Kennedy 1922) Elazari-Volcani 1957
Description: Bergey 8.

H. cutirubrum (Lochhead 1934) Elazari-Volcani 1957
Lochhead, A.G. 1934. Canadian Journal of Research **10**:275-286; Elazari-Volcani, B. 1957 in Breed, R.S., E.G.D. Murray and N.R. Smith (eds). Bergey's Manual of Determinative Bacteriology, 7th ed. The Williams and Wilkins Co., Baltimore. pp. 207-212. **Type strain:** NRC 34001
Description: Colwell, R.R., C.D. Litchfield, R.H. Vreeland, N.E. Gibbons and L.A. Kiefer. 1979. International Journal of Systematic Bacteriology **29**:379-399.

H. halobium (Petter 1931) Elazari-Volcani 1957
Petter, H.F.M. 1931. Proceedings Academy of Science, Amsterdam **34**:1417-1423; Elazari-Volcani, B. 1957 in Breed, R.S., E.G.D. Murray and N.R. Smith (eds). Bergey's Manual of Determinative Bacteriology, 7th ed. The Williams and Wilkins Co., Baltimore. pp. 207-212. **Type strain:** NCIB 8720
Description: Petter, H.F.M. 1931. Ibid.

H. saccharovorum Tomlinson and Hochstein 1977
Tomlinson, G.A. and L.I. Hochstein. 1977 in International Journal of Systematic Bacteriology **27**:306; Effective publication Tomlinson, G.A. and L.I. Hochstein. 1976. Canadian Journal of Microbiology **22**:587-591. **Type strain:** ATCC 29252
Description: Tomlinson, G.A. and L.I. Hochstein. 1976. Ibid.

H. salinarium (Harrison and Kennedy 1922) Elazari-Volcani 1957
Harrison, F.C. and M.E. Kennedy. 1922. Proceedings of the Royal Society of Canada, Section 5, **16**:101-152; Elazari-Volcani, B. 1957 in Breed, R.S., E.G.D. Murray and N.R. Smith (eds). Bergey's Manual of Determinative Bacteriology, 7th ed. The Williams and Wilkins Co., Baltimore. pp. 207-212. **Type strain:** NRC 34002
Description: Colwell, R.R., C.D. Litchfield, R.H. Vreeland, N.E. Gibbons and L. A. Kiefer. 1979. International Journal of Systematic Bacteriology **29**:379-399.

H. trapanicum (Petter 1931) Elazari-Volcani 1957
Petter, H.F.M. 1931. Proceedings Academy of Science, Amsterdam **34**:1417-1423; Elazari-Volcani, B. 1957 in Breed, R.S., E.G.D. Murray and N.R. Smith (eds). Bergey's Manual of Determinative Bacteriology, 7th ed. The Williams and Wilkins Co., Baltimore. pp. 207-212. **Type strain:** NRC 34021
Description: Petter, H.F.M. 1931. Ibid.

H. vallismortis Gonzalez, Gutierrez and Ramirez 1979
Gonzalez, C., C. Gutierrez and C. Ramirez. 1979 in International Journal of Systematic Bacteriology **29**:436; Effective publication Gonzalez, C., C. Gutierrez and C. Ramirez. 1978. Canadian Journal of Microbiology **24**:710-715. **Type strain:** ATCC 29715
Description: Gonzalez, C., C. Gutierrez and C. Ramirez. 1978. Ibid.

H. volcanii Mullakhanbhai and Larson 1975
Mullakhanbhai, M.F. and H. Larson.

1975. Archives of Microbiology **104**:207-214. **Type strain:** NCMB 2012
Description: Mullakhanbhai, M.F. and H. Larson. 1975. Ibid.

Halococcus Schoop 1935
Schoop, G. 1935. Deutsche Tierarztliche Wochenschrift **43**:817-820. **Type species:** *H. morrhuae* (Farlow 1880) Kocur and Hodgkiss 1973
Description: Bergey 8.

H. morrhuae (Farlow 1880) Kocur and Hodgkiss 1973
Farlow, W.G. 1880. United States Fisheries Commission Report for 1878 pp. 969-974; Kocur, M. and W. Hodgkiss. 1973. International Journal of Systematic Bacteriology **23**:151-156. **Type strain:** ATCC 17082
Description: Kocur, M. and W. Hodgkiss. 1973. Ibid.

Herpetosiphon Holt and Lewin 1968
Holt, J.G. and R.A. Lewin. 1968. Journal of Bacteriology **95**:2407-2408. **Type species:** *H. aurantiacus* Holt and Lewin 1968
Description: Bergey 8.

H. aurantiacus Holt and Lewin 1968
Holt, J.G. and R.A. Lewin. 1968. Journal of Bacteriology **95**:2407-2408. **Type strain:** ATCC 23779
Description: Holt, J.G. and R.A. Lewin. 1968. Ibid.

H. cohaerens Lewin 1970
Lewin, R.A. 1970. Canadian Journal of Microbiology **16**:517-520. **Type strain:** ATCC 23123
Description: Lewin, R.A. 1970. Ibid.

H. geysericola (Copeland 1936) Lewin 1970
Copeland, J.J. 1936. Annals of the New York Academy of Sciences **36**:1-232; Lewin, R.A. 1970. Canadian Journal of Microbiology **16**:517-520. **Type strain:** ATCC 23076
Description: Lewin, R.A. 1970. Ibid.

H. nigricans Lewin 1970
Lewin, R.A. 1970. Canadian Journal of Microbiology **16**:517-520. **Type strain:** ATCC 23147
Description: Lewin, R.A. 1970. Ibid.

H. persicus Lewin 1970
Lewin, R.A. 1970. Canadian Journal of Microbiology **16**:517-520. **Type strain:** ATCC 23167
Description: Lewin, R.A. 1970. Ibid.

Hyphomicrobium Stutzer and Hartleb 1899
Stutzer, A. and R. Hartleb. 1899. Mitteilungen des Landwirtschaftlichen Institutes der Kaiserlichen Universitat Breslau **1**:75-100. **Type species:** *H. vulgare* Stutzer and Hartleb 1899
Description: Bergey 8.

H. indicum Johnson and Weisrock 1969
Johnson, R.M. and W.P. Weisrock. 1969. International Journal of Systematic Bacteriology **19**:295-307. **Type strain:** ATCC 19614
Description: Bergey 8.

H. neptunium Leifson 1964
Leifson, E. 1964. Antonie van Leeuwenhoek Journal of Microbiology and Serology **30**:249-256. **Type strain:** ATCC 15444
Description: Bergey 8.

H. vulgare Stutzer and Hartleb 1899
Stutzer, A. and R. Hartleb. 1899. Mitteilungen des Landwirtschaftlichen Institutes der Kaiserlichen Universitat Breslau **1**:75-100. **Type strain:** NCIB 8698
Description: Bergey 8.

Intrasporangium Kalakoutskii, Kirillova and Krassilnikov 1967
Kalakoutskii, L.V., I.P. Kirillova and N.A. Krassilnikov. 1967. Journal of General Microbiology **48**:79-85. **Type species:** *I. calvum* Kalakoutskii, Kirillova and Krassilnikov 1967
Description: Kalakoutskii, L.V., I.P. Kirillova and N.A. Krassilnikov. 1976. Ibid.

I. calvum Kalakoutskii, Kirillova and Krassilnikov 1967
Kalakoutskii, L.V., I.P. Kirillova and N.A. Krassilnikov. 1967. Journal of General Microbiology **48**:79-85. **Type strain:** ATCC 23552
Description: Kalakoutskii, L.V., I.P. Kirillova and N.A. Krassilnikov. 1967. Ibid.

Janthinobacterium De Ley, Segers and Gillis 1978
De Ley, J., P. Segers and M. Gillis. 1978. International Journal of Systematic Bacteriology **28**:154-168. **Type species:** *J. livi-*

dum (Eisenberg 1891) De Ley, Segers and Gillis 1978
Description: De Ley, J., P. Segers and M. Gillis. 1978. Ibid.

J. lividum (Eisenberg 1891) De Ley, Segers and Willis 1978
Eisenberg, J. 1891. Bacteriologische Diagnostik, Hilfstabellen zum Gebrauche beim Praktischen Arbeiten, 3rd ed. Leopold Voss, Hamburg; De Ley, J., P. Segers and M. Gillis. 1978. International Journal of Systematic Bacteriology **28**:154-168. **Type strain;** NCTC 9796
Description: De Ley, J., P. segers and M. Gillis. 1978. Ibid.

Kineosporia Pagani and Parenti 1978
Pagani, H. and F. Parenti. 1978. International Journal of Systematic Bacteriology **28**:401-406. **Type species:** *K. aurantiaca* Pagani and Parenti 1978
Description: Pagani, H. and F. Parenti. 1978. Ibid.

K. aurantiaca Pagani and Parenti 1978
Pagani, H. and F. Parenti. 1978. International Journal of Systematic Bacteriology **28**:401-406. **Type strain:** ATCC 29727
Description: Pagani, H. and F. Parenti. 1978. Ibid.

Kingella Henriksen and Bovre 1976
Henriksen, S.D. and K. Bovre. 1976. International Journal of Systematic Bacteriology **26**:447-450. **Type species:** *K. kingae* (Henriksen and Bovre 1968) Henriksen and Bovre 1976
Description: Henriksen, S.D. and K. Bovre. 1976. Ibid.

K. denitrificans Snell and Lapage 1976
Snell, J.J.S. and S.P. Lapage. 1976. International Journal of Systematic Bacteriology **26**:451-458. **Type strain:** NCTC 10995
Description: Snell, J.J.S. and S.P. Lapage. 1976. Ibid.

K. indologenes Snell and Lapage 1976
Snell, J.J.S. and S.P. Lapage. 1976. International Journal of Systematic Bacteriology **26**:451-458. **Type strain:** ATCC 25869
Description: Snell, J.J.S. and S.P. Lapage 1976. Ibid.

K. kingae (Henriksen and Bovre 1968) Henriksen and Bovre 1976
Henriksen, S.D. and K. Bovre. 1968. Journal of General Microbiology **51**:377-385; Henriksen, S.D. and K. Bovre. 1976. International Journal of Systematic Bacteriology **26**:447-450. **Type strain:** ATCC 23330
Description: Henriksen, S.D. and K. Bovre. 1976. Ibid.

Kitasatoa Matsumae and Hata 1968
Matsumae, A.M. and T. Hata. 1968 in Matsumae, A.M., M. Ohtani, H. Takeshima and T. Hata. Journal of Antibiotics (Tokyo) **21**:616-625. **Type species:** *K. purpurea* Matsumae and Hata 1968
Description: Bergey 8.

K. diplospora Matsumae, Ohtani and Hata 1968
Matsumae, A.M., M. Ohtani and T. Hata. 1968 in Matsumae, A.M., M. Ohtani, H. Takeshima and T. Hata. Journal of Antibiotics (Tokyo) **21**:616-625. **Type strain:** ATCC 27788
Description: Bergey 8.

K. kauaiensis Matsumae, Ohtani and Hata 1968
Matsumae, A.M., M. Ohtani and T. Hata. 1968 in Matsumae, A.M., M. Ohtani, H. Takeshima and T. Hata. Journal of Antibiotics (Tokyo) **21**:616-625. **Type strain:** ATCC 21405
Description: Bergey 8.

K. nagasakiensis Matsumae and Hata 1968
Matsumae, A.M. and T. Hata. 1968 in Matsumae, A.M., M. Ohtani, H. Takeshima and T. Hata. Journal of Antibiotics (Tokyo) **21**:616-625. **Type strain:** ATCC 27789
Description: Bergey 8.

K. purpurea Matsumae and Hata 1968
Matsumae, A.M. and T. Hata. 1968 in Matsumae, A.M., M. Ohtani, H. Takeshima and T. Hata. Journal of Antibiotics (Tokyo) **21**:616-625. **Type strain:** ATCC 27787
Description: Bergey 8.

Klebsiella Trevisan 1885
Trevisan, V. 1885. Atti della Accademia Fisica-Medica-Statistica in Milano (ser 4) **3**:92-107. **Type species:** *K. pneumoniae* (Schroeter 1886) Trevisan 1887
Description: Bergey 8.

K. mobilis Bascomb, Lapage, Willcox and Curtis 1971
Bascomb, S., S.P. Lapage, W.R. Willcox and M.A. Curtis. 1971. Journal of General Microbiology **66**:279-295. **Type strain:** ATCC 13048
Description: Bascomb, S., S.P. Lapage, W.R. Willcox and M.A. Curtis. 1971. Ibid.

K. oxytoca (Flugge 1886) Lautrop 1956
Flugge, C. 1886. Die Mikroorganismen. F.C.W. Vogel, Leipzig; Lautrop, H. 1956. Acta Pathologica et Microbiologica Scandinavica **39**:375-384. **Type strain:** ATCC 13182
Description: Lautrop, H. 1956. Ibid.

K. ozaenae (Abel 1893) Bergey, Harrison, Breed, Hammer and Huntoon 1925
Abel, R. 1893. Zentralblatt fur Bakteriologie, Parasitenkunde, Infektionskrankheiten und Hygiene. Abteilung I. **13**:161-173; Bergey, D.H., F.C. Harrison, R.S. Breed, B.W. Hammer and F.M. Huntoon (eds). 1925. Bergey's Manual of Determinative Bacteriology, 2nd ed. The Williams and Wilkins Co., Baltimore. pp. 1-462. **Type strain:** ATCC 11296
Description: Bergey 8.

K. pneumoniae (Schroeter 1886) Trevisan 1887
Schroeter, J. 1885-1889 in Cohn, F. Kryptogamenflora von Schlesien. Band 3, Heft 3, Pilze J.U. Kern's Verlag, Breslau. pp. 1-814; Trevisan, V. 1887. Rendiconti dell'Instituto Lombardo di Scienze e Lettere (Ser 2) **20**:88-105. **Type strain:** ATCC 13883
Description: Bergey 8.

K. rhinoscleromatis Trevisan 1887
Trevisan, V. 1887. Rendiconti dell'Instituto Lombardo di Scienze e Lettere (Ser 2) **20**:88-105. **Type strain:** ATCC 13884
Description: Bergey 8.

Kurthia Trevisan 1885
Trevisan, V. 1885. Atti della Accademia Fisia-Medica-Statistica in Milano (Series 4) **3**:92-107. **Type species:** *K. zopfii* (Kurth 1883) Trevisan 1885
Description: Bergey 8.

K. zopfii (Kurth 1883) Trevisan 1885
Kurth, H. 1883. Berichte der Deutschen botanischen Gesellschaft **1**:97-100; Trevisan, V. 1885. Atti della Accademia Fisia-Medica-Statistica in Milano (Series 4) **3**:92-107. **Type strain:** NCIB 9878
Description: Bergey 8.

Lachnospira Bryant and Small 1956
Bryant, M.P. and N. Small. 1956. Journal of Bacteriology **72**:22-26. **Type species:** *L. multipara* Bryant and Small 1956
Description: Bergey 8.

L. multipara Bryant and Small 1956
Bryant, M.P. and N. Small. 1956. Journal of Bacteriology **72**:22-26. **Type strain:** ATCC 19207
Description: Holdeman, L.V., E.P. Cato and W.E.C. Moore (eds). 1977. Anaerobe Laboratory Manual, 4th ed. Virginia Polytechnic Institute and State University, Blacksburg, Virginia. pp. 1-156.

Lactobacillus Beijerinck 1901
Beijerinck, M.W. 1901. Archives Neerlandaises des Sciences Exactes et Naturelles (Section 2) **6**:212-243. **Type species:** *L. delbrueckii* (Leichmann 1896) Beijerinck 1901
Description: Bergey 8.

L. acidophilus (Moro 1900) Hansen and Mocquot 1970
Moro, E. 1900. Jahrbuch fur Kinderheilkunde und physiche Erziehung **52**:38-55; Hansen, P.A. and G. Mocquot. 1970. International Journal of Systematic Bacteriology **20**:325-327. **Type strain:** ATCC 4356
Description: Hansen, P.A. and G. Mocquot. 1970. Ibid.

L. brevis (Orla-Jensen 1919) Bergey, Breed, Hammer, Huntoon, Murray and Harrison 1934
Orla-Jensen, S. 1919. The lactic acid bacteria. Host, Copenhagen. pp. 1-196; Bergey, D.H., R.S. Breed, B.W. Hammer, F.M. Huntoon, E.G.D. Murray and F.C. Harrison. 1934. Bergey's Manual of Determinative Bacteriology, 4th ed. The Williams and Wilkins Co., Baltimore. pp. 1-664. **Type strain:** ATCC 14869
Description: Rogosa, M. and P.A. Hansen. 1971. International Journal of Systematic Bacteriology **21**:177-186.

L. buchneri (Henneberg 1903) Bergey, Harrison, Breed, Hammer, and Huntoon 1923

Henneberg, W. 1903. Zeitschrift fur Spiritusindustrie **26**:329-332; Bergey, D.H., F.C. Harrison, R.S. Breed, B.W. Hammer, and F.M. Huntoon. 1923. Bergey's Manual of Determinative Bacteriology, 1st ed. The Williams and Wilkins Co., Baltimore. pp. 1-442. **Type strain:** ATCC 4005
Description: Rogosa, M. and P.A. Hansen. 1971. Ibid.

L. bulgaricus (Orla-Jensen 1919) Rogosa and Hansen 1971
Orla-Jensen, S. 1919. The lactic acid bacteria. Host, Copenhagen. pp. 1-196; Rogosa, M. and P.A. Hansen. 1971. International Journal of Systematic Bacteriology **21**:177-186. **Type strain:** ATCC 11842
Description: Rogosa, M. and P.A. Hansen. 1971. Ibid.

L. casei (Orla-Jensen 1916) Hansen and Lessel 1971
Orla-Jensen, S. 1916. Maelkeri-Bakteriologi. Schonberske Forlag, Copenhagen; Hansen, P.A. and E.F. Lessel. 1971. International Journal of Systematic Bacteriology **21**:69-71. **Type strain:** ATCC 393
Description: Bergey 8.

L. casei subsp. *alactosus* Mills and Lessel 1973
Mills, C.K. and E.F. Lessel. 1973. International Journal of Systematic Bacteriology **23**:67-68. **Type strain:** ATCC 27216
Description: Mills, C.K. and E.F. Lessel. 1973. Ibid.

L. casei subsp. *casei* (Orla-Jensen 1916) Hansen and Lessel 1971
Orla-Jensen, S. 1916. Maelkeri-Bakteriologi. Schonberske Forlag, Copenhagen; Hansen, P.A. and E.F. Lessel. 1971. International Journal of Systematic Bacteriology **21**:69-71. **Type strain:** ATCC 393
Description: Bergey 8.

L. casei subsp. *pseudoplantarum* Abo-Elnaga and Kandler 1965
Abo-Elnaga, I.G. and O. Kandler. 1965. Zentralblatt fur Bakteriologie, Parasitenkunde, Infektionskrankheiten und Hygiene. Abteilung II. **119**:1-36. **Type strain:** ATCC 25598
Description: Bergey 8.

L. casei subsp. *rhamnosus* Hansen 1968
Hansen, P.A. 1968. A report by the Taxonomic Subcommittee on Lactobacilli and Closely Related Organisms. American Type Culture Collection, Rockville, Maryland. p. 76. **Type strain:** ATCC 7469
Description: Bergey 8.

L. casei subsp. *tolerans* Abo-Elnaga and Kandler 1965
Abo-Elnaga, I.G. and O. Kandler. 1965. Zentralblatt fur Bakteriologie, Parasitenkunde, Infektionskrankheiten und Hygiene. Abteilung II. **119**:1-39. **Type strain:** ATCC 25599
Description: Bergey 8.

L. catenaformis (Eggerth 1935) Moore and Holdeman 1970
Eggerth, A.H. 1935. Journal of Bacteriology **30**:277-299; Moore, W.E.C. and L.V. Holdeman. 1970. IN Cato, E.P., C.S. Cummins, L.V. Holdeman, J.L. Johnson, W.E.C. Moore, R.M. Smibert and L.DS. Smith (eds). Outline of clinical methods in anaerobic bacteriology, 2nd revision. Virginia Polytechnic Institute Anaerobe Laboratory, Blacksburg, Virginia pp. 15-22. **Type strain:** ATCC 25536
Description: Holdeman, L.V., E.P. Cato and W.E.C. Moore (eds). 1977. Anaerobe Laboratory Manual, 4th ed. Virginia Polytechnic Institute and State University, Blacksburg, Virginia. pp. 1-156.

L. cellobiosus Rogosa, Wiseman, Mitchell and Disraely 1953
Rogosa, M., R.F. Wiseman, J.A. Mitchell and M.N. Disraely. 1953. Journal of Bacteriology **65**:681-699. **Type strain:** ATCC 11739
Description: Bergey 8.

L. collinoides Carr and Davies 1972
Carr, J.G. and P.A. Davies. 1972. Journal of Applied Bacteriology **35**:463-471. **Type strain:** ATCC 27612
Description: Carr, J.G. and P.A. Davies. 1972. Ibid.

L. confusus (Holzapfel and Kandler 1969) Sharpe, Garvie and Tilbury 1972
Holzapfel, W. and O. Kandler. 1969. Zentralblatt fur Bakteriologie, Parasitenkunde, Infektionskrankheiten und Hygiene. Abteilung II. **123**:657-666; Sharpe, M.E., E.I. Garvie and R. Tilbury. 1972. Applied Microbiology **23**:389-397. **Type strain:** ATCC 10881
Description: Sharpe, M.E., E.I. Garvie

and R. Tilbury. 1972. Ibid.

L. coryniformis Abo-Elnaga and Kandler 1965
Abo-Elnaga, I.G. and O. Kandler. 1965. Zentralblatt fur Bakteriologie, Parasitenkunde, Infektionskrankheiten und Hygiene. Abteilung II. **119**:1-36. **Type strain:** DSM 20001
Description: Bergey 8.

L. coryniformis subsp. *coryniformis* Abo-Elnaga and Kandler 1965
Abo-Elnaga, I.G. and O. Kandler. 1965. Zentralblatt fur Bakteriologie, Parasitenkunde, Infektionskrankheiten und Hygiene. Abteilung II. **119**:1-36. **Type strain:** DSM 20001
Description: Bergey 8.

L. coryniformis subsp. *torquens* Abo-Elnaga and Kandler 1965
Abo-Elnaga, I.G. and O. Kandler. 1965. Zentralblatt fur Bakteriologie, Parasitenkunde, Infektionskrankheiten und Hygiene. Abteilung II. **119**:1-36. **Type strain:** ATCC 25600
Description: Bergey 8.

L. crispatus (Brygoo and Aladame 1953) Moore and Holdeman 1970
Brygoo, E.R. and N. Aladame. 1953. Annales de l'Institut Pasteur (Paris) **84**:640-641; Moore, W.E.C. and L.V. Holdeman. 1970 in Cato, E.P., C.S. Cummins, L.V. Holdeman, J.L. Johnson, W.E.C. Moore, R.M. Smibert and L.DS. Smith (eds). Outline of clinical methods in anaerobic bacteriology, 2nd revision. Virginia Polytechnic Institute Anaerobe Laboratory, Blacksburg, Virginia. pp. 15-21. **Type strain:** VPI 3199
Description: Holdeman, L.V., E.P. Cato and W.E.C. Moore (eds). 1977. Anaerobe Laboratory Manual, 4th ed. Virginia Polytechnic Institute and State University, Blacksburg, Virginia. pp. 1-156.

L. curvatus (Troili-Petersson 1903) Abo-Elnaga and Kandler 1965
Troili-Petersson, G. 1903. Zentralblatt fur Bakteriologie, Parasitenkunde, Infektionskrankheiten und Hygiene. Abteilung II. **11**:120-143; Abo-Elnaga, I.G. and O. Kandler. 1965. Zentralblatt fur Bakteriologie, Parasitenkunde, Infektionskrankheiten und Hygiene. Abteilung II. **119**:1-36. **Type strain:** ATCC 25601
Description: Bergey 8.

L. delbrueckii (Leichmann 1896) Beijerinck 1901
Leichmann, G. 1896. Zentralblatt fur Bakteriologie, Parasitenkunde, Infektionskrankheiten und Hygiene. Abteilung II. **2**:281-285; Beijerinck, M.W. 1901. Archives Neerlandaises des Sciences Exactes et Naturelles (Section 2) **6**:212-243. **Type strain:** ATCC 9649
Description: Rogosa, M. and P.A. Hansen. 1971. International Journal of Systematic Bacteriology **21**:177-186.

L. fermentum Beijerinck 1901
Beijerinck, M.W. 1901. Archives Neerlandaises des Sciences Exactes et Naturelles (Section 2) **6**:212-243. **Type strain:** ATCC 14931
Description: Rogosa, M. and P.A. Hansen. 1971. International Journal of Systematic Bacteriology **21**:177-186.

L. fructivorans Charlton, Nelson and Werkman 1934
Charlton, D.B., M.E. Nelson and C.H. Werkman. 1934. Iowa State Journal of Science **9**:1-ll. **Type strain:** ATCC 8288
Description: Bergey 8.

L. fructosus Kodama 1956
Kodama, R. 1956. Journal of the Agricultural Chemical Society of Japan **30**:705-708. **Type strain:** ATCC 13162
Description: Kodama, R. 1956. Ibid.

L. helveticus (Orla-Jensen 1919) Bergey, Harrison, Breed, Hammer and Huntoon 1925
Orla-Jensen, S. 1919. The lactic acid bacteria. Host, Copenhagen. pp. 1-196; Bergey, D.H., F.C. Harrison, R.S. Breed, B.W. Hammer and F.M. Huntoon (eds). 1925. Bergey's Manual of Determinative Bacteriology, 2nd ed. The Williams and Wilkins Co., Baltimore. pp. 1-462. **Type strain:** ATCC 15009
Description: Rogosa, M. and P.A. Hansen. 1971. International Journal of Systematic Bacteriology **21**:177-186.

L. heterohiochii Kitahara, Kaneko and Goto 1957
Kitahara, K., T. Kaneko and O. Goto. 1957. Journal of General and Applied Microbiology **3**:111-120. **Type strain:** ATCC 15435
Description: Bergey 8.

L. hilgardii Douglas and Cruess 1936
Douglas, H.C. and W.V. Cruess. 1936.

Food Research **1**:113-119. **Type strain:** ATCC 8290
Description: Bergey 8.

L. homohiochii Kitahara, Kaneko and Goto 1957
Kitahara, K., T. Kaneko and O. Goto. 1957. Journal of General and Applied Microbiology **3**:111-120. **Type strain:** ATCC 15434
Description: Bergey 8.

L. jensenii Gasser, Mandel and Rogosa 1970
Gasser, F., M. Mandel and M. Rogosa. 1970. Journal of General Microbiology **62**:219-222. **Type strain:** ATCC 25258
Description: Bergey 8.

L. lactis (Orla-Jensen 1919) Bergey, Breed, Hammer, Huntoon, Murray and Harrison 1934
Orla-Jensen, S. 1919. The lactic acid bacteria. Host, Copenhagen. pp. 1-196; Bergey, D.H., R.S. Breed, B.W. Hammer, F.M. Huntoon, E.G.D. Murray and F.C. Harrison (eds). 1934. Bergey's Manual of Determinative Bacteriology, 4th ed. The Williams and Wilkins Co., Baltimore. pp. 1-664. **Type strain:** ATCC 12315
Description: Rogosa, M. and P.A. Hansen. 1971. International Journal of Systematic Bacteriology **21**:177-186.

L. leichmannii (Henneberg 1903) Bergey, Harrison, Breed, Hammer and Huntoon 1923
Henneberg, W. 1903. Zeitschrift fur Spiritusindustrie **26**:329-332; Bergey, D.H., F.C. Harrison, R.S. Breed, B.W. Hammer and F.M. Huntoon. 1923. Bergey's Manual of Determinative Bacteriology, 1st ed. The Williams and Wilkins Co., Baltimore. pp. 1-442. **Type strain:** ATCC 4797
Description: Rogosa, M. and P.A. Hansen. 1971. International Journal of Systematic Bacteriology **21**:177-186.

L. mali Carr and Davies 1970
Carr, J.G. and P.A. Davies. 1970. Journal of Applied Bacteriology **33**:768-774. **Type strain:** ATCC 27053
Description: Carr, J.G. and P.A. Davis. 1970. Ibid.

L. maltaromicus Miller, Morgan and Libbey 1974
Miller III, A., M.E. Morgan and L.M. Libbey. 1974. International Journal of Systematic Bacteriology **24**:346-354. **Type strain:** ATCC 27865
Description: Miller III, A., M.E. Morgan and L.M. Libbey. 1974. Ibid.

L. minutus (Hauduroy, Ehringer, Urbain, Guillot and Magrou 1937) Moore and Holdeman 1972
Hauduroy, P., G. Ehringer, A. Urbain, G. Guillot and J. Magrou. 1937. Dictionnaire des bacteries pathogenes. Masson et Cie, Paris. pp. 1-597; Moore, W.E.C. and L.V. Holdeman. 1972 in Holdeman, L.V. and W.E.C. Moore (eds). Anaerobe Laboratory Manual. Virginia Polytechnic Institute Anaerobe Laboratory, Blacksburg, Virginia. **Type strain:** VPI 9428
Description: Holdeman, L.V., E.P. Cato, and W.E.C. Moore (eds). 1977. Anaerobe Laboratory Manual, 4th ed. Virginia Polytechnic Institute and State University, Blacksburg, Virginia. pp. 1-156.

L. plantarum (Orla-Jensen 1919) Bergey, Harrison, Breed, Hammer and Huntoon 1923
Orla-Jensen, S. 1919. The lactic acid bacteria. Host, Copenhagen. pp. 1-196; Bergey, D.H., F.C. Harrison, R.S. Breed, B.W. Hammer and F.M. Huntoon. 1923. Bergey's Manual of Determinative Bacteriology, lst ed. The Williams and Wilkins Co., Baltimore. pp. 1-442. **Type strain:** ATCC 14917
Description: Bergey 8.

L. rogosae Holdeman and Moore 1974
Holdeman, L.V. and W.E.C. Moore. 1974. International Journal of Systematic Bacteriology **24**:260-277. **Type strain:** VPI C37-38
Description: Holdeman, L.V. and W.E.C. Moore. 1974. Ibid.

L. ruminis Sharpe, Latham, Garvie, Zirngibl and Kandler 1973
Sharpe, M.E., M.J. Latham, E.I. Garvie, J. Zirngibl and O. Kandler. 1973. Journal of General Microbiology **77**:37-49. **Type strain:** ATCC 27780
Description: Sharpe, M.E., M.J. Latham, E.I. Garvie, J. Zirngibl and O. Kandler. 1973. Ibid.

L. sake Katagiri, Kitahara and Fukami 1934
Katagiri, H., K. Kitahara and K. Fukami. 1934. Bulletin of the Agricultural Chemical Society of Japan **10**:156-157. **Type strain:** ATCC 15521
Description: Katagiri, H., K. Kitahara

and K. Fukami. 1934. Ibid.

L. salivarius Rogosa, Wiseman, Mitchell and Disraely 1953
Rogosa, M., R.F. Wiseman, J.A. Mitchell and M.N. Disraely. 1953. Journal of Bacteriology **65**:681-699. **Type strain:** ATCC 11741
Description: Bergey 8.

L. salivarius subsp. *salicinius* Rogosa, Wiseman, Mitchell and Disraely 1953
Rogosa, M., R.F. Wiseman, J.A. Mitchell and M.N. Disraely. 1953. Journal of Bacteriology **65**:681-699. **Type strain:** ATCC 11742
Description: Bergey 8.

L. salivarius subsp. *salivarius* Rogosa, Wiseman, Mitchell and Disraely 1953
Rogosa, M., R.F. Wiseman, J.A. Mitchell and M.N. Disraely. 1953. Journal of Bacteriology **65**:681-699. **Type strain:** ATCC 11741
Description: Bergey 8.

L. trichodes Fornachon, Douglas and Vaughn 1949
Fornachon, J.C.M., H.C. Douglas and R.H. Vaughn. 1949. Hilgardia **19**:129-132. **Type strain:** ATCC 27394
Description: Bergey 8.

L. viridescens Niven and Evans 1957
Niven, C.F. and J.B. Evans. 1957. Journal of Bacteriology **73**:758-759. **Type strain:** ATCC 12706
Description: Bergey 8.

L. vitulinus Sharpe, Latham, Garvie, Zirngibl and Kandler 1973
Sharpe, M.E., M.J. Latham, E.I. Garvie, J. Zirngibl and O. Kandler. 1973. Journal of General Microbiology **77**:37-49. **Type strain:** ATCC 27783
Description: Sharpe, M.E., M.J. Latham, E.I. Garvie, J. Zirngibl and O. Kandler. 1973. Ibid.

L. xylosus Kitahara 1938
Kitahara, K. 1938. Journal of the Agricultural Chemical Society of Japan **14**:1449-1465. **Type strain:** ATCC 15577
Description: Bergey 8.

Lamprocystis Schroeter 1886
Schroeter, J. 1885-1889 in Cohn, F. Kryptogamenflora von Schlesien. Band 3, Heft 3, Pilze J.U. Kern's Verlag, Breslau. pp. 1-814. **Type species:** *L. roseopersicina* (Kutzing 1849) Schroeter 1886
Description: Bergey 8.

L. roseopersicina (Kutzing 1849) Schroeter 1886
Kutzing, F.T. 1849. Species Algarum. Lipsiae, pp. 1-922; Schroeter, J. 1885-1889 in Cohn, F. Kryptogamenflora von Schlesien. Band 3, Heft 3, Pilze J.U. Kern's Verlag. Breslau, pp. 1-814. **Type strain:** DSM 229
Description: Bergey 8.

Lampropedia Schroeter 1886
Schroeter, J. 1885-1889 in Cohn, F. Kryptogamenflora von Schlesien. Band 3, Heft 3, Pilze J.U. Kern's Verlag, Breslau. pp. 1-814. **Type species:** *L. hyalina* (Ehrenberg 1832) Schroeter 1886
Description: Bergey 8.

L. hyalina (Ehrenberg 1832) Schroeter 1886
Ehrenberg, C.G. 1832. Abhandlungen der Deutschen Akademie der Wissenschaften zu Berlin **1830**:1-88; Schroeter, J. 1885-1889 in Cohn, F. Kryptogamenflora von Schlesien. Band 3, Heft 3, Pilze J.U. Kern's Verlag, Breslau. pp. 1-814. **Type strain:** ATCC 11041
Description: Bergey 8.

Legionella Brenner, Steigerwalt and McDade 1979
Brenner, D.J., A.G. Steigerwalt and J.E. McDade. 1979. International Journal of Systematic Bacteriology **29**:436; Effective publication Brenner, D.J., A.G. Steigerwalt and J.E. McDade. 1979. Annals of Internal Medicine **90**:656-658. **Type species:** *L. pneumophila* Brenner, Steigerwalt and McDade 1979
Description: Brenner, D.J., A.G. Steigerwalt and J.E. McDade. 1979. Ibid.

L. pneumophila Brenner, Steigerwalt and McDade 1979
Brenner, D.J., A.G. Steigerwalt and J.E. McDade. 1979. International Journal of Systematic Bacteriology **29**:436; Effective pbulication Brenner, D.J., A.G. Steigerwalt and J.E. McDade. 1979. Annals of Internal Medicine **90**:656-658. **Type strain:** ATCC 33152
Description: Brenner, D.J., A.G. Steigerwalt and J.E. McDade. 1979. Ibid.

Leptospira Noguchi 1917

Noguchi, H. 1917. Journal of Experimental Medicine **25**:755-763. **Type species:** *L. interrogans* (Stimson 1907) Wenyon 1926
Description: Bergey 8.

L. biflexa (Wolbach and Binger 1914) Noguchi 1918
Wolbach, S.B. and C.A.L. Binger. 1914. Journal of Medical Research **30**:23-26; Noguchi, H. 1918. Journal of Experimental Medicine **27**:575-592. **Type strain:** ATCC 23582
Description: Bergey 8.

L. interrogans (Stimson 1907) Wenyon 1926
Stimson, A.M. 1907. Public Health Report **22**:541; Wenyon, C.M. 1926 in Wenyon, C.M. Protozoology, Vol. 2. Williams, Wood and Co., New York. pp. 1233-1288. **Type strain:** ATCC 23581
Description: Bergey 8.

Leptothrix Kutzing 1843
Kutzing, F.T. 1843. Phycologia Generalis. Leipzig. **Type species:** *L. ochracea* (Roth 1797) Kutzing 1843
Description: Bergey 8.

L. cholodnii Mulder and van Veen 1963
Mulder, E.G. and W.L. van Veen. 1963. Antonie van Leeuwenhoek Journal of Microbiology and Serology **29**:121-153. **Type strain:** LVMW 99
Description: Mulder, E.G. and W.L. van Veen. 1963. Ibid.

L. lopholea Dorff 1934
Dorff, P. 1934. Pflanzenforschung, Jena. **16**:1-62. **Type strain:** LVMW 124
Description: Mulder, E.G. and W.L. van Veen. 1963. Antonie van Leeuwenhoek Journal of Microbiology and Serology **29**: 121-153.

L. ochracea (Roth 1797) Kutzing 1843
Roth, A.W. 1797. Catalecta botanica quibus plantae novae et minus cognitae describuntur atque illustrantur. Lipsiae in Bibliopolio I.G. Mulleriano, Fasc. 1; Kutzing, F.T. 1843. Phycologia Generalis. Leipzig. **Type strain:** no culture available.
Description: Bergey 8.

Leptotrichia Trevisan 1879
Trevisan, V. 1879. R C Institut Lombardo (Seriale 2) **12**:133-151. **Type species:** *L. buccalis* (Robin 1853) Trevisan 1879
Description: Bergey 8.

L. buccalis (Robin 1853) Trevisan 1879
Robin, C. 1853. J.-B. Bailliere, Paris. pp. 1-702; Trevisan, V. 1879. R C Institut Lombardo (Seriale 2) **12**:133-151. **Type strain:** ATCC 14201
Description: Holdeman, L.V., E.P. Cato and W.E.C. Moore (eds). 1977. Anaerobe Laboratory Manual, 4th ed. Virginia Polytechnic Institute and State University, Blacksburg, Virginia. pp. 1-156.

Leuconostoc van Tieghem 1878
van Tiegham, P. 1878. Annales des Sciences Naturelles Botanique **7**:180-203. **Type species:** *L. mesenteroides* (Tsenkovskii 1878) van Tiegham 1878
Description: Bergey 8.

L. cremoris (Kundsen and Sorenson 1929) Garvie 1960
Kundsen, S. and A. Sorenson. 1929. Zentralblatt fur Bakteriologie, Parasitenkunde, Infektionskrankheiten und Hygiene. Abteilung II. **79**:75-85; Garvie, E.I. 1960. Journal of Dairy Research **27**:283-292. **Type strain:** NCDO 543
Description: Garvie, E.I. 1979. International Journal of Systematic Bacteriology **29**:149-151.

L. dextranicum (Beijerinck 1912) Hucker and Pederson 1930
Beijerinck, M.W. 1912. Folia Mikrobiologiya (Delft) **1**:4-100; Hucker, G.J. and C.S. Pederson. 1930. Technical Bulletin New York State Agricultural Experiment Station **167**:3-80. **Type strain:** NCDO 529
Description: Garvie, E.I. 1979. International Journal of Systematic Bacteriology **29**:149-151.

L. lactis Garvie 1960
Garvie, E.I. 1960. Journal of Dairy Research **27**:283-292. **Type strain:** NCDO 533
Description: Garvie, E.I. 1960. Ibid.

L. mesenteroides (Tsenkovskii 1878) van Tieghem 1878
Tsenkovskii, L. 1878. Proceedings of the Society of Natural Sciences of the Imperial University of Kharkov **12**:137-167; van Tieghem, P. 1878. Annales des Sciences Naturelles Botanique **7**:180-203. **Type strain:** ATCC 8293
Description: Garvie, E.I. 1979. International Journal of Systematic Bacteriology

29:149-151.

L. oenos Garvie 1967
Garvie, E.I. 1967. Journal of General Microbiology **48**:431-438. **Type strain:** NCDO 1674
Description: Garvie, E.I. 1967. Ibid.

L. paramesenteroides Garvie 1967
Garvie, E.I. 1967. Journal of General Microbiology **48**:439-447. **Type strain:** NCDO 803
Description: Garvie, E.I. 1967. Ibid.

Leucothrix Oersted 1844
Oersted, A.S. 1844. De Regionibus marinis. Elementa Topographiae Historiconaturalis Freti Oeresund. Inaugural Dissertation. J.C. Scharling, Copenhagen. **Type species:** *L. mucor* Oersted 1844
Description: Bergey 8.

L. mucor Oersted 1844
Oersted, A.S. 1844. De Regionibus Marinis. Elementa Topographiae Historiconaturalis Freti Oeresund. Inaugural Dissertation. J.C. Scharling, Copenhagen. **Type strain:** ATCC 25107
Description: Bergey 8.

Levinea Young, Kenton, Hobbs and Moody 1971
Young, V.M., D.M. Kenton, B.J. Hobbs and M.R. Moody. 1971. International Journal of Systematic Bacteriology **21**:58-63. **Type species:** *L. amalonatica* Young, Kenton, Hobbs and Moody 1971
Description: Young, V.M., D.M. Kenton, B.J. Hobbs and M.R. Moody. 1971. Ibid.

L. amalonatica Young, Kenton, Hobbs and Moody 1971
Young, V.M., D.M. Kenton, B.J. Hobbs and M.R. Moody. 1971. International Journal of Systematic Bacteriology **21**:58-63. **Type strain:** ATCC 25405
Description: Young, V.M., D.M. Kenton, B.J. Hobbs and M.R. Moody. 1971. Ibid.

L. malonatica Young, Kenton, Hobbs and Moody 1971
Young, V.M., D.M. Kenton, B.J. Hobbs and M.R. Moody. 1971. International Journal of Systematic Bacteriology **21**:58-63. **Type strain:** ATCC 25408
Description: Young, V.M., D.M. Kenton, B.J. Hobbs and M.R. Moody. 1971. Ibid.

Listeria Pirie 1940
Pirie, J.H.H. 1940. Science (Washington) **91**:383. **Type species:** *L. monocytogenes* (Murray, Webb and Swann 1926) Pirie 1940
Description: Bergey 8.

L. denitrificans Prevot 1961
Prevot, A.R. 1961. Traite de Systematique Bacterienne, Vol. 2, Dunod, Paris. pp. 1-771. **Type strain:** ATCC 14870
Description: Bergey 8.

L. grayi Errebo Larsen and Seeliger 1966
Errebo Larsen, H. and H.P.R. Seeliger. 1966. Proceedings of the Third International Symposium on Listeriosis July 13-16, Bilthoven. p. 35. **Type strain:** ATCC 19120
Description: Bergey 8.

L. monocytogenes (Murray, Webb and Swann 1926) Pirie 1940
Murray, E.G.D., A.A. Webb and M.B.R. Swann. 1926. Journal of Pathology and Bacteriology **29**:407-439; Pirie, J.H.H. 1940. Science (Washington) **91**:383. **Type strain:** ATCC 15313
Description: Bergey 8.

L. murrayi Welshimer and Meredith 1971
Welshimer, H.J. and A. Meredith. 1971. International Journal of Systematic Bacteriology **21**:3-7. **Type strain:** ATCC 25401
Description: Bergey 8.

Lucibacterium Hendrie, Hodgkiss and Shewan 1970
Hendrie, M.S., W. Hodgkiss and J.M. Shewan. 1970. Journal of General Microbiology **64**:151-169. **Type species:** *L. harveyi* (Johnson and Shunk 1936) Hendrie, Hodgkiss and Shewan 1970
Description: Bergey 8.

L. harveyi (Johnson and Shunk 1936) Hendrie, Hodgkiss and Shewan 1970 (See also Beneckea harveyi)
Johnson, F.H. and I.V. Shunk. 1936. Journal of Bacteriology **31**:585-592; Hendrie, M.S., W. Hodgkiss and J.M. Shewan. 1970. Journal of General Microbiology **64**:151-169. **Type strain:** ATCC 14126
Description: Johnson, F.H. and I.V. Shunk. 1936. Ibid.

Lysobacter Christensen and Cook 1978
Christensen, P. and F.D. Cook. 1978. International Journal of Systematic Bacteriology **28**:367-393. **Type species:** *L. enzymogenes* Christensen and Cook 1978
Description: Christensen, P. and F.D. Cook. 1978. Ibid.

L. antibioticus Christensen and Cook 1978
Christensen, P. and F.D. Cook. 1978. International Journal of Systematic Bacteriology **28**:367-393. **Type strain:** ATCC 29479
Description: Christensen, P. and F.D. Cook. 1978. Ibid.

L. brunescens Christensen and Cook 1978
Christensen, P. and F.D. Cook. 1978. International Journal of Systematic Bacteriology **28**:367-393. **Type strain:** ATCC 29482
Description: Christensen, P. and F.D. Cook. 1978. Ibid.

L. enzymogenes Christensen and Cook 1978
Christensen, P. and F.D. Cook. 1978. International Journal of Systematic Bacteriology **28**:367-393. **Type strain:** ATCC 29487
Description: Christensen, P. and F.D. Cook. 1978. Ibid.

L. gummosus Christensen and Cook 1978
Christensen, P. and F.D. Cook. 1978. International Journal of Systematic Bacteriology **28**:367-393. **Type strain:** ATCC 29489
Description: Christensen, P. and F.D. Cook. 1978. Ibid.

Macromonas Utermohl and Koppe 1924
Utermohl, H. and F. Koppe. 1924 in Koppe, F. Archiv fur Hydrobiologie **14**: 619-672. **Type species:** *M. mobilis* (Lauterborn 1915) Utermohl and Koppe 1924
Description: Bergey 8.

M. mobilis (Lauterborn 1915) Utermohl and Koppe 1924
Lauterborn, R. 1915. Verhandlungen des Naturhistorisch- Medizinischen Vereins zu Heidelberg **13**:395-481; Utermohl, H. and F. Koppe. 1924 in Koppe, F. Archiv fur Hydrobiologie **14**:619-672. **Type strain:** no culture available.
Description: Bergey 8.

Megasphaera Rogosa 1971
Rogosa, M. 1971. International Journal of Systematic Bacteriology **21**:187-189. **Type species:** *M. elsdenii* (Gutierrez, Davis, Lindahl and Warwick 1959) Rogosa 1971
Description: Bergey 8.

M. elsdenii (Gutierrez, Davis, Lindahl and Warwick 1959) Rogosa 1971
Gutierrez, J., R.E. Davis, I.L. Lindahl and E.J. Warwick. 1959. Applied Microbiology **7**:16-22; Rogosa, M. 1971. International Journal of Systematic Bacteriology **21**:187-189. **Type strain:** ATCC 25940
Description: Holdeman, L.V., E.P. Cato and W.E.C. Moore (eds). 1977. Anaerobe Laboratory Manual, 4th ed. Virginia Polytechnic Institute and State University, Blacksburg, Virginia. pp. 1-156.

Melittangium Jahn 1924
Jahn, E. 1924. Beitrage zur botanischen Protistologie. I. Die Polyangiden. Verlag von Gebruder Borntraeger, Leipzig. **Type species:** *M. boletus* Jahn 1924
Description: Bergey 8.

M. alboraceum (Peterson 1959) McCurdy 1971
Peterson, J.E. 1959. Mycologia **51**:163-172; McCurdy, H.D. 1971. International Journal of Systematic Bacteriology **21**:50-54. **Type strain:** UMH Slides Peterson 72
Description: Peterson, J.E. 1959. Ibid.

M. boletus Jahn 1924
Jahn, E. 1924. Beitrage zur botanischen Protistologie. I. Die Polyangiden. Verlag von Gebruder Borntraeger, Leipzig. **Type strain:** not cultivated.
Description: Jahn, E. 1924. Ibid.

M. lichenicola (Thaxter 1892) McCurdy 1971
Thaxter, R. 1892. Botanical Gazette **17**: 389-406; McCurdy, H.D. 1971. International Journal of Systematic Bacteriology **21**:50-54. **Type strain:** ATCC 25944
Description: McCurdy, H.D. 1971. Ibid.

Meniscus Irgens 1977
Irgens, R.L. 1977. International Journal of Systematic Bacteriology **27**:38-43. **Type species:** *M. glaucopis* Irgens 1977
Description: Irgens, R.L. 1977. Ibid.

M. glaucopis Irgens 1977
Irgens, R.L. 1977. International Journal of Systematic Bacteriology **27**:38-43. **Type strain:** ATCC 29398

Description: Irgens, R.L. 1977. Ibid.

Methanobacterium Kluyver and van Niel 1936
Kluyver, A.J. and C.B. van Niel. 1936. Zentralblatt fur Bakteriologie, Parasitenkunde, Infektionskrankheiten und Hygiene. Abteilung II. **94**:369-403. **Type species:** *M. formicicum* Schnellen 1947
Description: Bergey 8.

M. arbophilicum Zeikus and Henning 1975
Zeikus, J.G. and D.L. Henning. 1975. Antonie van Leeuwenhoek Journal of Microbiology and Serology **41**:543-552. **Type strain:** DSM 1125
Description: Zeikus, J.G. and D.L. Henning. 1975. Ibid.

M. formicicum Schnellen 1947
Schnellen, C.G.T.P. 1947. Onderzoekingen over de methaangisting. Thesis, Delft. pp. 1-137. **Type strain:** DSM 863
Description: Mylorie, R.L. and R.E. Hungate. 1954. Canadian Journal of Microbiology **1**:55-64.

M. mobile Paynter and Hungate 1968
Paynter, M.J.B. and R.E. Hungate. 1968. Journal of Bacteriology **95**:1943-1951. **Type strain:** DSM 1539
Description: Paynter, M.J.B. and R.E. Hungate. 1968. Ibid.

M. ruminantium Smith and Hungate 1958
Smith, P.H. and R.E. Hungate. 1958. Journal of Bacteriology **75**:713-718. **Type strain:** DSM 1093
Description: Bergey 8.

M. thermoautotrophicum Zeikus and Wolfe 1972
Zeikus, J.G. and R.S. Wolfe. 1972. Journal of Bacteriology **109**:707-713. **Type strain:** ATCC 29096
Description: Zeikus, J.G. and R.S. Wolfe. 1972. Ibid.

Methanococcus Kluyver and van Niel 1936 emend. Barker 1936
Kluyver, A.J. and C.B. van Niel. 1936. Zentralblatt fur Bakteriologie, Parasitenkunde, Infektionskrankheiten und Hygiene. Abteilung II. **94**:369-403; Barker, H.A. 1936. Archiv fur Mikrobiologie **7**: 420-438. **Type species:** *M. mazei* Barker 1936
Description: Bergey 8.

M. mazei Barker 1936
Barker, H.A. 1936. Archiv fur Mikrobiologie **7**:420-438. **Type strain:** not cultivated.
Description: Bergey 8.

M. vannielii Stadtman and Barker 1951
Stadtman, T.C. and H.A. Barker. 1951. Journal of Bacteriology **62**:269-280. **Type strain:** DSM 1224
Description: Stadtman, T.C. and H.A. Barker. 1951. Ibid.

Methanosarcina Kluyver and van Niel 1936 emend. Barker 1956
Kluyver, A.J. and C.B. van Niel. 1936. Zentralblatt fur Bakteriologie, Parasitenkunde, Infektionskrankheiten und Hygiene. Abteilung II. **94**:369-403; Barker, H.A. 1956. Bacterial fermentations. John Wiley and Sons, New York. pp. 1-95. **Type species:** *M. methanica* (Smit 1930) Kluyver and van Niel 1936
Description: Bergey 8.

M. barkeri Schnellen 1947
Schnellen, C.G.T.P. 1947. Onderzoekingen over de methaangisting. Thesis, Delft. pp. 1-137. **Type strain:** DSM 800
Description: Mah, R.A., M.R. Smith and L. Baresi. 1978. Applied and Environmental Microbiology **35**:1174-1184.

M. methanica (Smit 1930) Kluyver and van Niel 1936
Smit, J. 1930. Die Garungssarcinen. Eine Monographie. Pflanzenforschung Heft 14, pp. 1-59; Kluyver, A.J. and C.B. van Niel. 1936. Zentralblatt fur Bakteriologie, Parasitenkunde, Infektionskrankheiten und Hygiene. Abteilung II. **94**:369-403. **Type strain:** not cultivated.
Description: Bergey 8.

Methanospirillum Ferry, Smith and Wolfe 1974
Ferry, J.G., P.H. Smith and R.S. Wolfe. 1974. International Journal of Systematic Bacteriology **24**:465-469. **Type species:** *M. hungatii* Ferry, Smith and Wolfe 1974
Description: Ferry, J.G., P.H. Smith and R.S. Wolfe. 1974. Ibid.

M. hungatii Ferry, Smith and Wolfe 1974
Ferry, T.G., P.H. Smith and R.S. Wolfe. 1974. International Journal of Systematic Bacteriology **24**:465-469. **Type strain:** ATCC 27890
Description: Ferry, J.G., P.H. Smith and

R.S. Wolfe. 1974. Ibid.

Methylobacillus Yordy and Weaver 1977
Yordy, J.R. and T.L. Weaver. 1977. International Journal of Systematic Bacteriology **27**:247-255. **Type species:** *M. glycogenes* Yordy and Weaver 1977
Description: Yordy, J.R. and T.L. Weaver. 1977. Ibid.

M. glycogenes Yordy and Weaver 1977
Yordy, J.R. and T.L. Weaver. 1977. International Journal of Systematic Bacteriology **27**:247-255. **Type strain:** ATCC 29475
Description: Yordy, J.R. and T.L. Weaver. 1977. Ibid.

Methylobacterium Patt, Cole and Hanson 1976
Patt, T.E., G.C. Cole and R.S. Hanson. 1976. International Journal of Systematic Bacteriology **26**:226-229. **Type species:** *M. organophilum* Patt, Cole and Hanson 1976
Description: Patt, T.E., G.C. Cole and R.S. Hanson. 1976. Ibid.

M. organophilum Patt, Cole and Hanson 1976
Patt, T.E., G.C. Cole and R.S. Hanson. 1976. International Journal of Systematic Bacteriology **26**:226-229. **Type strain:** ATCC 27886
Description: Patt, T.E., G.C. Cole and R.S. Hanson. 1976. Ibid.

Methylococcus Foster and Davis 1966
Foster, J.W. and R.H. Davis. 1966. Journal of Bacteriology **91**:1924-1931. **Type species:** *M. capsulatus* Foster and Davis 1966
Description: Foster, J.W. and R.H. Davis. 1966. Ibid.

M. capsulatus Foster and Davis 1966
Foster, J.W. and R.H. Davis. 1966. Journal of Bacteriology **91**:1924-1931. **Type strain:** ATCC 19069
Description: Foster, J.W. and R.H. Davis. 1966. Ibid.

M. thermophilus Malashenko, Romanovskaya, Bogachenko and Stved 1975
Malashenko, Y.R., V.A. Romanovskaya, V.N. Bogachenko and A.D. Stved. 1975. Mikrobiologiya **44**:855-862 (English translation 773-779). **Type strain:** IMV-2Yu
Description: Malashenko, Y.R., V.A. Romanovskaya, V.N. Bogachenko and A.D. Stved. 1975. Ibid.

Microbacterium Orla-Jensen 1919
Orla-Jensen, S. 1919. The lactic acid bacteria. Host, Copenhagen. pp. 1-196. **Type species:** *M. lacticum* Orla-Jensen 1919
Description: Robinson, K. 1966. Journal of Applied Bacteriology **29**:607-615.

M. lacticum Orla-Jensen 1919
Orla-Jensen, S. 1919. The lactic acid bacteria. Host, Copenhagen. pp. 1-196. **Type strain:** ATCC 8180
Description: Robinson, K. 1966. Journal of Applied Bacteriology **29**:607-615.

Microbispora Nonomura and Ohara 1957
Nonomura, H. and Y. Ohara. 1957. Journal of Fermentation Technology **35**:307-311. **Type species:** *M. rosea* Nonomura and Ohara 1957
Description: Bergey 8.

M. aerata (Gerber and Lechevalier 1964) Cross 1974
Gerber, N.N. and M.P. Lechevalier. 1964. Biochemistry **3**:598-602; Cross, T. 1964. Bergey 8. **Type strain:** ATCC 15448
Description: Bergey 8.

M. amethystogenes Nonomura and Ohara 1960
Nonomura, H. and Y. Ohara. 1960. Journal of Fermentation Technology **38**:401-409. **Type strain:** DSM 43164
Description: Bergey 8.

M. bispora (Henssen 1957) Lechevalier 1965
Henssen, A. 1957. Archiv fur Mikrobiologie **26**:373-414; Lechevalier, H.A. 1965. International Bulletin of Bacteriological Nomenclature and Taxonomy **15**:139-142. **Type strain:** ATCC 19993
Description: Bergey 8.

M. chromogenes Nonomura and Ohara 1960
Nonomura, H. and Y. Ohara. 1960. Journal of Fermentation Technology **38**:401-409. **Type strain:** DSM 43165
Description: Bergey 8.

M. diastatica Nonomura and Ohara 1960
Nonomura, H. and Y. Ohara. 1960. Journal of Fermentation Technology **38**:401-409. **Type strain:** KCC A-0023
Description: Bergey 8.

M. echinospora Nonomura and Ohara 1971
Nonomura, H. and Y. Ohara. 1971. Journal of Fermentation Technology **49**:887-

894. **Type strain:** ATCC 27300
Description: Nonomura, H. and Y. Ohara. 1971. Ibid.

M. parva Nonomura and Ohara 1960
Nonomura, H. and Y. Ohara. 1960. Journal of Fermentation Technology **38**:401-409. **Type strain:** KCC A-0024
Description: Bergey 8.

M. rosea Nonomura and Ohara 1957
Nonomura, H. and Y. Ohara. 1957. Journal of Fermentation Technology **35**:307-311. **Type strain:** ATCC 12950
Description: Bergey 8.

M. thermodiastatica Nonomura and Ohara 1969
Nonomura, H. and Y. Ohara. 1969. Journal of Fermentation Technology **47**:701-709. **Type strain:** ATCC 27098
Description: Nonomura, H. and Y. Ohara. 1969. Ibid.

M. thermorosea Nonomura and Ohara 1969
Nonomura, H. and Y. Ohara. 1969. Journal of Fermentation Technology **47**:701-709. **Type strain:** ATCC 27099
Description: Nonomura, H. and Y. Ohara. 1969. Ibid.

Micrococcus Cohn 1872
Cohn, F. 1872. Beitrage zur Biologie der Pflanzen **1**:127-224. **Type species:** *M. luteus* (Schroeter 1872) Cohn 1872
Description: Bergey 8.

M. agilis Ali-Cohen 1889
Ali-Cohen, C.H. 1889. Zentralblatt fur Bakteriologie, Parasitenkunde, Infektionskrankheiten und Hygiene. Abteilung I. **6**: 33-36. **Type strain:** ATCC 966
Description: Kocur, M. and K.H. Schleifer. 1975. International Journal of Systematic Bacteriology **25**:294-297.

M. halobius Onishi and Kamekura 1972
Onishi, H. and M. Kamekura. 1972. International Journal of Systematic Bacteriology **22**:233-236. **Type strain:** ATCC 21727
Description: Onishi, H. and M. Kamekura. 1972. Ibid.

M. kristinae Kloos, Tornabene and Schleifer 1974
Kloos, W.E., T.G. Tornabene and K.H. Schleifer. 1974. International Journal of Systematic Bacteriology **24**:79-101. **Type strain:** ATCC 27570
Description: Kloos, W.E., T.G. Tornabene and K.H. Schleifer. 1974. Ibid.

M. luteus (Schroeter 1872) Cohn 1872
Schroeter, J. 1872 in Cohn, F. 1875. Beitrage zur Biologie der Pflanzen, J.U. Kern's Verlag, Breslau. pp. 109-126; Cohn, F. 1872. Beitrage zur Biologie der Pflanzen **1**:127-224. **Type strain:** ATCC 4698
Description: Kocur, M., Z. Pacova and T. Martinec. 1972. International Journal of Systematic Bacteriology **22**:218-223.

M. lylae Kloos, Tornabene and Schleifer 1974
Kloos, W.E., T.G. Tornabene and K. Schleifer. 1974. International Journal of Systematic Bacteriology **24**:79-101. **Type strain:** ATCC 17566
Description: Kloos, W.E., T.G. Tornabene and K. Schleifer. 1974. Ibid.

M. nishinomiyaensis Oda 1935
Oda, M. 1935. Journal of Fermentation and Technology **13**:1202-1228. **Type strain:** ATCC 29093
Description: Kocur, M., K.H. Schleifer and W.E. Kloos. 1975. International Journal of Systematic Bacteriology **25**: 290-293.

M. roseus Flugge 1886
Flugge, C. 1886. Die Mikroorganismen. F.C.W. Vogel, Leipzig. **Type strain:** ATCC 186
Description: Kocur, M. and Z. Pacova. 1970. International Journal of Systematic Bacteriology **20**:233-240.

M. sedentarius ZoBell and Upham 1944
ZoBell, C.E. and H.C. Upham. 1944. Bulletin Scripps Institution of Oceanography University of California (Technical Series) **5**:239-292. **Type strain:** ATCC 14392
Description: Kloos, W.E., T.G. Tornabene and K.H. Schleifer. 1974. International Journal of Systematic Bacteriology **24**:79-101.

M. varians Migula 1900
Migula, W. 1900. System der Bakterien. Vol. 2. Gustav Fischer, Jena. **Type strain:** ATCC 15306
Description: Kocur, M. and T. Martinec. 1972. International Journal of Systematic Bacteriology **22**:228-232.

Microcyclus Orskov 1928

Orskov, J. 1928. Zentralblatt fur Bakteriologie, Parasitenkunde, Infektionskrankheiten und Hygiene. Abteilung I. **107**: 180-184. **Type species:** *M. aquaticus* Orskov 1928
Description: Larkin, J.M., P.M. Williams and R. Taylor. 1977. International Journal of Systematic Bacteriology **27**:147-156.

M. aquaticus Orskov 1928
Orskov, J. 1928. Zentralblatt fur Bakteriologie, Parasitenkunde, Infektionskrankheiten und Hygiene. Abteilung I. **107**:180-184. **Type strain:** ATCC 25396
Description: Larkin, J.M., P.M. Williams and R. Taylor. 1977. International Journal of Systematic Bacteriology **27**:147-156.

Microellobosporia Cross, Lechevalier and Lechevalier 1963
Cross, T., M.P. Lechevalier and H.A. Lechevalier. 1963. Journal of General Microbiology **31**:421-429. **Type species:** *M. cinerea* Cross, Lechevalier and Lechevalier 1963
Description: Bergey 8.

M. cinerea Cross, Lechevalier and Lechevalier 1963
Cross, T., M.P. Lechevalier and H.A. Lechevalier. 1963. Journal of General Microbiology **31**:421-429. **Type strain:** ATCC 15840
Description: Cross, T., M.P. Lechevalier and H.A. Lechevalier. 1963. Ibid.

M. flavea Cross, Lechevalier and Lechevalier 1963
Cross, T., M.P. Lechevalier and H.A. Lechevalier. 1963. Journal of General Microbiology **31**:421-429. **Type strain:** ATCC 15332
Description: Cross, T., M.P. Lechevalier and H.A. Lechevalier. 1963. Ibid.

M. grisea (Konev, Tsyganov, Minbayev and Morozov 1967) Pridham 1970
Konev, I.E., V.A. Tsyganov, P. Minbayev and V.M. Morozov. 1967. Mikrobiologiya **36**:308-317; Pridham, T.G. 1970. Bulletin of the United States Department of Agriculture **1424**:1-55. **Type strain:** ATCC 19618
Description: Bergey 8.

M. violacea (Tsyganov, Zhukova and Timofeeva 1964) Pridham 1974
Tsyganov, V.A., R.A. Zhukova and K.A. Timofeeva. 1964. Mikrobiologiya **33**:863-869; Pridham, T.G. 1974. Bergey 8. **Type strain:** DSM 43167
Description: Bergey 8.

Micromonospora Orskov 1923
Orskov, J. 1923. Investigations into the morphology of the ray fungi. Levin and Munksgaard, Copenhagen, Denmark. **Type species:** *M. chalcea* (Foulerton 1905) Orskov 1923
Description: Bergey 8.

M. aurantiaca Sveshnikova, Maksimova and Kudrina 1969
Sveshnikova, M.A., T.S. Maksimova and E.S. Kudrina. 1969. Mikrobiologiya **38**: 754-763. **Type strain:** ATCC 27029
Description: Sveshnikova, M.A., T.S. Maksimova and E.S. Kudrina. 1969. Ibid.

M. brunnea Sveshnikova, Maksimova and Kudrina 1969
Sveshnikova, M.A., T.S. Maksimova and E.S. Kudrina. 1969. Mikrobiologiya **38**: 754-763. **Type strain:** ATCC 27334
Description: Sveshnikova, M.A., T.S. Maksimova and E.S. Kudrina. 1969. Ibid.

M. carbonacea Luedemann and Brodsky 1965
Luedemann, G.M. and B.C. Brodsky. 1965. Antimicrobial Agents and Chemotherapy **1964**:47-52. **Type strain:** ATCC 27114
Description: Bergey 8.

M. carbonacea subsp. *aurantiaca* Luedemann and Brodsky 1965
Luedemann, G.M. and B.C. Brodsky. 1965. Antimicrobial Agents and Chemotherapy **1964**:47-52. **Type strain:** ATCC 27115
Description: Bergey 8.

M. carbonacea subsp. *carbonacea* Luedemann and Brodsky 1965
Luedemann, G.M. and B.C. Brodsky. 1965. Antimicrobial Agents and Chemotherapy **1964**:47-52. **Type strain:** ATCC 27114
Description: Bergey 8.

M. chalcea (Foulerton 1905) Orskov 1923
Foulerton, A.G.R. 1905. Lancet **1**:1199-1200; Orskov, J. 1923. Investigations into the morphology of the ray fungi. Levin and Munksgaard, Copenhagen, Denmark. **Type strain:** ATCC 12452
Description: Bergey 8.

M. coerulea Jensen 1932

Jensen, H.L. 1932. Proceedings of the Linnean Society of New South Wales **57**: 173-180. **Type strain:** ATCC 27008
Description: Bergey 8.

M. echinospora Luedemann and Brodsky 1964
Luedemann, G.M. and B.C. Brodsky. 1964. Antimicrobial Agents and Chemotherapy **1963**:116-124. **Type strain:** ATCC 15837
Description: Bergey 8.

M. echinospora subsp. *echinospora* Luedemann and Brodsky 1964
Luedemann, G.M. and B.C. Brodsky. 1964. Antimicrobial Agents and Chemotherapy **1963**:116-124. **Type strain:** ATCC 15837
Description: Bergey 8.

M. echinospora subsp. *ferruginea* Luedemann and Brodsky 1964
Luedemann, G.M. and B.C. Brodsky. 1964. Antimicrobial Agents and Chemotherapy **1963**:116-124. **Type strain:** ATCC 15836
Description: Bergey 8.

M. echinospora subsp. *pallida* Luedemann and Brodsky 1964
Luedemann, G.M. and B.C. Brodsky. 1964. Antimicrobial Agents and Chemotherapy **1963**:116-124. **Type strain:** ATCC 15838
Description: Bergey 8.

M. gallica (Erikson 1935) Waksman 1961
Erickson, D. 1935. Medical Research Council (Great Britain) Special Report series **203**:5-61; Waksman, S.A. 1961. The Actinomycetes. Vol. 2. Classification, identification and descriptions of genera and species. The Williams and Wilkins Co., Baltimore. pp. 1-363. **Type strain:** NCTC 4582
Description: Bergey 8.

M. halophytica Weinstein, Luedemann, Oden and Wagman 1968
Weinstein, M.J., G.M. Luedemann, E.M. Oden and G.H. Wagman. 1968. Antimicrobial Agents and Chemotherapy **1967**: 435-441. **Type strain:** ATCC 27596
Description: Bergey 8.

M. halophytica subsp. *halophytica* Weinstein, Luedemann, Oden and Wagman 1968
Weinstein, M.J., G.M. Luedemann, E.M. Oden and G.H. Wagman. 1968. Antimicrobial Agents and Chemotherapy **1967**: 435-441. **Type strain:** ATCC 27596
Description: Bergey 8.

M. halophytica subsp. *nigra* Weinstein, Luedemann, Oden and Wagman 1968
Weinstein, M.J., G.M. Luedemann, E.M. Oden and G.H. Wagman. 1968. Antimicrobial Agents and Chemotherapy **1967**: 435-441. **Type strain:** ATCC 33088
Description: Bergey 8.

M. inositola Kawamoto, Okachi, Kato, Yamamoto, Takahashi, Takasawa and Nara 1974
Kawamoto, I., R. Okachi, H. Kato, S. Yamamoto, I. Takahashi, S. Takasawa and T. Nara. 1974. Journal of Antibiotics **27**:493-501. **Type strain:** ATCC 21773
Description: Bergey 8.

M. purpurea Luedemann and Brodsky 1964
Luedemann, G.M. and B.C. Brodsky. 1964. Antimicrobial Agents and Chemotherapy **1963**:116-124. **Type strain:** ATCC 15835
Description: Bergey 8.

M. purpureochromogenes (Waksman and Curtis 1916) Luedemann 1971
Waksman, S.A. and R.E. Curtis. 1916. Soil Science **1**:99-134; Luedemann, G.M. 1971. International Journal of Systematic Bacteriology **21**:240-247. **Type strain:** ATCC 27007
Description: Bergey 8.

M. rhodorangea Wagman, Testa, Marquez and Weinstein 1974
Wagman, G.H., R.T. Testa, J.A. Marquez and M.J. Weinstein. 1974. Antimicrobial Agents and Chemotherapy **6**:144-149. **Type strain:** ATCC 27932
Description: Wagman, G.H., R.T. Testa, J.A. Marquez and M.J. Weinstein. 1974. Ibid.

Micropolyspora Lechevalier, Solotorovsky and McDurmont 1961
Lechevalier, H.A., M. Solotorovsky and C.I. McDurmont. 1961. Journal of General Microbiology **26**:11-18. **Type species:** *M. brevicatena* Lechevalier, Solotorovsky and McDurmont 1961
Description: Bergey 8.

M. angiospora Zhukova, Tsyganov and Morozov 1968
Zhukova, R.A., V.A. Tsyganov and V.M. Morozov. 1968. Mikrobiologiya **37**:724-728. **Type strain:** LIA 3479-30
Description: Zhukova, R.A., V.A. Tsyga-

nov and V.M. Morozov. 1968. Ibid.

M. brevicatena Lechevalier, Solotorovsky and McDurmont 1961
Lechevalier, H.A., M. Solotorovsky and C.I. McDurmont. 1961. Journal of General Microbiology **26**:11-18. **Type strain:** ATCC 15333
Description: Bergey 8.

M. faeni Cross, Maciver and Lacey 1968
Cross, T., A.M. Maciver and J. Lacey. 1968. Journal of General Microbiology **50**:351-359. **Type strain:** ATCC 15347
Description: Bergey 8.

M. internatus Agre, Guzeva and Dorokhova 1974
Agre, N.S., L.N. Guzeva and L.A. Dorokhova. 1974. Mikrobiologiya **43**:577-583. **Type strain:** INMI 632
Description: Agre, N.S., L.N. Guzeva and L.A. Dorokhova. 1974. Ibid.

M. rectivirgula (Krassilnikov and Agre 1964) Prauser and Momirova 1970
Krassilnikov, N.A. and N.S. Agre. 1964. Hindustan Antibiotics Bulletin **6**:97-107; Prauser, H. and S. Momirova. 1970. Zeitschrift fur Allgemeine Mikrobiologie **10**: 219-222. **Type strain:** INMI 683
Description: Bergey 8.

Microscilla Pringsheim 1951
Pringsheim, E.G. 1951. Journal of General Microbiology **5**:124-149. **Type species:** *M. marina* Pringsheim 1951
Description: Pringsheim, E.G. 1951. Ibid.

M. marina (Pringsheim 1951) Lewin 1969
Pringsheim, E.G. 1951. Journal of General Microbiology **5**:124-149; Lewin, R.A. 1969. Journal of General Microbiology **58**:189-206. **Type strain:** ATCC 23134
Description: Lewin, R.A. 1969. Ibid.

Microtetraspora Thiemann, Pagani and Beretta 1968
Thiemann, J.E., H. Pagani and G. Beretta. 1968. Journal of General Microbiology **50**:295-303. **Type species:** *M. glauca* Thiemann, Pagani and Beretta 1968
Description: Thiemann, J.E., H. Pagani and G. Beretta. 1968. Ibid.

M. fusca Thiemann, Pagani and Beretta 1968
Thiemann, J.E., H. Pagani and G. Beretta. 1968. Journal of General Microbiology **50**:295-303. **Type strain:** ATCC 23058
Description: Thiemann, J.E., H. Pagani and G. Beretta. 1968. Ibid.

M. glauca Thiemann, Pagani and Beretta 1968
Thiemann, J.E., H. Pagani and G. Beretta. 1968. Journal of General Microbiology **50**:295-303. **Type strain:** ATCC 23057
Description: Thiemann, J.E., H. Pagani and G. Beretta. 1968. Ibid.

M. niveoalba Nonomura and Ohara 1971
Nonomura, H. and Y. Ohara. 1971. Journal of Fermentation Technology **49**:887-894. **Type strain:** ATCC 27301
Description: Thiemann, J.E., H. Pagani and G. Beretta. 1968. Journal of General Microbiology **50**:295-303.

M. viridis Nonomura and Ohara 1971
Nonomura, H. and Y. Ohara. 1971. Journal of Fermentation Technology **49**:887-894. **Type strain:** ATCC 27103
Description: Thiemann, J.E., H. Pagani and G. Beretta. 1968. Journal of General Microbiology **50**:295-303.

Moraxella Lwoff 1939
Lwoff, A. 1939. Annales de l'Institut Pasteur (Paris) **62**:168-176. **Type species:** *M. lacunata* (Eyre 1900) Lwoff 1939
Description: Bergey 8.

M. anatipestifer (Hendrickson and Hilbert 1932) Bruner and Fabricant 1954
Hendrickson, J.M. and K.F. Hilbert. 1932. Cornell Veterinarian **22**:239-252; Bruner, D.W. and J. Fabricant. 1954. Cornell Veterinarian **44**:461-464. **Type strain:** ATCC 11845
Description: Henriksen, S.D. 1973. Bacteriological Reviews **37**:522-561.

M. atlantae Bovre, Fuglesang, Hagen, Jantzen and Froholm 1976
Bovre, K., J.E. Fuglesang, N. Hagen, E. Jantzen and L.O. Froholm. 1976. International Journal of Systematic Bacteriology **26**:511-521. **Type strain:** ATCC 29525
Description: Bovre, K., J.E. Fuglesang, N. Hagen, E. Jantzen and L.O. Froholm. 1976. Ibid.

M. bovis (Hauduroy, Ehringer, Urbain, Guillot and Magrou 1937) Murray 1948
Hauduroy, P., G. Ehringer, A. Urbain, G. Guillot and J. Magrou. 1937. Dictionnaire des Bacteries Pathogenes. Masson et Cie, Paris. p. 247; Murray, E.G.D. 1948 in Breed, R.S., E.G.D. Murray and A.P. Hitchens (eds). Bergey's Manual of Determinative Bacteriology, 6th ed. The Wil-

liams and Wilkins Co., Baltimore. pp. 590-592. **Type strain:** ATCC 10900
Description: Henriksen, S.D. 1973. Bacteriological Reviews **37**:522-561.

M. catarrhalis (Frosch and Kolle 1896) Henriksen and Bovre 1968
Frosch, P. and W. Kolle. 1896 in Flugge, C. Die Mikroorganismen. 3rd ed. Verlag von Vogel, Leipzig. Part 2. p. 154; Henriksen, S.D. and K. Bovre. 1968. Journal of General Microbiology **51**:387-392. **Type strain:** ATCC 25238
Description: Catlin, B.W. 1970. International Journal of Systematic Bacteriology **20**:155-159.

M. caviae (Pelczar 1953) Henriksen and Bovre 1968
Pelczar, M.J. 1953. Journal of Bacteriology **65**:744; Henriksen, S.D. and K. Bovre. 1968. Journal of General Microbiology **51**:387-392. **Type strain:** ATCC 14659
Description: Berger, U. 1963. Ergebnisse der Mikrobiologie, Immunitatsforschung und experimentellen Therapie **36**:97-167.

M. equi Hughes and Pugh 1970
Hughes, D.E. and G.W. Pugh. 1970. American Journal of Veterinary Research **31**:457-462. **Type strain:** ATCC 25576
Description: Hughes, D.E. and G.W. Pugh. 1970. Ibid.

M. lacunata (Eyre 1900) Lwoff 1939
Eyre, J.W. 1900. Journal of Pathology and Bacteriology **6**:1-13; Lwoff, A. 1939. Annales de l'Institut Pasteur (Paris) **62**: 168-176. **Type strain:** ATCC 17967
Description: Henriksen, S.D. 1973. Bacteriological Reviews **37**:522-561.

M. nonliquefaciens (Scarlett 1916) Lwoff 1939
Scarlett, M. 1916. Annales d'Oculistique (Paris) **153**:100-111; Lwoff, A. 1939. Annales de l'Institut Pasteur (Paris) **62**:168-176. **Type strain:** ATCC 19975
Description: Henriksen, S.D. 1973. Bacteriological Reviews **37**:522-561.

M. osloensis Bovre and Henriksen 1967
Bovre, K. and S.D. Henriksen. 1967. International Journal of Systematic Bacteriology **17**:127-135. **Type strain:** ATCC 19976
Description: Henriksen, S.D. 1973. Bacteriological Reviews **37**:522-561.

M. ovis (Lindqvist 1960) Henriksen and Bovre 1968
Lindqvist, K. 1960. Journal of Infectious Diseases **106**:162-165; Henriksen, S.D. and K. Bovre. 1968. Journal of General Microbiology **51**:387-392. **Type strain:** ATCC 33078
Description: Berger, U. 1963. Ergebnisse der Mikrobiologie, Immunitatsforschung und experimentellen Therapie **36**:97-167.

M. phenylpyruvica Bovre and Henriksen 1967
Bovre, K. and S.D. Henriksen. 1967. International Journal of Systematic Bacteriology **17**:343-360. **Type strain:** ATCC 23333
Description: Henriksen, S.D. 1973. Bacteriological Reviews **37**:522-561.

M. saccharolytica Flamm 1956
Flamm, H. 1956. Zentralblatt fur Bakteriologie, Parasitenkunde, Infektionskrankheiten und Hygiene. Abteilung I. **166**: 498-502. **Type strain:** ATCC 19245
Description: Flamm, H. 1956. Ibid.

M. urethralis Lautrop, Bovre and Frederiksen 1970
Lautrop, H., K. Bovre and W. Frederiksen. 1970. Acta Pathologica et Microbiologica Scandinavica, Section B, **78**:255-256. **Type strain:** ATCC 17960
Description: Lautrop, H., K. Bovre and W. Frederiksen. 1970. Ibid.

Morganella Fulton 1943
Fulton, MacD. 1943. Journal of Bacteriology **46**:79-81. **Type species:** *M. morganii* (Winslow, Kligler and Rothberg 1919) Fulton 1943
Description: Fulton, MacD. 1943. Ibid.

M. morganii (Winslow, Kligler and Rothberg 1919) Fulton 1943
Winslow, C.E.A., I.J. Kligler and W. Rothberg. 1919. Journal of Bacteriology **4**:429-503; Fulton, MacD. 1943. Journal of Bacteriology **46**:79-81. **Type strain:** ATCC 25830
Description: Brenner, D.J., J.J. Farmer III, G.R. Fanning, A.G. Steigerwalt, P. Klykken, H.G. Wathen, F.W. Hickman and W.H. Ewing. 1978. Ibid.

Mycobacterium Lehmann and Neumann 1896
Lehmann, K.B. and R. Neumann. 1896. Atlas und Grundriss der Bacteriologie und Lehrbuch der speziellen bacteriologischen Diagnostik. 1st ed. J.F. Lehmann, Munchen. **Type species:** *M. tuberculosis* (Zopf 1883) Lehmann and Neumann 1896

Description: Bergey 8.

M. africanum Castets, Rist and Boisvert 1969
Castets, M., N. Rist and H. Boisvert. 1969. Medecin D'Afrique Noire **16**:321-322. **Type strain:** ATCC 25420
Description: Bergey 8.

M. asiaticum Weiszfeiler, Karasseva and Karczag 1971
Weiszfeiler, J., G.V. Karasseva and E. Karczag. 1971. Acta Microbiologica Academiae Scientiarum Hungaricae **18**:247-252. **Type strain:** ATCC 25276
Description: Weiszfeiler, J., G.V. Karasseva and E. Karczag. 1971. Ibid.

M. aurum Tsukamura 1966
Tsukamura, M. 1966. Journal of General Microbiology **45**:253-273. **Type strain:** ATCC 23366
Description: Tsukamura, M. 1966. Ibid.

M. avium Chester 1901
Chester, F.D. 1901. A manual of determinative bacteriology. The Macmillan Co., New York. **Type strain:** ATCC 25291
Description: Meissner, G., K.H. Schroder, G.E. Amadio, W. Anz, S. Chaparas, H.W.B. Engel, P.A. Jenkins, W. Kappler, H.H. Kleeberg, E. Kubala, M. Kubin, D. Lauterbach, A. Lind, M. Magnusson, Zd. Mikova, S.R. Pattyn, W.B. Schaefer, J.L. Stanford, M. Tsukamura, L.G. Wayne, I. Willers and E. Wolinsky. 1974. Journal of General Microbiology **83**:207-235.

M. bovis Karlson and Lessel 1970
Karlson, A.G. and E.F. Lessel. 1970. International Journal of Systematic Bacteriology **20**:273-282. **Type strain:** ATCC 19210
Description: Karlson, A.G. and E.F. Lessel. 1970. Ibid.

M. chelonae Bergey, Harrison, Breed, Hammer and Huntoon 1923
Bergey, D.H., F.C. Harrison, R.S. Breed, B.W. Hammer and F.M. Huntoon. 1923. Bergey's Manual of Determinative Bacteriology, 1st ed. The Williams and Wilkins Co., Baltimore. **Type strain:** NCTC 946
Description: Kubica, G.P., I. Baess, R.E. Gordon, P.A. Jenkins, J.B.G. Kwapinski, C. McDurmont, S.R. Pattyn, H. Saito, V. Silcox, J.L. Stanford, K. Takeya and M. Tsukamura. 1972. Journal of General Microbiology **73**:55-70.

M. chelonae subsp. *abscessus* (Moore and Frerichs 1953) Kubica, Baess, Gordon, Jenkins, Kwapinski, McDurmont, Pattyn, Saito, Silcox, Stanford, Takeya and Tsukamura 1972
Moore, M. and J.B. Frerichs. 1953. Journal of Investigative Dermatology **20**:133-169; Kubica, G.P., I. Baess, R.E. Gordon, P.A. Jenkins, J.B.G. Kwapinski, C. McDurmont, S.R. Pattyn, H. Saito, V. Silcox, J.L. Stanford, K. Takeya and M. Tsukamura. 1972. Journal of General Microbiology **73**:55-70. **Type strain:** ATCC 19977
Description: Kubica, G.P., I. Baess, R.E. Gordon, P.A. Jenkins, J.B.G. Kwapinski, G. McDurmont, S.R. Pattyn, H. Saito, V. Silcox, J.L. Stanford, K. Takeya and M. Tsukamura. 1972. Ibid.

M. chelonae subsp. *chelonae* (Bergey, Harrison, Breed, Hammer and Huntoon 1923) Kubica, Baess, Gordon, Jenkins, Kwapinski, McDurmont, Pattyn, Saito, Silcox, Stanford, Takeya and Tsukamura 1972
Bergey, D.H., F.C. Harrison, R.S. Breed, B.W. Hammer and F.M. Huntoon. 1923. Bergey's Manual of Determinative Bacteriology, 1st ed. The Williams and Wilkins Co., Baltimore; Kubica, G.P., I. Baess, R.E. Gordon, P.A. Jenkins, J.B.G. Kwapinski, C. McDurmont, S.R. Pattyn, H. Saito, V. Silcox, J.H. Stanford, K. Takeya and M. Tsukamura. 1972. Journal of General Microbiology **73**:55-70. **Type strain:** NCTC 946
Description: Kubica, G.P., I. Baess, R.E. Gordon, P.A. Jenkins, J.B.G. Kwapinski, C. McDurmont, S.R. Pattyn, H. Saito, V. Silcox, J.L. Stanford, K. Takeya and M. Tsukamura. 1972. Ibid.

M. chitae Tsukamura 1967
Tsukamura, M. 1967. Japanese Journal of Microbiology **11**:43-47. **Type strain:** ATCC 19627
Description: Tsukamura, M. 1967. Ibid.

M. duvalii Stanford and Gunthorpe 1971
Stanford, J.L. and W.J. Gunthorpe. 1971. British Journal of Experimental Pathology **52**:627-637. **Type strain:** NCTC 358
Description: Stanford, J.L. and W.J. Gunthorpe. 1971. Ibid.

M. farcinogenes Chamoiseau 1973

Chamoiseau, G. 1973. Annales de Microbiologie de l'Institut Pasteur (Paris) **124**: 215-222. **Type strain:** NCTC 10955
Description: Chamoiseau, G. 1973. Ibid.

M. flavescens Bojalil, Cerbon and Trujillo 1962
Bojalil, L.F., J. Cerbon and A. Trujillo. 1962. Journal of General Microbiology **28**:333-346. **Type strain:** ATCC 14474
Description: Kubica, G.P., I. Baess, R.E. Gordon, P.A. Jenkins, J.B.G. Kwapinski, C. McDurmont, S.R. Pattyn, H. Saito, V. Silcox, J.L. Stanford, K. Takeya and M. Tsukamura. 1972. Journal of General Microbiology **73**:55-70.

M. fortuitum da Costa Cruz 1938
da Costa Cruz, J.C. 1938. Acta Medica (Rio de Janeiro) **1**:297-301. **Type strain:** ATCC 6841
Description: Kubica, G.P., I. Baess, R.E. Gordon, P.A. Jenkins, J.B.G. Kwapinski, C. McDurmont, S.P. Pattyn, H. saito, V. Silcox, J.L. Stanford, K. Takeya and M. Tsukamura. 1972. Journal of General Microbiology **73**:55-70.

M. gadium Casal and Calero 1974
Casal, M. and J.R. Calero. 1974. Tubercle **55**:299-308. **Type strain:** ATCC 27726
Description: Casel, M. and J.R. Calero. 1974. Ibid.

M. gastri Wayne 1966
Wayne, L.G. 1966. American Review of Respiratory Diseases **93**:919-928. **Type strain:** ATCC 15754
Description: Wayne, L.G. 1966. Ibid.

M. gilvum Stanford and Gunthorpe 1971
Stanford, J.L. and W.J. Gunthorpe. 1971. British Journal of Experimental Pathology **52**:627-637. **Type strain:** NCTC 10742
Description: Stanford, J.L. and W.J. Gunthorpe. 1971. Ibid.

M. gordonae Bojalil, Cerbon and Trujillo 1962
Bojalil, L.F., J. Cerbon and A. Trujillo. 1962. Journal of General Microbiology **28**:333-346. **Type strain:** ATCC 14470
Description: Wayne, L.G., T.M. Dietz, C. Gernez-Rieux, P.A. Jenkins, W. Kappler, G.P. Kubica, J.B.G. Kwapinski, G. Meissner, S.R. Pattyn, E.H. Runyon, K.H. Schroder, V.A. Silcox, A. Tacquet, M. Tsukamura and E. Wolinsky. 1971. Journal of General Microbiology **66**:255-271.

M. haemophilum Sompolinsky, Lagziel, Naveh and Yankilevitz 1978
Sompolinsky, D., A. Lagziel, D. Naveh and L. Yankilevitz. 1978. International Journal of Systematic Bacteriology **28**:67-75. **Type strain:** ATCC 29548
Description: Sompolinsky, D., A. Lagziel, D. Naveh and T. Yankilevitz. 1978. Ibid.

M. intracellulare (Cuttino and McCabe 1949) Runyon 1965
Cuttino, J.T. and A.M. McCabe. 1949. American Journal of Clinical Pathology **25**:1-34; Runyon, E.H. 1965. Advances in Tuberculosis Research **14**:235-287. **Type strain:** ATCC 13950
Description: Meissner, G., K.H. Schroder, G.E. Amadio, W. Anz, S. Chaparas, H.W.B. Engel, P.A. Jenkins, W. Kappler, H.H. Kleeberg, E. Kubala, M. Kubin, D. Lauterbach, A. Lind, M. Magnusson, Zd. Mikova, S.R. Pattyn, W.B. Schaefer, J.L. Stanford, M. Tsukamura, L.G. Wayne, I. Willers and E. Wolinsky. 1974. Journal of General Microbiology **83**:207-235.

M. kansasii Hauduroy 1955
Hauduroy, P. 1955. Derniers aspects du monde des mycobacteries. Masson et Cie, Paris. **Type strain:** ATCC 12478
Description: Bergey 8.

M. komossense Kazda and Muller 1979
Kazda, J. and K. Muller. 1979. International Journal of Systematic Bacteriology **29**:361-365. **Type strain:** ATCC 33013
Description: Kazda, J. and K. Muller. 1979. Ibid.

M. leprae (Hansen 1880) Lehmann and Neumann 1896
Hansen, G.A. 1880. Virchows Archiv fur pathologische Anatomie und Physiologie und fur klinische Medizin **79**:32-42; Lehmann, K.B. and R. Neumann. 1896. Atlas und Grundriss der Bakteriologie und Lehrbuch der speziellen bakteriologischen Diagnostik. 1st ed. J.F. Lehmann, Munchen. **Type strain:** Has not been cultivated.
Description: Bergey 8.

M. lepraemurium Marchoux and Sorel 1912
Marchoux, E. and F. Sorel. 1912. Annales de l'Institut Pasteur (Paris) **26**:675-700. **Type strain:** Difficult to cultivate.
Description: Saito, H., K. Yamaoka and

K. Kiyotani. 1976. International Journal of Systematic Bacteriology **26**:111-115.

M. malmoense Schroder and Juhlin 1977
Schroder, K.H. and I. Juhlin. 1977. International Journal of Systematic Bacteriology **27**:241-246. **Type strain:** ATCC 29571
Description: Schroder, K.H. and I. Juhlin. 1977. Ibid.

M. marinum Aronson 1926
Aronson, J.D. 1926. Journal of Infectious Diseases **39**:314-320. **Type strain:** ATCC 927
Description: Bergey 8.

M. microti Reed 1957
Reed, G.B. 1957 in Breed, R.S., E.G.D. Murray and N.R. Smith (eds). Bergey's Manual of Determinative Bacteriology, 7th ed. The Williams and Wilkins Co., Baltimore. p. 703-704. **Type strain:** NCTC 8710
Description: Bergey 8.

M. nonchromogenicum Tsukamura 1965
Tsukamura, M. 1965. Medicine and Biology **71**:110-113. **Type strain:** ATCC 19530
Description: Meissner, G., K.H. Schroder, G.E. Amadio, W. Anz, S. Chaparas, H.W.B. Engel, P.A. Jenkins, W. Kappler, H.H. Kleeberg, E. Kubala, M. Kubin, D. Lauterbach, A. Lind, M. Magnusson, Zd. Mikova, S.R. Pattyn, W.B. Schaefer, J.L. Stanford, M. Tsukamura, L.G. Wayne, I. Willers and E. Wolinsky. 1974. Journal of General Microbiology **83**:207-235.

M. neoaurum Tsukamura 1972
Tsukamura, M. 1972. Medicine and Biology **85**:229-233. **Type strain:** ATCC 25795
Description: Tsukamura, M. 1972. Ibid.

M. parafortuitum Tsukamura, Toyama and Mizuno 1965
Tsukamura, M., H. Toyama and S. Mizuno. 1965. Medicine and Biology **70**:232-235. **Type strain:** ATCC 19686
Description: Saito, H., R.E. Gordon, I. Juhlin, W. Kappler, J.B.G. Kwapinski, C. McDurmont, S.R. Pattyn, E.H. Runyon, J.L. Stanford, I. Tarnok, H. Tasaka, M. Tsukamura and J. Weiszfeiler. 1977. International Journal of Systematic Bacteriology **27**:75-85.

M. paratuberculosis Bergey, Harrison, Breed, Hammer and Huntoon 1923
Bergey, D.H., F.C. Harrison, R.S. Breed, B.W. Hammer and F.M. Huntoon. 1923. Bergey's Manual of Determinative Bacteriology, 1st ed. The Williams and Wilkins Co., Baltimore. p. 374. **Type strain:** ATCC 19698
Description: Thorel, M.F. and L. Valette. 1976. Annales Recherches Veterinaire **7**: 207-213.

M. phlei Lehmann and Neumann 1899
Lehmann, K.B. and R. Neumann. 1899. Lehmann's Medizin, Handatlanten X. Atlas und Grundriss der Bakteriologie und Lehrbuch der speziellen bakteriologischen Diagnostik. **2**:1-497. **Type strain:** ATCC 11758
Description: Kubica, G.P., I. Baess, R.E. Gordon, P.A. Jenkins, J.B.G. Kwapinski, C. McDurmont, S.R. Pattyn, H. Saito, V. Silcox, J.L. Stanford, K. Takeya and M. Tsukamura. 1972. Journal of General Microbiology **73**:55-70.

M. scrofulaceum Prissick and Masson 1956
Prissick, F.H. and A.M. Masson. 1956. Canadian Medical Association Journal **75**: 798-803. **Type strain:** ATCC 19981
Description: Wayne, L.G., T.M. Dietz, C. Gernez-Rieux, P.A. Jenkins, W. Kappler, G.P. Kubica, J.B.G. Kwapinski, G. Meissner, S.R. Pattyn, E.H. Runyon, K.H. Schroder, V.A. Silcox, A. Tacquet, M. Tsukamura and E. Wolinsky. 1971. Journal of General Microbiology **66**:255-271.

M. senegalense (Chamoiseau 1973) Chamoiseau 1979
Chamoiseau, G. 1973. Annales de Microbiologie de l'Institut Pasteur (Paris) **124**: 215-222; Chamoiseau, G. 1979. International Journal of Systematic Bacteriology **29**:407-410. **Type strain:** NCTC 10956
Description: Chamoiseau, G. 1973. Ibid.

M. simiae Karassova, Weiszfeiler and Krasznay 1965
Karassova, V., J. Weiszfeiler and E. Krasznay. 1965. Acta Microbiologica Academiae Scientiarum Hungaricae **12**:275-282. **Type strain:** ATCC 25275
Description: Bergey 8.

M. smegmatis (Trevisan 1889) Lehmann and Neumann 1899
Trevisan, V. 1889. I Generi e le Specie delle Bacteriacee. Zanaboni and Gabuzzi, Milano; Lehmann, K.B. and R. Neu-

mann. 1899. Lehmann's Medizin Handatlanten X. Atlas und Grundriss der Bakteriologie und Lehrbuch der speziellen bakteriologischen Diagnostik. **2**:1-497. **Type strain:** ATCC 19420
Description: Kubica, G.P., I. Baess, R.E. Gordon, P.A. Jenkins, J.B.G. Kwapinski, C. McDurmont, S.R. Pattyn, H. Saito, V. Silcox, J.L. Stanford, K. Takeya and M. Tsukamura. 1972. Journal of General Microbiology **73**:55-70.

M. szulgai Marks, Jenkins and Tsukamura 1972
Marks, J., P.A. Jenkins and M. Tsukamura. 1972. Tubercle **53**:210-214. **Type strain:** NCTC 10831
Description: Marks, J., P.A. Jenkins and M. Tsukamura. 1972. Ibid.

M. terrae Wayne 1966
Wayne, L.G. 1966. American Review of Respiratory Diseases **93**:919-928. **Type strain:** ATCC 15755
Description: Meissner, G., K.H. Schroder, G.E. Amadio, W. Anz, S. Chaparas, H.W.B. Engel, P.A. Jenkins, W. Kappler, H.H. Kleeberg, E. Kubala, M. Kubin, D. Lauterbach, A. Lind, M. Magnusson, Zd. Mikova, S.R. Pattyn, W.B. Schaefer, J.L. Stanford, M. Tsukamura, L.G. Wayne, I. Willers and E. Wolinsky. 1974. Journal of General Microbiology **83**:207-235.

M. thermoresistibile Tsukamura 1966
Tsukamura, M. 1966. Journal of General Microbiology **45**:253-273. **Type strain:** ATCC 19527
Description: Tsukamura, M. 1966. Ibid.

M. triviale Kubica 1970
Kubica, G.P. 1970 in Kubica, G.P., V.A. Silcox, J.O. Kilburn, R.W. Smithwick, R.E. Beam, W.D. Jones, Jr. and K.D. Stottmeier. International Journal of Systematic Bacteriology **20**:161-174. **Type strain:** ATCC 23292
Description: Kubica, G.P. 1970. Ibid.

M. tuberculosis (Zopf 1883) Lehmann and Neumann 1896
Zopf, W. 1883. Die Spaltpilze, Edward Trewendt, Breslau. pp. 1-100; Lehmann, K.B. and R. Neumann. 1896. Atlas und Grundriss der Bakteriologie und Lehrbuch der speziellen bakteriologischen Diagnostik. 1st ed. J.F. Lehmann, Munchen. **Type strain:** ATCC 27294
Description: Bergey 8.

M. ulcerans MacCallum, Tolhurst and Buckle 1950
MacCallum, P., J.C. Tolhurst and G. Buckle. 1950 in Fenner, F. Medical Journal of Australia **2**:813-818. **Type strain:** ATCC 19423
Description: Boisvert, H. 1977. Bulletin de la Societe de Pathologie Exotique **70**: 125-131.

M. vaccae Bonicke and Juhasz 1964
Bonicke, R. and S.E. Juhasz. 1964. Zentralblatt fur Bakteriologie, Parasitenkunde, Infektionskrankheiten und Hygiene. Abteilung I. **192**:133-135. **Type strain:** ATCC 15483
Description: Saito, H., R.E. Gordon, I. Juhlin, W. Kappler, J.B.G. Kwapinski, C. McDurmont, S.R. Pattyn, E.H. Runyon, J.L. Stanford, I. Tarnok, H. Tasaka, M. Tsukamura and J. Weiszfeiler. 1977. International Journal of Systematic Bacteriology **27**:75-85.

M. xenopi Schwabacher 1959
Schwabacher, H. 1959. Journal of Hygiene **57**:57-67. **Type strain:** NCTC 10042
Description: Meissner, G., K.H. Schroder, G.E. Amadio, W. Anz, S. Chaparas, H.W.B. Engel, P.A. Jenkins, W. Kappler, H.H. Kleeberg, E. Kubala, M. Kubin, D. Lauterbach, A. Lind, M. Magnusson, Zd. Mikova, S.R. Pattyn, W.B. Schaefer, J.L. Stanford, M. Tsukamura, L.G. Wayne, I. Willers and E. Wolinsky. 1974. Journal of General Microbiology **83**:207-235.

Mycoplana Gray and Thornton 1928
Gray, P.H.H. and H.G. Thornton. 1928. Zentralblatt fur Bakteriologie, Parasitenkunde, Infektionskrankheiten und Hygiene. Abteilung II. **73**:74-96. **Type species:** *M. dimorpha* Gray and Thornton 1928
Description: Gray, P.H.H. and H.G. Thornton. 1928. Ibid.

M. bullata Gray and Thornton 1928
Gray, P.H.H. and H.G. Thornton. 1928. Zentralblatt fur Bakteriologie, Parasitenkunde, Infektionskrankheiten und Hygiene. Abteilung II. **73**:74-96. **Type strain:** ATCC 4278
Description: Gray, P.H.H. and H.G.

Thornton. 1928. Ibid.

M. dimorpha Gray and Thornton 1928
Gray, P.H.H. and H.G. Thornton. 1928. Zentralblatt fur Bakteriologie, Parasitenkunde, Infektionskrankheiten und Hygiene. Abteilung II. **73**:74-96. **Type strain:** ATCC 4279
Description: Gray, P.H.H. and H.G. Thornton. 1928. Ibid.

Mycoplasma Nowak 1929
Nowak, J. 1929. Annales de l'Institut Pasteur (Paris) **43**:1330-1352. **Type species:** *M. mycoides* (Borrel, Dujardin-Beaumetz, Jeantet and Jouan 1910) Freundt 1955
Description: Bergey 8.

M. agalactiae (Wroblewski 1931) Freundt 1955
Wroblewski, W. 1931. Annales de l'Institut Pasteur (Paris) **47**:94-115; Freundt, E.A. 1955. International Bulletin of Bacteriological Nomenclature and Taxonomy **5**:67-78. **Type strain:** NCTC 10123
Description: Cottew, G.S. 1979 in Barile, M.F., S. Razin, J.G. Tully and R.F. Whitcomb (eds). The Mycoplasmas. Academic Press, New York, Volume II. pp. 103-132.

M. alkalescens Leach 1973
Leach, R.H. 1973. Journal of General Microbiology **75**:135-153. **Type strain:** ATCC 29103
Description: Leach, R.H. 1973. Ibid.

M. alvi Gourlay, Wyld and Leach 1977
Gourlay, R.N., S.G. Wyld and R.H. Leach. 1977. International Journal of Systematic Bacteriology **27**:86-96. **Type strain:** ATCC 29626
Description: Gourlay, R.N., S.G. Wyld and R.H. Leach. 1977. Ibid.

M. anatis Roberts 1964
Roberts, D.H. 1964. Veterinary Record **76**:470-473. **Type strain:** ATCC 25524
Description: Bergey 8.

M. arginini Barile, Del Giudice, Carski, Gibbs and Morris 1968
Barile, M.F., R.A. Del Giudice, T.R. Carski, C.J. Gibbs and J.A. Morris. 1968. Proceedings of the Society for Experimental Biology and Medicine **129**:489-494. **Type strain:** ATCC 23838
Description: Gourlay, R.N. and C.J. Howard. 1979 in Barile, M.F., S. Razin, J.G. Tully and R.F. Whitcomb (eds). The Mycoplasmas. Academic Press, New York, Volume II. pp. 49-102.

M. arthritidis (Sabin 1941) Freundt 1955
Sabin, A.B. 1941. Bacteriological Reviews **5**:1-66; Freundt, E.A. 1955. International Bulletin of Bacteriological Nomenclature and Taxonomy **5**:67-78. **Type strain:** ATCC 19611
Description: Cassell, G.H. and A. Hill. 1979 in Barile, M.F., S. Razin, J.G. Tully and R.F. Whitcomb (eds). The Mycoplasmas. Academic Press, New York, Volume II. pp. 235-273.

M. bovigenitalium Freundt 1955
Freundt, E.A. 1955. International Bulletin of Bacteriological Nomenclature and Taxonomy **5**:67-78. **Type strain:** ATCC 19852
Description: Gourlay, R.N. and C.J. Howard. 1979 in Barile, M.F., S. Razin, J.G. Tully and R.F. Whitcomb (eds). The Mycoplasmas. Academic Press, New York, Volume II. pp. 49-102.

M. bovirhinis Leach 1967
Leach, R.H. 1967. Annals of the New York Academy of Sciences **143**:305-316. **Type strain:** ATCC 27748
Description: Gourlay, R.N. and C.J. Howard. 1979 in Barile, M.F., S. Razin, J.G. Tully and R.F. Whitcomb (eds). The Mycoplasmas. Academic Press, New York, Volume II. pp. 49-102.

M. bovis (Hale, Helmboldt, Plastridge and Stula 1962) Askaa and Erno 1976
Hale, H.H., C.F. Helmboldt, W.N. Plastridge and E.F. Stula. 1962. Cornell Veterinarian **52**:582-591; Askaa, G. and H. Erno. 1976. International Journal of Systematic Bacteriology **26**:323-325. **Type strain:** ATCC 25523
Description: Askaa, G. and H. Erno. 1976. Ibid.

M. bovoculi Langford and Leach 1973
Langford, E.V. and R.H. Leach. 1973. Canadian Journal of Microbiology **19**:1435-1444. **Type strain:** ATCC 29104
Description: Langford, E.V. and R.H. Leach. 1973. Ibid.

M. buccale Freundt, Taylor-Robinson, Purcell, Chanock and Black 1974
Freundt, E.A., D. Taylor-Robinson, R.H. Purcell, R.M. Chanock and F.T. Black. 1974. International Journal of Systematic Bacteriology **24**:252-255. **Type strain:**

ATCC 23636
Description: Freundt, E.A., D. Taylor-Robinson, R.H. Purcell, R.M. Chanock and F.T. Black. 1974. Ibid.

M. canadense Langford, Ruhnke and Onoviran 1976
Langford, E.V., H.L. Ruhnke and O. Onoviran. 1976. International Journal of Systematic Bacteriology **26**:212-219. **Type strain:** ATCC 29418
Description: Langford, E.V., H.L. Ruhnke and O. Onoviran. 1976. Ibid.

M. canis Edward 1955
Edward, D.G. ff. 1955. International Bulletin of Bacteriological Nomenclature and Taxonomy **5**:85-93. **Type strain:** ATCC 19525
Description: Rosendal, S. 1979 in Barile, M.F., S. Razin, J.G. Tully and R.F. Whitcomb (eds). The Mycoplasmas. Academic Press, New York, Volume II. pp. 217-234.

M. capricolum Tully, Barile, Edward, Theodore and Erno 1974
Tully, J.G., M.F. Barile, D.G. ff. Edward, T.S. Theodore and H. Erno. 1974. Journal of General Microbiology **85**:102-120. **Type strain:** ATCC 27343
Description: Tully, J.G., M.F. Barile, D.G. ff. Edward, T.S. Theodore and H. Erno. 1974. Ibid.

M. caviae Hill 1971
Hill, A. 1971. Journal of General Microbiology **65**:109-113. **Type strain:** ATCC 27108
Description: Cassel, G.H. and A. Hill. 1979 in Barile, M.F., S. Razin, J.G. Tully and R.F. Whitcomb (eds). The Mycoplasmas. Academic Press, New York, Volume II. pp. 235-273.

M. citelli Rose, Tully and Langford 1978
Rose, D.L., J.G. Tully and E.V. Langford. 1978. International Journal of Systematic Bacteriology **28**:567-572. **Type strain:** ATCC 29760
Description: Rose, D.L., J.G. Tully and E.V. Langford. 1978. Ibid.

M. columbinum Shimizu, Erno and Nagatomo 1978
Shimizu, T., H. Erno and H. Nagatomo. 1978. International Journal of Systematic Bacteriology **28**:538-546. **Type strain:** ATCC 29257
Description: Shimizu, T., H. Erno and H. Nagatomo. 1978. Ibid.

M. columborale Shimizu, Erno and Nagatomo 1978
Shimizu, T., H. Erno and H. Nagatomo. 1978. International Journal of Systematic Bacteriology **28**:538-546. **Type strain:** ATCC 29258
Description: Shimizu, T., H. Erno and H. Nagatomo. 1978. Ibid.

M. conjunctivae Barile, Del Giudice and Tully 1972
Barile, M.F., R.A. Del Giudice and J.G. Tully. 1972. Infection and Immunity **5**:70-76. **Type strain:** ATCC 25834
Description: Barile, M.F., R.A. Del Giudice and J.G. Tully. 1972. Ibid.

M. cynos Rosendal 1973
Rosendal, S. 1973. International Journal of Systematic Bacteriology **23**:49-54. **Type strain:** ATCC 27544
Description: Rosendal, S. 1973. Ibid.

M. dispar Gourlay and Leach 1970
Gourlay, R.N. and R.H. Leach. 1970. Journal of Medical Microbiology **3**:111-123. **Type strain:** ATCC 27140
Description: Gourlay, R.N. and C.J. Howard. 1979 in Barile, M.F., S. Razin, J.G. Tully and R.F. Whitcomb (eds). The Mycoplasmas. Academic Press, New York, Volume II. pp. 49-102.

M. edwardii Tully, Barile, Del Giudice, Carski, Armstrong and Razin 1970
Tully, J.G., M.F. Barile, R.A. Del Giudice, T.R. Carski, D. Armstrong and S. Razin. 1970. Journal of Bacteriology **101**: 346-349. **Type strain:** ATCC 23462
Description: Rosendal, S. 1979 in Barile, M.F., S. Razin, J.G. Tully and R.F. Whitcomb (eds). The Mycoplasmas. Academic Press, New York, Volume II. pp. 217-234.

M. equigenitalium Kirchhoff 1978
Kirchhoff, H. 1978. International Journal of Systematic Bacteriology **28**:496-502. **Type strain:** ATCC 29869
Description: Kirchhoff, H. 1978. Ibid.

M. equirhinis Allam and Lemcke 1975
Allam, N.M. and R.M. Lemcke. 1975. Journal of Hygiene Cambridge **74**:385-408. **Type strain:** ATCC 29420
Description: Allam, N.M. and R.M. Lemcke. 1975. Ibid.

M. faucium Freundt, Taylor-Robinson, Purcell, Chanock and Black 1974
Freundt, E.A., D. Taylor-Robinson, R.H. Purcell, R.M. Chanock and F.T. Black 1974. International Journal of Systematic Bacteriology **24**:252-255. **Type strain:** ATCC 25293
Description: Freundt, E.A., D. Taylor-Robinson, R.H. Purcell, R.M. Chanock and F.T. Black. 1974. Ibid.

M. feliminutum Heyward, Sabry and Dowdle 1969
Heyward, J.T., M.Z. Sabry and W.R. Dowdle. 1969. American Journal of Veterinary Research **30**:615-622. **Type strain:** ATCC 25749
Description: Rosendal, S. 1979 in Barile, M.F., S. Razin, J.G. Tully and R.F. Whitcomb (eds). The Mycoplasmas. Academic Press, New York, Volume II. pp. 217-234.

M. felis Cole, Golightly and Ward 1967
Cole, B.C., L. Golightly and J.R. Ward. 1967. Journal of Bacteriology **94**:1451-1458. **Type strain:** ATCC 23391
Description: Bergey 8.

M. fermentans Edward 1955
Edward, D.G. ff. 1955. International Bulletin of Bacteriological Nomenclature and Taxonomy **5**:85-93. **Type strain:** ATCC 19989
Description: Bergey 8.

M. flocculare Meyling and Friss 1972
Meyling, A. and N.F. Friis. 1972. Acta Veterinaria Scandinavica **13**:287-289. **Type strain:** ATCC 27399
Description: Rose, D.L. J.G. Tully and R.G. Wittler. 1979. International Journal of Systematic Bacteriology **29**:83-91.

M. gallinarum Freundt 1955
Freundt, E.A. 1955. International Bulletin of Bacteriological Nomenclature and Taxonomy **5**:67-78. **Type strain:** ATCC 19708
Description: Bergey 8.

M. gallisepticum Edward and Kanarek 1960
Edward, D.G. ff. and A.D. Kanarek. 1960. Annals of the New York Academy of Sciences **79**:696-702. **Type strain:** ATCC 19610
Description: Bergey 8.

M. gateae Cole, Golightly and Ward 1967
Cole, B.C., L. Golightly and J.R. Ward. 1967. Journal of Bacteriology **94**:1451-1458. **Type strain:** ATCC 23392
Description: Bergey 8.

M. hominis (Freundt 1953) Edward 1955
Freundt, E.A. 1953. Acta Pathologica et Microbiologica Scandinavica **32**:468-480; Edward, D.G. ff. 1955. International Bulletin of Bacteriological Nomenclature and Taxonomy **5**:85-93. **Type strain:** ATCC 23114
Description: Bergey 8.

M. hyopneumoniae Mare and Switzer 1965
Mare, C.J. and W.P. Switzer. 1965. Veterinary Medicine **60**:841-846. **Type strain:** ATCC 25934
Description: Rose, D.L., J.G. Tully and R.G. Wittler. 1979. International Journal of Systematic Bacteriology **29**:83-91.

M. hyorhinis Switzer 1955
Switzer, W.P. 1955. American Journal of Veterinary Research **16**:540-544. **Type strain:** ATCC 17981
Description: Whittlestone, P. 1979 in Barile, M.F., S. Razin, J.G. Tully and R.F. Whitcomb (eds). The Mycoplasmas. Academic Press, New York, Volume II. pp. 133-176.

M. hyosynoviae Ross and Karmon 1970
Ross, R.F. and J.A. Karmon. 1970. Journal of Bacteriology **103**:707-713. **Type strain:** ATCC 25591
Description: Whittlestone, P. 1979 in Barile, M.F., S. Razin, J.G. Tully and R.F. Whitcomb (eds). The Mycoplasmas. Academic Press, New York, Volume II. pp. 133-176.

M. iners Edward and Kanarek 1960
Edward, D.G. ff. and A.D. Kanarek. 1960. Annals of the New York Academy of Sciences **79**:696-702. **Type strain:** ATCC 19705
Description: Bergey 8.

M. lipophilum Del Giudice, Purcell, Carski and Chanock 1974
Del Giudice, R.A., R.H. Purcell, T.R. Carski and R.M. Chanock 1974. International Journal of Systematic Bacteriology **24**:147-153. **Type strain:** ATCC 27104
Description: Del Giudice, R.A., R.H. Purcell, T.R. Carski and R.M. Chanock. 1974. Ibid.

M. maculosum Edward 1955
Edward, D.G. ff. 1955. International Bulletin of Bacteriological Nomenclature and

Taxonomy **5**:85-93. **Type strain:** ATCC 19327
Description: Rosendal, S. 1979 in Barile, M.F., S. Razin, J.G. Tully and R.F. Whitcomb (eds). The Mycoplasmas. Academic Press, New York, Volume II. pp. 217-234.

M. meleagridis Yamamoto, Bigland and Ortmayer 1965
Yamamoto, R., C.H. Bigland and H.B. Ortmayer. 1965. Journal of Bacteriology **90**:47-49. **Type strain:** ATCC 25294
Description: Bergey 8.

M. moatsii Madden, Moats, London, Matthew and Sever 1974
Madden, D.L., K.E. Moats, W.T. London, E.B. Matthew and J.L. Sever. 1974. International Journal of Systematic Bacteriology **24**:459-464. **Type strain:** ATCC 27625
Description: Madden, D.L., K.E. Moats, W.T. London, E.B. Matthew and J.L. Sever. 1974. Ibid.

M. molare Rosendal 1974
Rosendal, S. 1974. International Journal of Systematic Bacteriology **24**:125-130. **Type strain:** ATCC 27746
Description: Rosendal, S. 1974. Ibid.

M. mycoides (Borrel, Dujardin-Beaumetz, Jeantet and Jouan 1910) Freundt 1955
Borrel, A., E. Dujardin-Beaumetz, Jeantet and Jouan. 1910. Annales de l'Institut Pasteur (Paris) **24**:168-179; Freundt, E.A. 1955. International Bulletin of Bacteriological Nomenclature and Taxonomy **5**: 67-78. **Type strain:** NCTC 10114
Description: Gourlay, R.N. and C.J. Howard. 1979 in Barile, M.F., S. Razin, J.G. Tully and R.F. Whitcomb (eds). The Mycoplasmas. Academic Press, New York, Volume II. pp. 103-132.

M. mycoides subsp. *capri* (Edward 1953) Freundt 1955
Edward, D.G. ff. 1953. Veterinary Record **65**:873-874; Freundt, E.A. 1955. International Bulletin of Bacteriological Nomenclature and Taxonomy **5**:67-78. **Type strain:** NCTC 10137
Description: Cottew, G.S. 1979 in Barile, M.F., S. Razin, J.G. Tully and R.F. Whitcomb (eds). The Mycoplasmas. Academic Press, New York, Volume II. pp. 103-132.

M. mycoides subsp. *mycoides* (Borrel, Dujardin-Beaumetz, Jeantet and Jouan 1910) Freundt 1955
Borrel, A., E. Dujardin-Beaumetz, Jeantet and Jouan. 1910. Annales de l'Institut Pasteur (Paris) **24**:168-179; Freundt, E.A. 1955. International Bulletin of Bacteriological Nomenclature and Taxonomy **5**: 67-78. **Type strain:** NCTC 10114
Description: Gourlay, R.N. and C.J. Howard. 1979 in Barile, M.F., S. Razin, J.G. Tully and R.F. Whitcomb (eds). The Mycoplasmas. Academic Press, New York, Volume II. pp. 103-132.

M. neurolyticum (Sabin 1941) Freundt 1955
Sabin, A.B. 1941. Bacteriological Reviews **5**:1-66; Freundt, E.A. 1955. International Bulletin of Bacteriological Nomenclature and Taxonomy **5**:67-78. **Type strain:** ATCC 19988
Description: Cassell, G.H. and A. Hill. 1979 in Barile, M.F., S. Razin, J.G. Tully and R.F. Whitcomb (eds). The Mycoplasmas. Academic Press, New York, Volume II. pp. 235-273.

M. opalescens Rosendal 1975
Rosendal, S. 1975. Acta Pathologica et Microbiologica Scandinavica Section B **83**:463-470. **Type strain:** ATCC 27921
Description: Rosendal, S. 1975. Ibid.

M. orale Taylor-Robinson, Canchola, Fox and Chanock 1964
Taylor-Robinson, D., J. Canchola, H. Fox and R.M. Chanock. 1964. American Journal of Hygiene **80**:135-148. **Type strain:** ATCC 23714
Description: Bergey 8.

M. ovipneumoniae Carmichael, St. George, Sullivan and Horsfall 1972
Carmichael, L.E., T.D. St. George, N.D. Sullivan and N. Horsfall. 1972. Cornell Veterinarian **62**:654-679. **Type strain:** ATCC 29419
Description: Cottew, G.S. 1979 in Barile, M.F., S. Razin, J.G. Tully and R.F. Whitcomb (eds). The Mycoplasmas. Academic Press, New York, Volume II. pp. 103-132.

M. pneumoniae Somerson, Taylor-Robinson and Chanock 1963
Somerson, N.L., D. Taylor-Robinson and R.M. Chanock. 1963. American Journal of Hygiene **77**:122-128. **Type strain:**

ATCC 15531
Description: Bergey 8.

M. primatum Del Giudice, Carski, Barile, Lemcke and Tully 1971
Del Giudice, R.A., T.R. Carski, M.F. Barile, R.M. Lemcke and J. Tully. 1971. Journal of Bacteriology **108**:439-445. **Type strain:** ATCC 25948
Description: Del Giudice, R.A., T.R. Carski, M.F. Barile, R.M. Lemcke and J.G. Tully. 1971. Ibid.

M. pulmonis (Sabin 1941) Freundt 1955
Sabin, A.B. 1941. Bacteriological Reviews **5**:1-66; Freundt, E.A. 1955. International Bulletin of Bacteriological Nomenclature and Taxonomy **5**:67-78. **Type strain:** ATCC 19612
Description: Cassell, G.H. and A. Hill. 1979 in Barile, M.F., S. Razin, J.G. Tully and R.F. Whitcomb (eds). The Mycoplasmas. Academic Press, New York, Volume II. pp. 235-273.

M. putrefaciens Tully, Barile, Edward, Theodore and Erno 1974
Tully, J.G., M.F. Barile, D.G. ff. Edward, T.S. Theodore and H. Erno. 1974. Journal of General Microbiology **85**:102-120. **Type strain:** ATCC 15718
Description: Tully, J.G., M.F. Barile, D.G. ff. Edward, T.S. Theodore and H. Erno. 1974. Ibid.

M. salivarium Edward 1955
Edward, D.G. ff. 1955. International Bulletin of Bacteriological Nomenclature and Taxonomy **5**:85-93. **Type strain:** ATCC 23064
Description: Bergey 8.

M. spumans Edward 1955
Edward, D.G. ff. 1955. International Bulletin of Bacteriological Nomenclature and Taxonomy **5**:85-93. **Type strain:** ATCC 19526
Description: Rosendal, S. 1979 in Barile, M.F., S. Razin, J.G. Tully and R.F. Whitcomb (eds). The Mycoplasmas. Academic Press, New York, Volume II. pp. 217-234.

M. sualvi Gourlay, Wyld and Leach 1978
Gourlay, R.N., S.G. Wyld and R.H. Leach. 1978. International Journal of Systematic Bacteriology **28**:289-292. **Type strain:** NCTC 10170
Description: Gourlay, R.N., S.G. Wyld and R.H. Leach. 1978. Ibid.

M. subdolum Lemcke and Kirchhoff 1979
Lemcke, R.M. and H. Kirchhoff. 1979. International Journal of Systematic Bacteriology **29**:42-50. **Type strain:** ATCC 29870
Description: Lemcke, R.M. and H. Kirchhoff. 1979. Ibid.

M. synoviae Olson, Kerr and Campbell 1964
Olson, N.O., K.M. Kerr and A. Campbell. 1964. Avian Diseases **8**:209-214. **Type strain:** ATCC 25204
Description: Bergey 8.

M. verecundum Gourlay, Leach and Howard 1974
Gourlay, R.N., R.H. Leach and C.J. Howard. 1974. Journal of General Microbiology **81**:475-484. **Type strain:** ATCC 27862
Description: Gourlay, R.N., R.H. Leach and C.J. Howard. 1974. Ibid.

Myxococcus Thaxter 1892
Thaxter, R. 1892. Botanical Gazette **17**: 389-406. **Type species:** *M. fulvus* (Cohn 1875) Jahn 1911
Description: Bergey 8.

M. coralloides Thaxter 1892
Thaxter, R. 1892. Botanical Gazette **17**: 389-406. **Type strain:** ATCC 25202
Description: Bergey 8.

M. disciformis Thaxter 1904
Thaxter, R. 1904. Botanical Gazette **37**: 405-416. **Type strain:** ATCC 33172
Description: Thaxter, R. 1904. Ibid.

M. fulvus (Cohn 1875) Jahn 1911
Cohn, F. 1875. Beitrage zur Biologie der Pflanzen **1**:141-207; Jahn, E. 1911. Kryptogamenflora der Mark Brandenburg **5**: 187-206. **Type strain:** ATCC 25199
Description: Bergey 8.

M. macrosporus (Krzemieniewska and Krzemieniewski 1926) Zahler and McCurdy 1974
Krzemieniewska, H. and S. Krzemieniewski. 1926. Acta Societatis Botanicorum Poloniae **4**:1-54; Zahler, S.A. and H.D. McCurdy. 1974. Bergey 8. **Type strain:** Windsor M271
Description: Bergey 8.

M. stipitatus Thaxter 1897
Thaxter, R. 1897. Botanical Gazette **23**: 395-411. **Type strain:** Windsor M78
Description: Bergey 8.

M. virescens Thaxter 1892
Thaxter, R. 1892. Botanical Gazette **17**: 389-406. **Type strain:** ATCC 25203
Description: Bergey 8.

M. xanthus Beebe 1941
Beebe, J.M. 1941. Journal of Bacteriology **42**:193-223. **Type strain:** ATCC 25232
Description: Bergey 8.

Nannocystis Reichenbach 1970
Reichenbach, H. 1970. Archiv fur Mikrobiologie **70**:119-138. **Type species:** *N. exedens* Reichenbach 1970
Description: Bergey 8.

N. exedens Reichenbach 1970
Reichenbach, H. 1970. Archiv fur Mikrobiologie **70**:119-138. **Type strain:** DSM 71
Description: Reichenbach, H. 1970. Ibid.

Neisseria Trevisan 1885
Trevisan, V. 1885. Atti della Accademia Fisica-Medica-Statistica in Milano (Ser 4) **3**:92-107. **Type species:** *N. gonorrhoeae* (Zopf 1885) Trevisan 1885
Description: Bergey 8.

N. animalis Berger 1960
Berger, U. 1960. Zeitschrift fur Hygiene und Infektionskrankheiten **147**:158-161. **Type strain:** NCTC 10212
Description: Berger, U. 1960. Ibid.

N. canis Berger 1962
Berger, U. 1962. Zeitschrift fur Hygiene und Infektionskrankheiten **148**:445-457. **Type strain:** ATCC 14678
Description: Berger, U. 1962. Ibid.

N. caviae Pelczar 1953
Pelczar, M.J. 1953. Journal of Bacteriology **65**:744. **Type strain:** ATCC 14659
Description: Branham, S.E. 1957 in Breed, R.S., E.G.D. Murray and N.R. Smith (eds). Bergey's Manual of Determinative Bacteriology, 7th ed. The Williams and Wilkins Co., Baltimore. pp. 480-485.

N. cinerea (von Lingelsheim 1906) Murray 1939
von Lingelsheim, W. 1906. Klinisches Jahrbuch **15**:373-489; Murray, E.G.D. 1939 in Bergey, D.H., R.S. Breed, E.G.D. Murray and A.P. Hitchens (eds). Bergey's Manual of Determinative Bacteriology, 5th ed. The Williams and Wilkins Co., Baltimore. pp. 278-288. **Type strain:** ATCC 14685
Description: Murray, E.G.D. 1939. Ibid.

N. cuniculi Berger 1962
Berger, U. 1962. Zeitschrift fur Hygiene und Infektionskrankheiten **148**:445-457. **Type strain:** ATCC 14688
Description: Berger, U. 1962. Ibid.

N. denitrificans Berger 1962
Berger, U. 1962. Zeitschrift fur Hygiene und Infektionskrankheiten **148**:445-457. **Type strain:** ATCC 14686
Description: Berger, U. 1962. Ibid.

N. elongata Bovre and Holten 1970
Bovre, K. and E. Holten. 1970. Journal of General Microbiology **60**:67-75. **Type strain:** ATCC 25295
Description: Bovre, K. and E. Holten. 1970. Ibid.

N. elongata subsp. *elongata* Bovre and Holten 1970
Bovre, K. and E. Holten. 1970. Journal of General Microbiology **60**:67-75. **Type strain:** ATCC 25295
Description: Bovre, K. and E. Holten. 1970. Ibid.

N. elongata subsp. *glycolytica* Henriksen and Holten 1976
Henriksen, S.D. and E. Holten. 1976. International Journal of Systematic Bacteriology **26**:478-481. **Type strain:** ATCC 29315
Description: Henriksen, S.D. and E. Holten. 1976. Ibid.

N. flava Bergey, Harrison, Breed, Hammer and Huntoon 1923
Bergey, D.H., F.C. Harrison, R.S. Breed, B.W. Hammer and F.M. Huntoon. 1923. Bergey's Manual of Determinative Bacteriology, 1st ed. The Williams and Wilkins Co., Baltimore. pp. 1-442. **Type strain:** NRL 30,008
Description: Branham, S.E. 1957 in Breed, R.S., E.G.D. Murray and N.R. Smith (eds). Bergey's Manual of Determinative Bacteriology, 7th ed. The Williams and Wilkins Co., Baltimore. pp. 480-485.

N. flavescens Branham 1930
Branham, S.E. 1930. United States Public Health Services Public Health Report **45**: 845-849. **Type strain:** ATCC 13120
Description: Bergey 8.

N. gonorrhoeae (Zopf 1885) Trevisan 1885
Zopf, W. 1885. Die Spaltpilze, 3rd ed. Edward Trewendt, Breslau. pp. 1-127; Trevisan, V. 1885. Atti della Accademia

Fisica-Medica-Statistica in Milano (Ser 4) **3**:92-107. **Type strain:** ATCC 19424
Description: Bergey 8.

N. lactamica Hollis, Wiggins and Weaver 1969
Hollis, D.G., G.L. Wiggins and R.E. Weaver. 1969. Applied Microbiology **17**: 71-77. **Type strain:** ATCC 23970
Description: Hollis, D.G., G.L. Wiggins and R.E. Weaver. 1969. Ibid.

N. meningitidis (Albrecht and Ghon 1901) Murray 1929
Albrecht, H. and A. Ghon. 1901. Wiener klinische Wochenschrift **14**:984-996; Murray, E.G.D. 1929. Medical Research Council (Great Britain) Special Report Series **124**:7-142. **Type strain:** ATCC 13077
Description: Bergey 8.

N. mucosa Veron, Thibault and Second 1959
Veron, M., P. Thibault and L. Second. 1959. Annales de l'Institut Pasteur (Paris) **97**:497-510. **Type strain:** ATCC 19696
Description: Bergey 8.

N. ovis Lindqvist 1960
Lindqvist, K. 1960. Journal of Infectious Diseases **106**:162-165. **Type strain:** ATCC 33078
Description: Lindqvist, K. 1960. Ibid.

N. perflava Bergey, Harrison, Breed, Hammer and Huntoon 1923
Bergey, D.H., F.C. Harrison, R.S. Breed, B.W. Hammer and F.M. Huntoon. 1923. Bergey's Manual of Determinative Bacteriology, 1st ed. The Williams and Wilkins Co., Baltimore. pp. 1-442. **Type strain:** ATCC 10555
Description: Branham, S.E. 1957 in Breed, R.S., E.G.D. Murray and N.R. Smith (eds). Bergey's Manual of Determinative Bacteriology, 7th ed. The Williams and Wilkins Co., Baltimore. pp. 480-485.

N. sicca (von Lingelsheim 1908) Bergey, Harrison, Breed, Hammer and Huntoon 1923
von Lingelsheim, W. 1908. Zentralblatt fur Bakteriologie, Parasitenkunde, Infektionskrankheiten und Hygiene. Abteilung I. **59**:457-476; Bergey, D.H., F.C. Harrison, R.S. Breed, B.W. Hammer and F.M. Huntoon. 1923. Bergey's Manual of Determinative Bacteriology, 1st ed. The Williams and Wilkins Co., Baltimore. pp. 1-442. **Type strain:** NRL 30,016
Description: Bergey 8.

N. subflava (Flugge 1886) Trevisan 1889
Flugge, 1886. Die Mikroorganismen. F.C.W. Vogel, Leipzig; Trevisan, V. 1889. I. Generi e le Specie delle Bactieriacee. Zanaboni and Gabuzzi, Milano. **Type strain:** NRL 30,017
Description: Bergey 8.

Neorickettsia Philip, Hadlow and Hughes 1953
Philip, C.B., W.J. Hadlow and L.E. Hughes. 1953. Riassunti della Communicazione VI Congresso Internazionale di Microbiologia, Roma **2**:256-257. **Type species:** *N. helminthoeca* Philip, Hadlow and Hughes 1953
Description: Bergey 8.

N. helminthoeca Philip, Hadlow and Hughes 1953
Philip, C.B., W.J. Hadlow and L.E. Hughes. 1953. Riassunti della Communicazione VI Congresso Internazionale di Microbiologia, Roma **2**:256-257. **Type strain:** no culture isolated.
Description: Bergey 8.

Nevskia Famintzin 1892
Famintzin, A. 1892. Bulletin of the Academy of Science St. Petersbourg New Series 2 **34**:481-486. **Type species:** *N. ramosa* Famintzin 1892
Description: Bergey 8.

N. ramosa Famintzin 1892
Famintzin, A. 1892. Bulletin of the Academy of Science St. Petersbourg New Series 2 **34**:481-486. **Type strain:** no culture isolated.
Description: Bergey 8.

Nitrobacter Winogradsky 1892
Winogradsky, S. 1892. Arkhiv Biologicheskikh Nauk (St. Petersbourg) **1**:87-137. **Type species:** *N. winogradskyi* Winslow, Broadhurst, Buchanan, Krumwiede, Rogers and Smith 1917
Description: Bergey 8.

N. winogradskyi Winslow, Broadhurst, Buchanan, Krumwiede, Rogers and Smith 1917
Winslow, C.E.A., J. Broadhurst, R.E. Buchanan, C. Krumwiede Jr., L.A. Rogers and G.H. Smith. 1917. Journal of Bacteriology **2**:505-566. **Type strain:** ATCC 25391
Description: Watson, S.W. 1971. International Journal of Systematic Bacteriology

21:254-270.

Nitrococcus Watson and Waterbury 1971
Watson, S.W. and J.B. Waterbury. 1971. Archiv fur Mikrobiologie **77**:203-230. **Type species:** *N. mobilis* Watson and Waterbury 1971
Description: Bergey 8.

N. mobilis Watson and Waterbury 1971
Watson, S.W. and J.B. Waterbury. 1971. Archiv fur Mikrobiologie **77**:203-230. **Type strain:** ATCC 25380
Description: Watson, S.W. and J.B. Waterbury. 1971. Ibid.

Nitrosococcus Winogradsky 1892
Winogradsky, S. 1892. Arkhiv Biologicheskikh Nauk (St. Petersbourg) **1**:87-137. **Type species:** *N. nitrosus* (Migula 1900) Buchanan 1925
Description: Bergey 8.

N. nitrosus (Migula 1900) Buchanan 1925
Migula, W. 1900. System der Bakterien, Vol. 2. Gustav Fischer, Jena; Buchanan, R.E. 1925. General Systematic Bacteriology. The Williams and Wilkins Co., Baltimore. **Type strain:** no culture isolated.
Description: Bergey 8.

N. oceanus (Watson 1965) Watson 1971
Watson, S.W. 1965. Limnology and Oceanography **10**:R274-R289; Watson, S.W. 1971. International Journal of Systematic Bacteriology **21**:254-270. **Type strain:** ATCC 19707
Description: Watson, S.W. 1971. Ibid.

Nitrosolobus Watson, Graham, Remsen and Valois 1971
Watson, S.W., L.B. Graham, C.C. Remsen and F.W. Valois. 1971. Archiv fur Mikrobiologie **76**:83-203. **Type species:** *N. multiformis* Watson, Graham, Remsen and Valois 1971
Description: Bergey 8.

N. multiformis Watson, Graham, Remsen and Valois 1971
Watson, S.W., L.B. Graham, C.C. Remsen and F.W. Valois. 1971. Archiv fur Mikrobiologie **76**:183-203. **Type strain:** ATCC 25196
Description: Watson, S.W. 1971. International Journal of Systematic Bacteriology **21**:254-270.

Nitrosomonas Winogradsky 1892
Winogradsky, S. 1892. Arkhiv Biologicheskikh Nauk (St. Petersbourg) **1**:87-137. **Type species:** *N. europaea* Winogradsky 1892
Description: Bergey 8.

N. europaea Winogradsky 1892
Winogradsky, S. 1892. Arkhiv Biologicheskikh Nauk (St. Petersbourg) **1**:87-137. **Type strain:** ATCC 25978
Description: Watson, S.W. 1971. International Journal of Systematic Bacteriology **21**:254-270.

Nitrosospira Winogradsky and Winogradsky 1933
Winogradsky, S. and H. Winogradsky. 1933. Annales de l'Institut Pasteur (Paris) **50**:350-434. **Type species:** *N. briensis* Winogradsky and Winogradsky 1933
Description: Bergey 8.

N. briensis Winogradsky and Winogradsky 1933
Winogradsky, S. and H. Winogradsky. 1933. Annales de l'Institut Pasteur (Paris) **50**:350-434. **Type strain:** no culture available.
Description: Watson, S.W. 1971. International Journal of Systematic Bacteriology **21**:254-270.

Nitrospina Watson and Waterbury 1971
Watson, S.W. and J.B. Waterbury. 1971. Archiv fur Mikrobiologie **77**:203-230. **Type species:** *N. gracilis* Watson and Waterbury 1971
Description: Bergey 8.

N. gracilis Watson and Waterbury 1971
Watson, S.W. and J.B. Waterbury. 1971. Archiv fur Mikrobiologie **77**:203-230. **Type strain:** no culture available.
Description: Watson, S.W. and J.B. Waterbury. 1971. Ibid.

Nocardia Trevisan 1889
Trevisan, V. 1889. I Generi e le Specie delle Battieriacee. Zanaboni and Gabuzzi, Milano. **Type species:** *N. asteroides* (Eppinger 1891) Blanchard 1896
Description: Bergey 8.

N. amarae Lechevalier and Lechevalier 1974
Lechevalier, M.P. and H.A. Lechevalier. 1974. International Journal of Systematic

Bacteriology **24**:278-288. **Type strain:** ATCC 27808
Description: Lechevalier, M.P. and H.A. Lechevalier. 1974. Ibid.

N. asteroides (Eppinger 1891) Blanchard 1896
Eppinger, H. 1891. Beitrage zur pathologischen Anatomie **9**:287-328; Blanchard, R. 1896 in Bouchard, C. Traite de Pathologie Generale, Vol. 2, G. Masson, Paris. pp. 1-932. **Type strain:** ATCC 19247
Description: Gordon, R.E. and J.M. Mihm. 1957. Journal of Bacteriology **73**: 15-27.

N. autotrophica (Takamiya and Tubaki 1956) Hirsch 1961
Takamiya, A. and K. Tubaki. 1956. Archiv fur Mikrobiologie **25**:58-64; Hirsch, P. 1961. Archiv fur Mikrobiologie **39**:360-373. **Type strain:** ATCC 19727
Description: Shirling, E.B. and D. Gottlieb. 1968. International Journal of Systematic Bacteriology **18**:69-189.

N. brasiliensis (Lindenberg 1909) Pinoy 1913
Lindenberg, A. 1909. Archives de Parasitologie **13**:265-282; Pinoy, E. 1913. Bulletin de l'Institut Pasteur **11**:929-938. **Type strain:** ATCC 19296
Description: Gordon, R.E. and J.M. Mihm. 1959. Journal of General Microbiology **20**:129-135.

N. calcarea Metcalf and Brown 1957
Metcalf, G. and M. Brown. 1957. Journal of General Microbiology **17**:567-572. **Type strain:** ATCC 19369
Description: Metcalf, G. and M. Brown. 1957. Ibid.

N. carnea (Rossi Doria 1891) Castellani and Chalmers 1913
Rossi Doria, T. 1891. Annali dell' Istituto d'Igiene Sperimentale, Universita Roma **1**: 399-438; Castellani, A. and A.J. Chalmers. 1913. Manual of Tropical Medicine, 2nd ed. Bailliere, Tindall and Cox, London. **Type strain:** ATCC 6847
Description: Gordon, R.E., S.K. Mishra and D.A. Barnett. 1978. Journal of General Microbiology **109**:69-78.

N. cellulans Metcalf and Brown 1957
Metcalf, G. and M. Brown. 1957. Journal of General Microbiology **17**:567-572. **Type strain:** ATCC 12830
Description: Metcalf, G. and M. Brown. 1957. Ibid.

N. coeliaca (Gray and Thornton 1928) Waksman and Henrici 1948
Gray, P.H.H. and H.G. Thornton. 1928. Zentralblatt fur Bakteriologie, Parasitenkunde, Infektionskrankheiten und Hygiene. Abteilung II. **73**:74-96; Waksman, S.A. and A.T. Henrici. 1948 in Breed, R.S., E.G.D. Murray and A.P. Hitchens (eds). Bergey's Manual of Determinative Bacteriology, 6th ed. The Williams and Wilkins Co., Baltimore. pp. 892-928. **Type strain:** ATCC 17041
Description: Gordon, R.E., D.A. Barnett, J.E. Handerhan and C. Hor-Nay Pang. 1974. International Journal of Systematic Bacteriology **24**:54-63.

N. corynebacterioides Serrano, Tablante, Serrano, San Blas and Imaeda 1972
Serrano, J.A., R.V. Tablante, A.A. Serrano, G.C. San Blas and T. Imaeda. 1972. Journal of General Microbiology **70**:339-349. **Type strain:** ATCC 14898
Description: Serrano, J.A., R.V. Tablante, A.A. Serrano, G.C. San Blas and T. Imaeda. 1972. Ibid.

N. farcinica Trevisan 1889
Trevisan, V. 1889. I. Generi e le Specie delle Battieriacee. Zanaboni and Gabuzzi, Milano. **Type strain:** ATCC 3318
Description: Trevisan, V. 1889. Ibid.

N. globerula (Gray 1928) Waksman and Henrici 1948
Gray, P.H.H. 1928. Proceedings of the Royal Society Series B **102**:263-280; Waksman, S.A. and A.T. Henrici. 1948 in Breed, R.S., E.G.D. Murray and A.P. Hitchens (eds). Bergey's Manual of Determinative Bacteriology, 6th ed. The Williams and Wilkins Co., Baltimore. pp. 892-928. **Type strain:** ATCC 19370
Description: Bergey 8.

N. hydrocarbonoxydans Nolof and Hirsch 1962
Nolof, G. and P. Hirsch. 1962. Archiv fur Mikrobiologie **44**:266-277. **Type strain:** ATCC 15104
Description: Nolof, G. and P. Hirsch. 1962. Ibid.

N. mediterranei (Margalith and Beretta 1960) Thiemann, Zucco and Pelizza 1969
Margalith, P. and G. Beretta. 1960. Mycopathologia et Mycologia Applicata **12**: 189-195; Thiemann, J.E., G. Zucco and G. Pelizza. 1969. Archiv fur Mikrobiolo-

gie **67**:147-155. **Type strain:** ATCC 13685
Description: Thiemann, J.E., G. Zucco and G. Pelizza. 1969. Ibid.

N. orientalis (Pittenger and Brigham 1956) Pridham and Lyons 1969
Pittenger, R.C. and R.B. Brigham. 1956. Antibiotics and Chemotherapy **6**:642-647; Pridham, T.G. and A.J. Lyons. 1969. Developments in Industrial Microbiology **10**: 183-221. **Type strain:** ATCC 19795
Description: Shirling, E.B. and D. Gottlieb. 1968. International Journal of Systematic Bacteriology **18**:69-189.

N. otitidiscaviarum Snijders 1924
Snijders, E.P. 1924. Geneeskundig Tijdschrift voor Nederlandsch-Indie **64**:85-87. **Type strain:** ATCC 14629
Description: Bergey 8.

N. petroleophila Hirsch and Engel 1956
Hirsch, P. and H. Engel. 1956. Berichte der Deutschen Botanischen Gesellschaft **69**:441-454. **Type strain:** ATCC 15777
Description: Hirsch, P. and H. Engel. 1956. Ibid.

N. restricta (Turfitt 1944) McClung 1974
Turfitt, G.E. 1944. Journal of Bacteriology **47**:487-493; McClung, N.M. 1974 Bergey 8. **Type strain:** DSM 43199
Description: Bergey 8.

N. saturnea Hirsch 1960
Hirsch, P. 1960. Archiv fur Mikrobiologie **35**:391-414. **Type strain:** ATCC 15809
Description: Hirsch, P. 1960. Ibid.

N. transvalensis Pijper and Pullinger 1927
Pijper, A. and B.D. Pullinger. 1927. Journal of Tropical Medicine and Hygiene **30**:153-156. **Type strain:** ATCC 6865
Description: Gordon, R.E., S.K. Mishra and D.A. Barnett. 1978. Journal of General Microbiology **109**:69-78.

N. vaccinii Demaree and Smith 1952
Demaree, J.B. and N.R. Smith. 1952. Phytopathology **42**:249-252. **Type strain:** ATCC 11092
Description: Demaree, J.B. and N.R. Smith. 1952. Ibid.

Nocardioides Prauser 1976
Prauser, H. 1976. International Journal of Systematic Bacteriology **26**:58-65. **Type species:** *N. albus* Prauser 1976
Description: Prauser, H. 1976. Ibid.

N. albus Prauser 1976
Prauser, H. 1976. International Journal of Systematic Bacteriology **26**:58-65. **Type strain:** ATCC 27980
Description: Prauser, H. 1976. Ibid.

Nocardiopsis (Brocq-Rousseau 1904) Meyer 1976
Brocq-Rousseau, D. 1904. Revue Generale de Botanique **16**:219-230; Meyer, J. 1976. International Journal of Systematic Bacteriology **26**:487-493. **Type species:** *N. dassonvillei* (Brocq-Rousseau 1904) Meyer 1976
Description: Meyer, J. 1976. Ibid.

N. dassonvillei (Brocq-Rousseau 1904) Meyer 1976
Brocq-Rousseau, D. 1904. Revue Generale de Botanique **16**:219-230; Meyer, J. 1976. International Journal of Systematic Bacteriology **26**:487-493. **Type strain:** ATCC 23218
Description: Meyer, J. 1976. Ibid.

Obesumbacterium Shimwell 1963
Shimwell, J.L. 1963. Brewers Journal **99**: 759-760. **Type species:** *O. proteus* Shimwell 1963
Description: Shimwell, J.L. 1963. Ibid.

O. proteus Shimwell 1963
Shimwell, J.L. 1963. Brewers Journal **99**: 759-760. **Type strain:** NCIB 8771
Description: Shimwell, J.L. 1963. Ibid.

Oceanospirillum Hylemon, Wells, Krieg and Jannasch 1973
Hylemon, P.B., J.S. Wells Jr., N.R. Krieg and H.W. Jannasch. 1973. International Journal of Systematic Bacteriology **23**:340-380. **Type species:** *O. linum* (Williams and Rittenberg 1957) Hylemon, Wells, Krieg and Jannasch 1973
Description: Hylemon, P.B., J.S. Wells Jr., N.R. Krieg and H.W. Jannasch. 1973. Ibid.

O. beijerinckii (Williams and Rittenberg 1957) Hylemon, Wells, Krieg and Jannasch 1973
Williams, M.A. and S.C. Rittenberg. 1957. International Bulletin of Bacteriological Nomenclature and Taxonomy **7**: 49-110; Hylemon, P.B., J.S. Wells Jr., N.R. Krieg and H.W. Jannasch. 1973. International Journal of Systematic Bacteriology **23**:340-380. **Type strain:** ATCC

12754
Description: Hylemon, P.B., J.S. Wells Jr., N.R. Krieg and H.W. Jannasch. 1973. Ibid.

O. hiroshimense (Terasaki 1973) Terasaki 1979
Terasaki, Y. 1973. Bulletin of the Suzugamine Women's College, Natural Science **17**:1-71; Terasaki, Y. 1979. International Journal of Systematic Bacteriology **29**: 130-144. **Type strain:** IFO 13616
Description: Terasaki, Y. 1979. Ibid.

O. japonicum (Watanabe 1959) Hylemon, Wells, Krieg and Jannasch 1973
Watanabe, N. 1959. Botanical Magazine (Tokyo) **72**:77-86; Hylemon, P.B., J.S. Wells Jr., N.R. Krieg and H.W. Jannasch. 1973. International Journal of Systematic Bacteriology **23**:340-380. **Type strain:** ATCC 19191
Description: Hylemon, P.B., J.S. Wells Jr., N.R. Krieg and H.W. Jannasch. 1973. Ibid.

O. linum (Williams and Rittenberg 1957) Hylemon, Wells, Krieg and Jannasch 1973
Williams, M.A. and S.C. Rittenberg. 1957. International Bulletin of Bacteriological Nomenclature and Taxonomy **7**: 49-110; Hylemon, P.B., J.S. Wells Jr., N.R. Krieg and H.W. Jannasch. 1973. International Journal of Systematic Bacteriology **23**:340-380. **Type strain:** ATCC 11336
Description: Hylemon, P.B., J.S. Wells Jr., N.R. Krieg and H.W. Jannasch. 1973. Ibid.

O. maris Hylemon, Wells, Krieg and Jannasch 1973
Hylemon, P.B., J.S. Wells Jr., N.R. Krieg and H.W. Jannasch. 1973. International Journal of Systematic Bacteriology **23**:340-380. **Type strain:** ATCC 27509
Description: Hylemon, P.B., J.S. Wells Jr., N.R. Krieg and H.W. Jannasch. 1973. Ibid.

O. minutulum (Watanabe 1959) Hylemon, Wells, Krieg and Jannasch 1973
Watanabe, N. 1959. Botanical Magazine (Tokyo) **72**:77-86; Hylemon, P.B., J.S. Wells Jr., N.R. Krieg and H.W. Jannasch. 1973. International Journal of Systematic Bacteriology **23**:340-380. **Type strain:** ATCC 19193
Description: Hylemon, P.B., J.S. Wells Jr., N.R. Krieg and H.W. Jannasch. 1973. Ibid.

O. multiglobuliferum (Terasaki 1973) Terasaki 1979
Terasaki, Y. 1973. Bulletin of the Suzugamine Women's College, Natural Science **17**:1-71; Terasaki, Y. 1979. International Journal of Systematic Bacteriology **29**: 130-144. **Type strain:** IFO 13614
Description: Terasaki, Y. 1979. Ibid.

O. pelagicum (Terasaki 1973) Terasaki 1979
Terasaki, Y. 1973. Bulletin of the Suzugamine Women's College, Natural Science **17**:1-71; Terasaki, Y. 1979. International Journal of Systematic Bacteriology **29**: 130-144. **Type strain:** IFO 13612
Description: Terasaki, Y. 1979. Ibid.

O. pusillum (Terasaki 1973) Terasaki 1979
Terasaki, Y. 1973. Bulletin of the Suzugamine Women's College, Natural Science **17**:1-71; Terasaki, Y. 1979. International Journal of Systematic Bacteriology **29**: 130-144. **Type strain:** IFO 13613
Description: Terasaki, Y. 1979. Ibid.

Oerskovia Prauser, Lechevalier and Lechevalier 1970
Prauser, H., M.P. Lechevalier and H.A. Lechevalier. 1970. Applied Microbiology **19**:534. **Type species:** *O. turbata* (Erikson 1954) Prauser, Lechevalier and Lechevalier 1970
Description: Prauser, H., M.P. Lechevalier and H.A. Lechevalier. 1970. Ibid.

O. turbata (Erikson 1954) Prauser, Lechevalier and Lechevalier 1970
Erikson, D. 1954. Journal of General Microbiology **11**:198-208; Prauser, H., M.P. Lechevalier and H.A. Lechevalier. 1970. Applied Microbiology **19**:534. **Type strain:** ATCC 25835
Description: Prauser, H., M.P. Lechevalier and H.A. Lechevalier. 1970. Ibid.

O. xanthineolytica Lechevalier 1972
Lechevalier, M.P. 1972. International Journal of Systematic Bacteriology **22**: 260-264. **Type strain:** ATCC 27402
Description: Lechevalier, M.P. 1972. Ibid.

Oscillospira Chatton and Perard 1913
Chatton, E. and C. Perard. 1913. Comptes Rendus des Seances de la Societe de Biologie (Paris) **74**:1159-1162. **Type**

species: *O. guilliermondii* Chatton and Perard 1913
Description: Bergey 8.

O. guilliermondii Chatton and Perard 1913
Chatton, E. and C. Perard. 1913. Comptes Rendus des Seances de la Societe de Biologie (Paris) **74**:1159-1162. **Type strain:** no culture isolated.
Description: Bergey 8.

Paracoccus Davis 1969
Davis, D.H. 1969 in Davis, D.H., M. Doudoroff, R.Y. Stanier and M. Mandel. International Journal of Systematic Bacteriology **19**:375-390. **Type species:** *P. denitrificans* (Beijerinck and Minkman 1910) Davis 1969
Description: Bergey 8.

P. denitrificans (Beijerinck and Minkman 1910) Davis 1969
Beijerinck, M.W. and D.C.J. Minkman. 1910. Zentralblatt fur Bakteriologie, Parasitenkunde, Infektionskrankheiten und Hygiene. Abteilung II. **25**:30-63; Davis, D.H. 1969 in Davis, D.H., M. Doudoroff, R.Y. Stanier and M. Mandel. International Journal of Systematic Bacteriology **19**:375-390. **Type strain:** ATCC 17741
Description: Davis, D.H. 1969. Ibid.

P. halodenitrificans (Robinson and Gibbons 1952) Davis 1969
Robinson, J. and N.E. Gibbons. 1952. Canadian Journal of Botany **30**:147-154; Davis, D.H. 1969 in Davis, D.H., M. Doudoroff, R.Y. Stanier and M. Mandel. International Journal of Systematic Bacteriology **19**:375-390. **Type strain:** ATCC 13511
Description: Bergey 8.

Pasteurella Trevisan 1887
Trevisan, V. 1887. Rendiconti dell' Istituto Lombardo di Scienze e Lettere (Ser 2) **20**:88-105. **Type species:** *P. multocida* (Lehmann and Neumann 1899) Rosenbusch and Merchant 1939
Description: Bergey 8.

P. aerogenes McAllister and Carter 1974
McAllister, H.A. and G.R. Carter. 1974. American Journal of Veterinary Research **35**:917-922. **Type strain:** ATCC 27883
Description: McAllister, H.A. and G.R. Carter. 1974. Ibid.

P. gallicida (Burrill 1883) Buchanan 1925
Burrill, T.J. 1883. The American Naturalist **17**:319-320; Buchanan, R.E. 1925. General systematic bacteriology. The Williams and Wilkins Co., Baltimore. pp. 412-415. **Type strain:** NCTC 10322
Description: Buchanan, R.E. 1925. Ibid.

P. gallinarum Hall, Heddleston, Legenhausen and Hughes 1955
Hall, W.J., K.L. Heddleston, D.H. Legenhausen and R.W. Hughes. 1955. American Journal of Veterinary Research **16**:598-604. **Type strain:** ATCC 13361
Description: Mraz, O., P. Jelen and J. Bohacek. 1977. Acta Veterinaria Brno **46**: 135-147.

P. haemolytica Newsome and Cross 1932
Newsome, I.E. and F. Cross. 1932. Journal of the American Veterinary Medical Association **80**:711-719. **Type strain:** NCTC 9380
Description: Bergey 8.

P. multocida (Lehmann and Neumann 1899) Rosenbusch and Merchant 1939
Lehmann, K.B. and R. Neumann. 1899. Lehmann's Medizin, Handatlanten Band X. Atlas und Grundriss der Bakteriologie und Lehrbuch der speziellen bakteriologischen Diagnostik. 2 Auflage, Munchen. p. 196; Rosenbusch, C.T. and I.A. Merchant. 1939. Journal of Bacteriology **37**: 69-89. **Type strain:** NCTC 10322
Description: Bergey 8.

P. pneumotropica Jawetz 1950
Jawetz, E. 1950. Journal of Infectious Diseases **86**:172-183. **Type strain:** NCTC 8141
Description: Bergey 8.

P. ureae Jones 1962
Jones, D.M. 1962. Journal of Pathology and Bacteriology **83**:143-151. **Type strain:** ATCC 25976
Description: Jones, D.M. 1962. Ibid.

Pasteuria Metchnikoff 1888
Metchnikoff, E. 1888. Annales de l'Institut Pasteur (Paris) **2**:165-170. **Type species:** *P. ramosa* Metchnikoff 1888
Description: Bergey 8.

P. ramosa Metchnikoff 1888
Metchnikoff, E. 1888. Annales de l'Institut Pasteur (Paris) **2**:165-170. **Type strain:** ATCC 27377

Description: Staley, J.T. 1973. Canadian Journal of Microbiology **19**:609-614.

Pectinatus Lee, Mabee and Jangaard 1978
Lee, S.Y., M.S. Mabee and N.O. Jangaard. 1978. International Journal of Systematic Bacteriology **28**:582-594. **Type species:** *P. cerevisiiphilus* Lee, Mabee and Jangaard 1978
Description: Lee, S.Y., M.S. Mabee and N.O. Jangaard. 1978. Ibid.

P. cerevisiiphilus Lee, Mabee and Jangaard 1978
Lee, S.Y., M.S. Mabee and N.O. Jangaard. 1978. International Journal of Systematic Bacteriology **28**:582-594. **Type strain:** ATCC 29359
Description: Lee, S.Y., M.S. Mabee and N.O. Jangaard. 1978. Ibid.

Pectobacterium Waldee 1945
Waldee, E.L. 1945. Iowa State Journal of Science **19**:435-484. **Type species:** *P. carotovorum* (Jones 1901) Waldee 1945
Description: Waldee, E.L. 1945. Ibid.

P. carnegieana (Standring 1942) Brenner, Steigerwalt, Miklos and Fanning 1973 (See also Erwinia carnegieana)
Standring, E.T. 1942 in Lightle, P.C., E.T. Standring and J.G. Brown. Phytopathology **32**:303-313; Brenner, D.J., A.G. Steigerwalt, G.V. Miklos and G.R. Fanning. 1973. International Journal of Systematic Bacteriology **23**:205-216. **Type strain:** NCPPB 439
Description: Bergey 8.

P. carotovorum (Jones 1901) Waldee 1945 (See also Erwinia carotovora)
Jones, L.R. 1901. Zentralblatt fur Bakteriologie, Parasitenkunde, Infektionskrankheiten und Hygiene. Abteilung II. **7**:12-21; Waldee, E.L. 1945. Iowa State Journal of Science **19**:435-484. **Type strain:** ATCC 15713
Description: Bergey 8.

P. chrysanthemi (Burkholder, McFadden and Dimock 1953) Brenner, Steigerwalt, Miklos and Fanning 1973 (See also Erwinia chrysanthemi)
Burkholder, W.H., L.A. McFadden and A.V. Dimock. 1953. Phytopathology **43**:522-526; Brenner, D.J., A.G. Steigerwalt, G.V. Miklos and G.R. Fanning. 1973. International Journal of Systematic Bacteriology **23**:205-216. **Type strain:** ATCC 11663
Description: Bergey 8.

P. cypripedii (Hori 1911) Brenner, Steigerwalt, Miklos and Fanning 1973 (See also Erwinia cypripedii)
Hori, S. 1911. Zentralblatt fur Bakteriologie, Parasitenkunde, Infektionskrankheiten und Hygiene. Abteilung II. **31**:85-92; Brenner, D.J., A.G. Steigerwalt, G.V. Miklos and G.R. Fanning. 1973. International Journal of Systematic Bacteriology **23**:205-216. **Type strain:** PDDCC 1591
Description: Bergey 8.

P. rhapontici (Millard 1924) Patel and Kulkarni 1951 (See also Erwinia rhapontici)
Millard, W.A. 1924. Bulletin of the Univeristy of Leeds and the Yorkshire Counsil for Agricultural Education **134**:1-28; Patel, M.K. and Y.S. Kulkarni. 1951. Indian Phytopathology **4**:74-84. **Type strain:** ATCC 29283
Description: Bergey 8.

Pediococcus Balcke 1884
Balcke, J. 1884. Wochenschrift fur Brauerei **1**:257 **Type species:** *P. damnosus* Claussen 1903
Description: Bergey 8.

P. acidilactici Lindner 1887
Lindner, P. 1887. Wochenschrift fur Brauerei **4**:437-440. **Type strain:** NCDO 1859
Description: Garvie, E.I. 1974. International Journal of Systematic Bacteriology **24**:301-306.

P. damnosus Claussen 1903
Claussen, N.H. 1903. Compte Rendu des Travaux du Laboratoire de Carlsberg **6**:64-83. **Type strain:** NCDO 1832
Description: Garvie, E.I. 1974. International Journal of Systematic Bacteriology **24**:301-306.

P. dextrinicus (Coster and White 1964) Back 1978
Coster, E. and H.R. White. 1964. Journal of General Microbiology **37**:15-31; Back, W. 1978. International Journal of Systematic Bacteriology **28**:523-527. **Type strain:** DSM 20335
Description: Back, W. 1978. Ibid.

P. halophilus Mees 1934
Mees, R.H. 1934. Thesis. Technical University Delft, Holland. pp. 1-110. **Type strain:** NCDO 1635
Description: Gunther, H.L. and H.R. White. 1961. Journal of General Microbiology **26**:185-197.

P. parvulus Gunther and White 1961
Gunther, H.L. and H.R. White. 1961. Journal of General Microbiology **26**:185-197. **Type strain:** NCDO 1634
Description: Gunther, H.L., E. Coster and H.R. White. 1962. International Bulletin of Bacteriological Nomenclature and Taxonomy **12**:189-190.

P. pentosaceus Mees 1934
Mees, R.H. 1934. Thesis. Technical University, Delft, Holland. pp. 1-110. **Type strain:** NCDO 990
Description: Garvie, E.I. 1974. International Journal of Systematic Bacteriology **24**:301-306.

Pedomicrobium Aristovskaya 1961
Aristovakaya, T.V. 1961. Doklady Akademii Nauk SSSR **136**:954-957. **Type species:** *P. ferrugineum* Aristovskaya 1961
Description: Bergey 8.

P. ferrugineum Aristovskaya 1961
Aristovskaya, T.V. 1961. Doklady Akademii Nauk SSSR **136**:954-957. **Type strain:** IFAM S-1290
Description: Bergey 8.

P. manganicum Aristovskaya 1961
Aristovskaya, T.V. 1961. Doklady Akademii Nauk SSSR **136**:954-957. **Type strain:** IFAM E-1129
Description: Bergey 8.

Pelodictyon Lauterborn 1913
Lauterborn, R. 1913. Allgemeine Botanische Zeitschrift **19**:97-100. **Type species:** *P. clathratiforme* (Szafer 1911) Lauterborn 1913
Description: Bergey 8.

P. clathratiforme (Szafer 1911) Lauterborn 1913
Szafer, W. 1911. Bulletin de l'Academie des Sciences de Cracovie, Serie B. pp. 160-167; Lauterborn, R. 1913. Allgemeine Botanische Zeitschrift **19**:97-100. **Type strain:** no pure culture.
Description: Bergey 8.

P. luteolum (Schmidle 1901) Pfennig and Truper 1971
Schmidle, W. 1901. Beihefte zum Botanischen Zentralblatt **10**:179-180; Pfennig, N. and H.G. Truper. 1971. International Journal of Systematic Bacteriology **21**:11-14. **Type strain:** DSM 273
Description: Bergey 8.

P. phaeum Gorlenko 1972
Gorlenko, V.M. 1972. Mikrobiologiya **41**: 370-372. **Type strain:** DSM 728
Description: Gorlenko, V.M. 1972. Ibid.

Peptococcus Kluyver and van Niel 1936
Kluyver, A.J. and C.B. van Niel. 1936. Zentralblatt fur Bakteriologie, Parasitenkunde, Infektionskrankheiten und Hygiene. Abteilung II. **94**:369-403. **Type species:** *P. niger* (Hall 1930) Kluyver and van Niel 1936
Description: Bergey 8.

P. asaccharolyticus (Distaso 1912) Douglas 1957
Distaso, A. 1912. Zentralblatt fur Bakteriologie, Parasitenkunde, Infektionskrankheiten und Hygiene. Abteilung I. **62**:433-468; Douglas, H.C. 1957 in Breed, R.S., E.G.D. Murray and N.R. Smith (eds). Bergey's Manual of Determinative Bacteriology, 7th ed. The Williams and Wilkins Co., Baltimore. pp. 474-480. **Type strain:** ATCC 14963
Description: Holdeman, L.V., E.P. Cato and W.E.C. Moore (eds). 1977. Anaerobe Laboratory Manual, 4th ed. Virginia Polytechnic Institute and State University, Blacksburg, Virginia. pp. 1-156.

P. glycinophilus (Cardon and Barker 1946) Douglas 1957
Cardon, B.P. and H.A. Barker. 1946. Journal of Bacteriology **52**:629-634; Douglas, H.C. 1957 in Breed, R.S., E.G.D. Murray and N.R. Smith (eds). Bergey's Manual of Determinative Bacteriology, 7th ed. The Williams and Wilkins Co., Baltimore. pp. 474-480. **Type strain:** ATCC 23195
Description: Douglas, H.C. 1957. Ibid.

P. indolicus (Christiansen 1934) Sorensen 1975
Christiansen, M. 1934. Acta Pathologica et Microbiologica Scandinavica, Supplement **18**:42-63; Sorensen, G. 1975. Acta Veterinaria Scandinavica **16**:218-225. **Type strain:** ATCC 29427

Description: Sorensen, G. 1973. Acta Veterinaria Scandinavica **14**:301-326.

P. magnus (Prevot 1933) Holdeman and Moore 1972
Prevot, A.R. 1933. Annales Societe Nature **15**:23-260; Holdeman, L.V. and W.E.C. Moore (eds). 1972. Anaerobe Laboratory Manual. Virginia Polytechnic Institute Anaerobe Laboratory, Blacksburg, Virginia. **Type strain:** ATCC 15794
Description: Holdeman, L.V., E.P. Cato and W.E.C. Moore (eds). 1977. Anaerobe Laboratory Manual, 4th ed. Virginia Polytechnic Institute and State University, Blacksburg, Virginia. pp. 1-156.

P. niger (Hall 1930) Kluyver and van Niel 1936
Hall, I.C. 1930. Journal of Bacteriology **20**:407-415; Kluyver, A.J. and C.B. van Niel. 1936. Zentralblatt fur Bakteriologie, Parasitenkunde, Infektionskrankheiten und Hygiene. Abteilung II. **94**:369-403. **Type strain:** ATCC 27731
Description: Wilkins, T.D., W.E.C. Moore, S.E.H. West and L.V. Holdeman. 1975. International Journal of Systematic Bacteriology **25**:47-49.

P. prevotii (Foubert and Douglas 1948) Douglas 1957
Foubert Jr., E.L. and H.C. Douglas. 1948. Journal of Bacteriology **56**:25-34; Douglas, H.C. 1957 in Breed, R.S., E.G.D. Murray and N.R. Smith (eds). Bergey's Manual of Determinative Bacteriology, 7th ed. The Williams and Wilkins Co., Baltimore. pp. 474-480. **Type strain:** ATCC 9321
Description: Holdeman, L.V., E.P. Cato and W.E.C. Moore (eds). 1977. Anaerobe Laboratory Manual, 4th ed. Virginia Polytechnic Institute and State University, Blacksburg, Virginia. pp. 1-156.

P. saccharolyticus (Foubert and Douglas 1948) Douglas 1957
Foubert Jr., E.L. and H.C. Douglas. 1948. Journal of Bacteriology **56**:25-34; Douglas, H.C. 1957 in Breed, R.S., E.G.D. Murray and N.R. Smith (eds). Bergey's Manual of Determinative Bacteriology, 7th ed. The Williams and Wilkins Co., Baltimore. pp. 474-480. **Type strain:** ATCC 14953
Description: Holdeman, L.V., E.P. Cato and W.E.C. Moore (eds). 1977. Anaerobe Laboratory Manual, 4th ed. Virginia Polytechnic Institute and State University, Blacksburg, Virginia. pp. 1-156.

Peptostreptococcus Kluyver and van Niel 1936
Kluyver, A.J. and C.B. van Niel. 1936. Zentralblatt fur Bakteriologie, Parasitenkunde, Infektionskrankheiten und Hygiene. Abteilung II. **94**:369-403. **Type species:** *P. anaerobius* (Natvig 1905) Kluyver and van Niel 1936
Description: Bergey 8.

P. anaerobius (Natvig 1905) Kluyver and van Niel 1936
Natvig, H. 1905. Archiv fur Gynakologie (Berlin) **76**:701-858; Kluyver, A.J. and C.B. van Niel. 1936. Zentralblatt fur Bakteriologie, Parasitenkunde, Infektionskrankheiten und Hygiene. Abteilung II. **94**:369-403. **Type strain:** ATCC 27337
Description: Holdeman, L.V., E.P. Cato and W.E.C. Moore (eds). 1977. Anaerobe Laboratory Manual, 4th ed. Virginia Polytechnic Institute and State University, Blacksburg, Virginia. pp. 1-156.

P. micros (Prevot 1933) Smith 1957
Prevot, A.R. 1933. Annales des Sciences Naturelles **15**:23-260; Smith, L.DS. 1957 in Breed, R.S., E.G.D. Murray and N.R. Smith (eds). Bergey's Manual of Determinative Bacteriology, 7th ed. The Williams and Wilkins Co., Baltimore. pp. 533-541. **Type strain:** VPI 5464
Description: Holdeman, L.V., E.P. Cato and W.E.C. Moore (eds). 1977. Anaerobe Laboratory Manual, 4th ed. Virginia Polytechnic Institute and State University, Blacksburg, Virginia. pp. 1-156.

P. parvulus (Weinberg, Nativelle and Prevot 1937) Smith 1957
Weinberg, M., R. Nativelle and A.R. Prevot. 1937. Les Microbes Anaerobies. Masson et Cie, Paris. pp. 1-1186; Smith, L.DS. 1957 in Breed, R.S., E.G.D. Murray and N.R. Smith (eds). Bergey's Manual of Determinative Bacteriology, 7th ed. The Williams and Wilkins Co., Baltimore. pp. 533-541. **Type strain:** VPI 0546
Description: Holdeman, L.V., E.P. Cato and W.E.C. Moore (eds). 1977. Anaerobe Laboratory Manual, 4th ed. Virginia Polytechnic Institute and State University,

Blacksburg, Virginia. pp. 1-156.

P. productus (Prevot 1941) Smith 1957
Prevot, A.R. 1941. Comptes Rendus des Seances de la Societe de Biologie (Paris) **135**:105-107; Smith, L.DS. 1957 in Breed, R.S., E.G.D. Murray and N.R. Smith (eds). Bergey's Manual of Determinative Bacteriology, 7th ed. The Williams and Wilkins Co., Baltimore. pp. 533-541. **Type strain:** ATCC 27340
Description: Holdeman, L.V., E.P. Cato and W.E.C. Moore (eds). 1977. Anaerobe Laboratory Manual, 4th ed. Virginia Polytechnic Institute and State University, Blacksburg, Virginia. pp. 1-156.

Photobacterium Beijerinck 1889
Beijerinck, M.W. 1889. Archives Neerlandaises des Science Exactes et Naturelles **23**:401-415. **Type species:** *P. phosphoreum* (Cohn 1878) Beijerinck 1889
Description: Bergey 8.

P. angustum Reichelt, Baumann and Baumann 1979
Reichelt, J.L., P. Baumann and L. Baumann. 1979 in International Journal of Systematic Bacteriology **29**:79-80; Effective publication Reichelt, J.L., P. Baumann and L. Baumann. 1976. Archives of Microbiology **110**:101-120. **Type strain:** ATCC 25915
Description: Baumann, P., L. Baumann and M. Mandel. 1971. Journal of Bacteriology **107**:268-294.

P. fischeri (Beijerinck 1889) Reichelt and Baumann 1973 (See also Vibrio fischeri)
Beijerinck, M.W. 1889. Archives Neerlandaises des Science Exactes et Naturelles **23**:401-415; Reichelt, J.L. and P. Baumann. 1973. Archiv fur Mikrobiologie **94**: 283-330. **Type strain:** ATCC 7744
Description: Reichelt, J.L. and P. Baumann. 1973. Ibid.

P. leiognathi Boisvert, Chatelain and Bassot 1967
Boisvert, H., R. Chatelain and J.-M. Bassot. 1967. Annales de l'Institut Pasteur (Paris) **112**:520-524. **Type strain:** ATCC 25521
Description: Reichelt, J.L. and P. Baumann. 1975. International Journal of Systematic Bacteriology **25**:208-209.

P. phosphoreum (Cohn 1878) Beijerinck 1889
Cohn, F. 1878. Letter to J. Penn which describes Micrococcus phosphoreum. Versameling van stucken betreffende het geneeskundig staats toerzich, 126-130; Beijerinck, M.W. 1889. Archives Neerlandaises des Science Exactes et Naturelles **23**:401-415. **Type strain:** ATCC 11040
Description: Reichelt, J.L. and P. Baumann. 1973. Archiv fur Mikrobiologie **94**: 283-330.

Pilimelia Kane 1966
Kane, W.D. 1966. Journal of the Elisha Mitchell Scientific Society **82**:220-230. **Type species:** *P. terevasa* Kane 1966
Description: Bergey 8.

P. anulata Kane 1966
Kane, W.D. 1966. Journal of the Elisha Mitchell Scientific Society **82**:220-230. **Type strain:** ATCC 25604
Description: Bergey 8.

P. terevasa Kane 1966
Kane, W.D. 1966. Journal of the Elisha Mitchell Scientific Society **82**:220-230. **Type strain:** ATCC 25603
Description: Bergey 8.

Planctomyces Gimesi 1924
Gimesi, N. 1924. Budapest, Kiadja a Magyar Ciszterci. Rend, pp. 1-8. **Type species:** *P. bekefii* Gimesi 1924
Description: Bergey 8.

P. bekefii Gimesi 1924
Gimesi, N. 1924. Budapest, Kiadja a Magyar Ciszterci. Rend, pp. 1-8. **Type strain:** no culture isolated.
Description: Bergey 8.

Planobispora Thiemann and Beretta 1968
Thiemann, J.E. and G. Beretta. 1968. Archiv fur Mikrobiologie **62**:157-166. **Type species:** *P. longispora* Thiemann and Beretta 1968
Description: Bergey 8.

P. longispora Thiemann and Beretta 1968
Thiemann, J.E. and G. Beretta. 1968. Archiv fur Mikrobiologie **62**:157-166. **Type strain:** ATCC 23867
Description: Bergey 8.

P. rosea Thiemann 1967
Thiemann, J.E. 1967 in Prauser, H. (ed). The Actinomycetales. The Jena International Symposium on Taxonomy, Sept.

1968. VEB Gustav Fischer Verlag Jena. pp. 245-257. **Type strain:** ATCC 23866
Description: Bergey 8.

Planococcus Migula 1894
Migula, W. 1894. Arbeiten aus dem Bakteriologischen Institut der Technischen Hochschule zu Karlsruhe **1**:235-238. **Type species:** *P. citreus* Migula 1894
Description: Bergey 8.

P. citreus Migula 1894
Migula, W. 1894. Arbeiten aus dem Bakteriologischen Institut der Technischen Hochschule zu Karlsruhe **1**:235-238. **Type strain:** ATCC 14404
Description: Kocur, M., Z. Pacova, W. Hodgkiss and T. Martinec. 1970. International Journal of Systematic Bacteriology **20**:241-248.

P. halophilus Novitsky and Kushner 1976
Novitsky, T.J. and D.J. Kushner. 1976. International Journal of Systematic Bacteriology **26**:53-57. **Type strain:** ATCC 27964
Description: Novitsky, T.J. and D.J. Kushner. 1976. Ibid.

Planomonospora Thiemann, Pagani and Beretta 1967
Thiemann, J.E., H. Pagani and G. Beretta. 1967. Giornale di Microbiologia **15**:27-38. **Type species:** *P. parontospora* Thiemann, Pagani and Beretta 1967
Description: Bergey 8.

P. parontospora Thiemann, Pagani and Beretta 1967
Thiemann, J.E., H. Pagani and G. Beretta. 1967. Giornale di Microbiologia **15**:27-38. **Type strain:** ATCC 23863
Description: Bergey 8.

P. parontospora subsp. *antibiotica* Thiemann, Coronelli, Pagani, Beretta, Tamoni and Arioli 1968
Thiemann, J.E., C. Coronelli, H. Pagani, G. Beretta, G. Tamoni and V. Arioli. 1968. Journal of Antibiotics (Tokyo) Series A **21**:525-531. **Type strain:** ATCC 23864
Description: Bergey 8.

P. parontospora subsp. *parontospora* Thiemann, Pagani and Beretta 1967
Thiemann, J.E., H. Pagani and G. Beretta. 1967. Giornale di Microbiologia **15**:27-38. **Type strain:** ATCC 23863
Description: Bergey 8.

P. venezuelensis Thiemann 1967
Thiemann, J.E. 1967 in Prauser, H. (ed). The Actinomycetales. The Jena International Symposium on Taxonomy, Sept. 1968. VEB Gustav Fischer Verlag Jena. pp. 245-257. **Type strain:** ATCC 23865
Description: Thiemann, J.E. 1970. Ibid.

Plesiomonas Habs and Schubert 1962
Habs, H. and R.H.W. Schubert. 1962. Zentralblatt fur Bakteriologie, Parasitenkunde, Infektionskrankheiten und Hygiene. Abteilung I. **186**:316-327. **Type species:** *P. shigelloides* (Bader 1954) Habs and Schubert 1962
Description: Bergey 8.

P. shigelloides (Bader 1954) Habs and Schubert 1962
Bader, R.E. 1954. Zentralblatt fur Bakteriologie, Parasitenkunde, Infektionskrankheiten und Hygiene. Abteilung I. **140**: 450-456; Habs, H. and R.H.W. Schubert. 1962. Zentralblatt fur Bakteriologie, Parasitenkunde, Infektionskrankheiten und Hygiene. Abteilung I. **186**:316-327. **Type strain:** ATCC 14029
Description: Bergey 8.

Polyangium Link 1809
Link, H.F. 1809. Gesellschaft fur Natur (Berlin) **3**:3-42. **Type species:** *P. vitellinum* Link 1809
Description: Bergey 8.

P. fumosum Krzemieniewska and Krzemieniewski 1930
Krzemieniewska, H. and S. Krzemieniewski. 1930. Acta Societatis Botanicorum Poloniae **7**:250-273. **Type strain:** Windsor M257
Description: Bergey 8.

P. minor (Peterson 1959) McCurdy 1970
Peterson, J.E. 1959. Mycologia **51**:1-8; McCurdy, H.D. 1970. International Journal of Systematic Bacteriology **20**:283-296. **Type strain:** UMH Peterson 44
Description: Peterson, J.E. 1959. Ibid.

P. rugiseptum (Peterson 1959) McCurdy 1970
Peterson, J.E. 1959. Mycologia **51**:1-8; McCurdy, H.D. 1970. International Journal of Systematic Bacteriology **20**:283-296. **Type strain:** UMH Peterson 51

Description: Peterson, J.E. 1959. Ibid.

P. vitellinum Link 1809
Link, H.F. 1809. Gesellschaft fur Natur (Berlin) **3**:3-42. **Type strain:** TC 4564
Description: Bergey 8.

Promicromonospora Krassilnikov, Kalakoutskii and Kirillova 1961
Krassilnikov, N.A., L.V. Kalakoutskii and N.F. Kirillova. 1961. Bulletin of the Academy of Sciences USSR (Series Biology) **1**:107-112. **Type species:** *P. citrea* Krassilnikov, Kalakoutskii and Kirillova 1961
Description: Krassilnikov, N.A., L.V. Kalakoutskii and N.F. Kirillova. 1961. Ibid.

P. citrea Krassilnikov, Kalakoutskii and Kirillova 1961
Krassilnikov, N.A., L.V. Kalakoutskii and N.F. Kirillova. 1961. Bulletin of the Academy of Sciences USSR (Series Biology) **1**:107-112. **Type strain:** ATCC 15908
Description: Krassilnikov, N.A., L.V. Kalakoutskii and N.F. Kirillova. 1961. Ibid.

Propionibacterium Orla-Jensen 1909
Orla-Jensen, S. 1909. Zentralblatt fur Bakteriologie, Parasitenkunde, Infektionskrankheiten und Hygiene. Abteilung II. **22**:305-346. **Type species:** *P. freudenreichii* van Niel 1928
Description: Bergey 8.

P. acidi-propionici Orla-Jensen 1909
Orla-Jensen, S. 1909. Zentralblatt fur Bakteriologie Parasitenkunde, Infektionskrankheiten und Hygiene. Abteilung II. **22**:305-346. **Type strain:** ATCC 25562
Description: Holdeman, L.V., E.P. Cato and W.E.C. Moore (eds). 1977. Anaerobe Laboratory Manual, 4th ed. Virginia Polytechnic Institute and State University, Blacksburg, Virginia. pp. 1-156.

P. acnes (Gilchrist 1900) Douglas and Gunter 1946
Gilchrist, T.C. 1900. Johns Hopkins Hospital Report **9**:409-430; Douglas, H.D. and S.E. Gunter. 1946. Journal of Bacteriology **52**:15-23. **Type strain:** ATCC 6919
Description: Holdeman, L.V., E.P. Cato and W.E.C. Moore (eds). 1977. Anaerobe Laboratory Manual, 4th ed. Virginia Polytechnic Institute and State University, Blacksburg, Virginia. pp. 1-156.

P. avidum (Eggerth 1935) Moore and Holdeman 1969
Eggerth, A.H. 1935. Journal of Bacteriology **30**:277-299; Moore, W.E.C. and L.V. Holdeman. 1969 in Cato, E.P., C.S. Cummins, L.V. Holdeman, J.L. Johnson, W.E.C. Moore, R.M. Smibert and L.DS. Smith (eds). Outline of clinical methods in anaerobic bacteriology. Virginia Polytechnic Institute Anaerobe Laboratory, Blacksburg, Virginia. **Type strain:** ATCC 25577
Description: Holdeman, L.V., E.P. Cato and W.E.C. Moore (eds). 1977. Anaerobe Laboratory Manual, 4th ed. Virginia Polytechnic Institute and State University, Blacksburg, Virginia. pp. 1-156.

P. freudenreichii van Niel 1928
van Niel, C.B. 1928. The propionic acid bacteria. Uitgeverszaak and Boissevain and Co., Haarlem, Holland. pp. 1-187. **Type strain:** ATCC 6207
Description: Holdeman, L.V., E.P. Cato and W.E.C. Moore (eds). 1977. Anaerobe Laboratory Manual, 4th ed. Virginia Polytechnic Institute and State University, Blacksburg, Virginia. pp. 1-156.

P. freudenreichii subsp. *freudenreichii* (van Niel 1928) Moore and Holdeman 1970
van Niel, C.B. 1928. The propionic acid bacteria. Uitgeverszaak and Boissevain and Co., Haarlem, Holland. pp. 1-187; Moore, W.E.C. and L.V. Holdeman. 1970 in Cato, E.P., C.S. Cummins, L.V. Holdeman, J.L. Johnson, W.E.C. Moore, R.M. Smibert and L.DS. Smith (eds). Outline of clinical methods in anaerobic bacteriology, 2nd revision. Virginia Polytechnic Institute Anaerobe Laboratory, Blacksburg, Virginia. pp. 15-21. **Type strain:** ATCC 6207
Description: Holdeman, L.V., E.P. Cato and W.E.C. Moore (eds). 1977. Anaerobe Laboratory Manual, 4th ed. Virginia Polytechnic Institute and State University, Blacksburg, Virginia. pp. 1-156.

P. freudenreichii subsp. *shermanii* (van Niel 1928) Holdeman and Moore 1970
van Niel, C.B. 1928. The propionic acid bacteria. N.V. Uitgeverszaak and J.W. Boissevain and Co., Haarlem, Holland. pp. 1-187; Moore, W.E.C. and L.V. Holdeman. 1970 in Cato, E.P., C.S.

Cummins, L.V. Holdeman, J.L. Johnson, W.E.C. Moore, R.M. Smibert and L.DS. Smith (eds). Outline of clinical methods in anaerobic bacteriology, 2nd revision. Virginia Polytechnic Institute Anaerobe Laboratory, Blacksburg, Virginia. pp. 15-21. **Type strain:** ATCC 9614
Description: Holdeman, L.V., E.P. Cato and W.E.C. Moore (eds). 1977. Anaerobe Laboratory Manual, 4th ed. Virginia Polytechnic Institute and State University, Blacksburg, Virginia. pp. 1-156.

P. granulosum (Prevot 1938) Moore and Holdeman 1970
Prevot, A.R. 1938. Annales de l'Institut Pasteur (Paris) **60**:285-307; Moore, W.E.C. and L.V. Holdeman. 1970 in Cato, E.P., C.S. Cummins, L.V. Holdeman, J.L. Johnson, W.E.C. Moore, R.M. Smibert and L.DS. Smith (eds). Outline of clinical methods in anaerobic bacteriology, 2nd revision. Virginia Polytechnic Institute Anaerobe Laboratory, Blacksburg, Virginia. pp. 15-21. **Type strain:** ATCC 25564
Description: Holdeman, L.V., E.P. Cato and W.E.C. Moore (eds). 1977. Anaerobe Laboratory Manual, 4th ed. Virginia Polytechnic Institute and State University, Blacksburg, Virginia. pp. 1-156.

P. jensenii van Niel 1928
van Niel, C.B. 1928. The propionic acid bacteria. Uitgeverszaak and Boissevain and Co., Haarlem, Holland. pp. 1-187. **Type strain:** ATCC 4868
Description: Holdeman, L.V., E.P. Cato and W.E.C. Moore (eds). 1977. Anaerobe Laboratory Manual, 4th ed. Virginia Polytechnic Institute and State University, Blacksburg, Virginia. pp. 1-156.

P. lymphophilum (Torrey 1916) Johnson and Cummins 1972
Torrey, J.C. 1916. Journal of Medical Research **34**:65-80; Johnson, J.L. and C.S. Cummins. 1972. Journal of Bacteriology **109**:1047-1066. **Type strain:** ATCC 27520
Description: Holdeman, L.V., E.P. Cato and W.E.C. Moore (eds). 1977. Anaerobe Laboratory Manual, 4th ed. Virginia Polytechnic Institute and State University, Blacksburg, Virginia. pp. 1-156.

P. thoenii van Niel 1928
van Niel, C.B. 1928. The propionic acid bacteria. Uitgeverszaak and Boissevain and Co., Haarlem, Holland. pp. 1-187. **Type strain:** ATCC 4874
Description: Holdeman, L.V., E.P. Cato and W.E.C. Moore (eds). 1977. Anaerobe Laboratory Manual, 4th ed. Virginia Polytechnic Institute and State University, Blacksburg, Virginia. pp. 1-156.

Prosthecochloris Gorlenko 1970
Gorlenko, V.M. 1970. Zeitschrift fur Allgemeine Mikrobiologie **10**:147-149. **Type species:** *P. aestuarii* Gorlenko 1970
Description: Bergey 8.

P. aestuarii Gorlenko 1970
Gorlenko, V.M. 1970. Zeitschrift fur Allgemeine Mikrobiologie **10**:147-149. **Type strain:** DSM 271
Description: Bergey 8.

Prosthecomicrobium Staley 1968
Staley, J.T. 1968. Journal of Bacteriology **95**:1921-1942. **Type species:** *P. pneumaticum* Staley 1968
Description: Bergey 8.

P. enhydrum Staley 1968
Staley, J.T. 1968. Journal of Bacteriology **95**:1921-1942. **Type strain:** ATCC 23634
Description: Staley, J.T. 1968. Ibid.

P. pneumaticum Staley 1968
Staley, J.T. 1968. Journal of Bacteriology **95**:1921-1942. **Type strain:** ATCC 23633
Description: Staley, J.T. 1968. Ibid.

Proteus Hauser 1885
Hauser, G. 1885. Uber Faulnissbakterien und deren Beziehungen zur Septicamie. Ein Beitrag zur Morphologie der Spaltpilze. Vogel, Leipzig. **Type species:** *P. vulgaris* Hauser 1885
Description: Bergey 8.

P. inconstans (Orstein 1920) Shaw and Clarke 1955
Orstein, M. 1920. Zeitschrift fur Hygiene **91**:152-178; Shaw, C. and P.H. Clarke. 1955. Journal of General Microbiology **13**:155-161. **Type strain:** ATCC 9886
Description: Shaw, C. and P.H. Clarke. 1955. Ibid.

P. mirabilis Hauser 1885
Hauser, G. 1885. Uber Faulnissbakterien und deren Beziehungen zur Septicamie. Ein Beitrag zur Morphologie der Spalt-

pilze. Vogel, Leipzig. p. 34. **Type strain:** ATCC 29906
Description: Rustigian, R. and C.A. Stuart. 1943. Proceedings of the Society for Experimental Biology and Medicine **53**: 241-243.

P. morganii (Winslow, Kligler and Rothberg 1919) Yale 1939 (See also Morganella morganii)
Winslow, C.E.A., I.J. Kligler and W. Rothberg. 1919. Journal of Bacteriology **4**:429-503; Yale, M.W. 1939 in Bergey, D.H., R.S. Breed, E.G.D. Murray and A.P. Hitchens (eds). Bergey's Manual of Determinative Bacteriology, 5th ed. The Williams and Wilkins Co., Baltimore. pp. 430-436. **Type strain:** ATCC 25830
Description: Rustigian, R. and C.A. Stuart. 1943. Proceedings of the Society for Experimental Biology and Medicine **53**: 241-243.

P. myxofaciens Cosenza and Podgwaite 1966
Cosenza, B.J. and J.D. Podgwaite. 1966. Antonie van Leeuwenhoek Journal of Microbiology and Serology **32**:187-191. **Type strain:** ATCC 19692
Description: Cosenza, B.J. and J.D. Podgwaite. 1966. Ibid.

P. rettgeri (Hadley, Elkins and Caldwell 1918) Rustigian and Stuart 1943 (See also Providencia rettgeri)
Hadley, P., M.W. Elkins and D.W. Caldwell. 1918. Bulletin of the Rhode Island Agricultural Experimental Station **174**:1-216; Rustigian, R. and C.A. Stuart. 1943. Proceedings of the Society for Experimental Biology and Medicine **53**:241-243. **Type strain:** ATCC 29944
Description: Rustigian, R. and C.A. Stuart. 1943. Ibid.

P. vulgaris Hauser 1885
Hauser, G. 1885. Uber Faulnissbakterien und deren Beziehungen zue Septicamie. Ein Beitrag zur Morphologie der Spaltpilze. Vogel. Leipzig. p. 12. **Type strain:** ATCC 13315
Description: Rustigian, R. and C.A. Stuart. 1943. Proceedings of the Society for Experimental Biology and Medicine **53**: 241-243.

Providencia Ewing 1962
Ewing, W.H. 1962. International Bulletin of Bacteriological Nomenclature and Taxonomy **12**:93-102. **Type species:** *P. alcalifaciens* (de Salles Gomes 1944) Ewing 1962
Description: Ewing, W.H. 1962. Ibid.

P. alcalifaciens (de Salles Gomes 1944) Ewing 1962
de Salles Gomes, L. 1944. Revista do Instituto Adolfo Lutz **14**:182-195; Ewing, W.H. 1962. International Bulletin of Bacteriological Nomenclature and Taxonomy **12**:93-102. **Type strain:** ATCC 9886
Description: Ewing, W.H. 1962. Ibid.

P. rettgeri (Hadley, Elkins and Caldwell 1918) Brenner, Farmer, Fanning, Steigerwalt, Klykken, Wathen, Hickman and Ewing 1978 (See also Proteus rettgeri)
Hadley, P., M.W. Elkins and D.W. Caldwell. 1918. Bulletin of the Rhode Island Agricultural Experimental Station **174**:1-216; Brenner, D.J., J.J. Farmer III, G.R. Fanning, A.G. Steigerwalt, P. Klykken, H.G. Wathen, F.W. Hickman and W.H. Ewing. 1978. International Journal of Systematic Bacteriology **28**:269-282. **Type strain:** ATCC 29944
Description: Rustigan, R. and C.A. Stuart. 1943. Proceedings of the Society for Experimental Biology and Medicine **53**: 241-243.

P. stuartii (Buttiaux, Osteux, Fresnoy and Moriamez 1954) Ewing 1962
Buttiaux, R., R. Osteux, R. Fresnoy and J. Moriamez. 1954. Annales de l'Institut Pasteur (Paris) **87**:375-386; Ewing, W.H. 1962. International Bulletin of Bacteriological Nomenclature and Taxonomy **12**: 93-102. **Type strain:** ATCC 29914
Description: Ewing, W.H. 1962. Ibid.

Pseudomonas Migula 1894
Migula, W. 1894. Arbeiten aus dem Bakteriologischen Institut der Technischen Hochschule zu Karlsruhe **1**:235-238. **Type species:** *P. aeruginosa* (Schroeter 1872) Migula 1900
Description: Bergey 8.

P. acidovorans den Dooren de Jong 1926
den Dooren de Jong, L.E. 1926. Bijdrage tot de kennis van het Mineralisatieproces. Thesis. Technische Hogeschool, Delft. pp. 1-199. **Type strain:** ATCC 15668
Description: Bergey 8.

P. aeruginosa (Schroeter 1872) Migula 1900
Schroeter, J. 1872 in Cohn, F. 1875. Beitrage zur Biologie der Pflanzen, J.U. Kern's Verlag, Breslau. pp. 109-126; Migula, W. 1900. System der Bakterien, Vol. 2. Gustav Fischer, Jena. **Type strain:** ATCC 10145
Description: Bergey 8.

P. agarici Young 1970
Young, J.M. 1970. New Zealand Journal of Agricultural Research **13**:977-990. **Type strain:** ATCC 25941
Description: Young, J.M. 1970. Ibid.

P. alcaligenes Monias 1928
Monias, B.L. 1928. Journal of Infectious Diseases **43**:330-334. **Type strain:** ATCC 14909
Description: Bergey 8.

P. aminovorans den Dooren de Jong 1926
den Dooren de Jong, L.E. 1926. Bijdrage tot de kennis van het mineralisatieproces. Thesis. Technische Hogeschool, Delft. pp. 1-199. **Type strain:** ATCC 23314
Description: den Dooren de Jong, L.E. 1926. Ibid.

P. amygdali Psallidas and Panagopoulos 1975
Psallidas, P.G. and C.G. Panagopoulos. 1975. Annales de l'Institut Phytopathologique Benaki N.S. **11**:94-108. **Type strain:** NCPPB 2607
Description: Psallidas, P.G. and C.G. Panagopoulos. 1975. Ibid.

P. andropogonis (Smith 1911) Stapp 1928
Smith, E.F. 1911. Carnegie Institute, Washington **2**:1-368; Stapp, C. 1928 in Sorauer, Handbuch der Pflanzenkrankheiten, 5th ed. Paul Parey, Berlin. pp. 758. **Type strain:** ATCC 23061
Description: Goto, M. and M.P. Starr. 1971. Annals of the Phytopathological Society of Japan **37**:233-241.

P. anguilliseptica Wakabayashi and Egusa 1972
Wakabayashi, H. and S. Egusa. 1972. Bulletin of the Japanese Society of Scientific Fisheries **38**:577-587. **Type strain:** NCMB 1949
Description: Wakabayashi, H. and S. Egusa. 1972. Ibid.

P. asplenii (Ark and Tompkins 1946) Savulescu 1947
Ark, P.A. and C.M. Tompkins. 1946. Phytopathology **36**:758-761; Savulescu, T. 1947. Analele Academiei Romane Series III Tom 22, Memoire **4**:1-26. **Type strain:** ATCC 23835
Description: Ark, P.A. and C.M. Tompkins. 1946. Ibid.

P. aurantiaca Nakhimovskaya 1948
Nakhimovskaya, M.I. 1948. Mikrobiologiya **17**:58-65. **Type strain:** NCIB 10068
Description: Nakhimovskaya, M.I. 1948. Ibid.

P. aureofaciens Kluyver 1956
Kluyver, A.J. 1956. Journal of Bacteriology **72**:406-411. **Type strain:** ATCC 13985
Description: Bergey 8.

P. avenae Manns 1909
Manns, T.F. 1909. Bulletin of the Ohio Agriculture Experiment Station **210**:91-167. **Type strain:** NCPPB 1011
Description: Schaad, N.W., C.I. Kado and D.R. Sumner. 1975. International Journal of Systematic Bacteriology **25**: 133-137.

P. azotoformans Iizuka and Komagata 1963
Iizuka, H. and K. Komagata. 1963. Journal of the Agricultural Chemical Society of Japan **37**:137-141. **Type strain:** IAM 1603
Description: Iizuka, H. and K. Komagata. 1963. Ibid.

P. beijerinckii Hof 1935
Hof, T. 1935. Recueil des Travaux Botaniques Neerlandais **32**:151-173. **Type strain:** ATCC 19372
Description: Haynes, W.C. 1957 in Breed, R.S., E.G.D. Murray and N.R. Smith (eds). Bergey's Manual of Determinative Bacteriology, 7th ed. The Williams and Wilkins Co., Baltimore. pp. 89-183.

P. betle (Ragunathan 1928) Savulescu 1947
Ragunathan, C. 1928. Annals of the Royal Botanical Garden of Peradeniya, Ceylon **11**:51-61; Savulescu, T. 1947. Analele Academiei Romane Series III Tom 22, Memoire **4**:1-26. **Type strain:** ATCC 19861
Description: Ragunathan, C. 1928. Ibid.

P. boreopolis Gray and Thornton 1928
Gray, P.H.H. and H.G. Thornton. 1928. Zentralblatt fur Bakteriologie, Parasitenkunde, Infektionskrankheiten und Hygiene. Abteilung II. **73**:74-96. **Type strain:** NCIB 9401
Description: Haynes, W.C. 1957 in Breed, R.S., E.G.D. Murray and N.R. Smith

(eds). Bergey's Manual of Determinative Bacteriology, 7th ed. The Williams and Wilkins Co., Baltimore. p. 101.

P. caricapapayae Robbs 1956
Robbs, C.F. 1956. Revista da Sociedade Brasileira de Agronomia **12**:73-76. **Type strain:** NCPPB 1873
Description: Robbs, C.F. 1956. Ibid.

P. caryophylli (Burkholder 1942) Starr and Burkholder 1942
Burkholder, W.H. 1942. Phytopathology **32**:141-149; Starr, M.P. and W.H. Burkholder. 1942. Phytopathology **32**:598-604. **Type strain:** ATCC 25418
Description: Burkholder, W.J. 1942. Ibid.

P. cattleyae (Pavarino 1911) Savulescu 1947
Pavarino, G.L. 1911. Atti della Reale Accademia dei Lincei **20**:233-237; Savulescu, T. 1947. Analele Academiei Romane Series III Tom 22, Memoire **4**:1-26. **Type strain:** NCPPB 961
Description: Ark, P.A. and H.E. Thomas. 1946. Phytopathology **36**:695-698.

P. chlororaphis (Guignard and Sauvageau 1894) Bergey, Harrison, Breed, Hammer and Huntoon 1930
Guignard, L. and C. Sauvageau. 1894. Comptes Rendus des Seances de la Societe de Biologie, Paris, Series 10, **1**:841-843; Bergey, D.H., F.C. Harrison, R.S. Breed, B.W. Hammer and F.M. Huntoon (eds). 1930. Bergey's Manual of Determinative Bacteriology, 3rd ed. The Williams and Wilkins Co., Baltimore. pp. 1-589. **Type strain:** ATCC 9446
Description: Bergey 8.

P. cichorii (Swingle 1925) Stapp 1928
Swingle, D.B. 1925. Phytopathology **15**: 730; Stapp, C. 1928. In Sorauer, Handbuch der Pflanzenkrankheiten, 5th ed. Paul Parey, Berlin. pp. 758. **Type strain:** ATCC 10857
Description: Bergey 8.

P. cissicola (Takimoto 1939) Burkholder 1948
Takimoto, S. 1939. Annals of the Phytopathological Society, Japan **9**:41-43; Burkholder, W.H. 1948 in Breed, R.S., E.G.D. Murray and A.P. Hitchens (eds). Bergey's Manual of Determinative Bacteriology, 6th ed. The Williams and Wilkins Co., Baltimore. pp. 82-150. **Type strain:** NCPPB 2982
Description: Goto, M. and T. Makino. 1977. Annals of the Phytopathological Society, Japan **43**:40-45.

P. citronellolis Seubert 1960
Seubert, W. 1960. Journal of Bacteriology **79**:426-434. **Type strain:** ATCC 13674
Description: Seubert, W. 1960. Ibid.

P. cocovenenans van Damme, Johannes, Cox and Berends 1960
van Damme, P.A., A.G. Johannes, H.C. Cox and W. Berends. 1960. Recueil des Travaux Chimiques des Pays-Bas **79**:255-267. **Type strain:** NCIB 9450
Description: van Damme, P.A., A.G. Johannes, H.C. Cox and W. Berends. 1960. Ibid.

P. delafieldii Davis 1970
Davis, D.H. 1970 in Davis, D.H., R.Y. Stanier, M. Doudoroff and M. Mandel. Archiv fur Mikrobiologie **70**:1-13. **Type strain:** ATCC 17505
Description: Bergey 8.

P. diminuta Leifson and Hugh 1954
Leifson, E. and R. Hugh. 1954. Journal of General Microbiology **10**:68-70. **Type strain:** ATCC 11568
Description: Leifson, E. and R. Hugh. 1954. Ibid.

P. doudoroffii Baumann, Baumann, Mandel and Allen 1972
Baumann, L., P. Baumann, M. Mandel and R.D. Allen. 1972. Journal of Bacteriology **110**:402-429. **Type strain:** ATCC 27123
Description. Baumann, L., P. Baumann, M. Mandel and R.D. Allen. 1972. Ibid.

P. echinoides Heumann 1962
Heumann, W. 1962. Biologisches Zentralblatt **81**:341-354. **Type strain:** ATCC 14820
Description: Heumann, W. 1962. Ibid.

P. elongata Humm 1946
Humm, H.J. 1946. Bulletin of Duke University Marine Station **3**:45-75. **Type strain:** ATCC 10144
Description: Humm, H.J. 1946. Ibid.

P. facilis (Schatz and Bovell 1952) Davis 1969
Schatz, A. and C.R. Bovell. 1952. Journal of Bacteriology **63**:87-98; Davis, D.H. 1969 in Davis, D.H., R.Y. Stanier, M. Doudoroff and M. Mandel. International Journal of Systematic Bacteriology **19**: 375-390. **Type strain:** ATCC 11228
Description: Bergey 8.

P. flava (Niklewski 1910) Davis 1969
Niklewski, B. 1910. Jahrbuch fur Wissenschaftliche Botanik **48**:113-142; Davis, D.H. 1969 in Davis, D.H., R.Y. Stanier, M. Doudoroff and M. Mandel. International Journal of Systematic Bacteriology **19**:375-390. **Type strain:** DSM 619
Description: Bergey 8.

P. flectens Johnson 1956
Johnson, J.C. 1956. Queensland Journal of Agricultural Science **13**:127-158. **Type strain:** ATCC 12775
Description: Johnson, J.C. 1956. Ibid.

P. fluorescens Migula 1895
Migula, W. 1895 in Engler, A. and K. Prantl (eds). Die Naturlichen Pfanzenfamilien, W. Engelmann, Leipzig, Teil I, Abteilung Ia. pp. 20-30. **Type strain:** ATCC 13525
Description: Bergey 8.

P. fragi (Eichholz 1902) Gruber 1905
Eichholz, W. 1902. Zentralblatt fur Bakteriologie, Parasitenkunde, Infektionskrankheiten und Hygiene. Abteilung II. **9**:425-428; Gruber, T. 1905. Zentralblatt fur Bakteriologie, Parasitenkunde, Infektionskrankheiten und Hygiene. Abteilung II. **14**:122-123. **Type strain:** ATCC 4973
Description: Haynes, W.C. 1957 in Breed, R.S., E.G.D. Murray and N.R. Smith (eds). Bergey's Manual of Determinative Bacteriology, 7th ed. The Williams and Wilkins Co., Baltimore. p. 110.

P. fulva Iizuka and Komagata 1963
Iizuka, H. and K. Komagata. 1963. Journal of the Agricultural Chemical Society of Japan **37**:137-141. **Type strain:** IAM 1529
Description: Iizuka, H. and K. Komagata. 1963. Ibid.

P. gelidicola Kadota 1951
Kadota, H. 1951. Memoirs of the College of Science, Kyoto University **59**:54-67. **Type strain:** IAM 1127
Description: Kadota, H. 1951. Ibid.

P. geniculata (Wright 1895) Chester 1901
Wright, J.H. 1895. Memoirs of the National Academy of Science **7**:422-484; Chester, F.D. 1901. A manual of determinative bacteriology. The Macmillan Co., New York. pp. 1-401. **Type strain:** ATCC 19374
Description: Haynes, W.C. 1957 in Breed, R.S., E.G.D. Murray and N.R. Smith (eds). Bergey's Manual of Determinative Bacteriology, 7th ed. The Williams and Wilkins Co., Baltimore. p. 105.

P. gladioli Severini 1913
Severini, G. 1913. Annali di Botanica (Roma) **11**:413-424. **Type strain:** NCPPB 1891
Description: Hildebrand, D.C., N.J. Palleroni and M. Doudoroff. 1973. International Journal of Systematic Bacteriology **23**:433-437.

P. glathei Zolg and Ottow 1975
Zolg, W. and J.C.G. Ottow. 1975. Zeitschrift fur Allgemeine Mikrobiologie **15**: 287-299. **Type strain:** ATCC 29195
Description: Zolg, W. and J.C.G. Ottow. 1975. Ibid.

P. glumae Kurita and Tabei 1967
Kurita, T. and H. Tabei. 1967. Annals of the Phytopathological Society, Japan **33**: 111. **Type strain:** NCPPB 2981
Description: Kurita, T. and H. Tabei. 1967. Ibid.

P. hibiscicola Moniz 1963
Moniz, L. 1963. Current Science **32**:177. **Type strain:** ATCC 19867
Description: Moniz, L. 1963. Ibid.

P. huttiensis Leifson 1962
Leifson, E. 1962. International Bulletin of Bacteriological Nomenclature and Taxonomy **12**:161-170. **Type strain:** ATCC 14670
Description: Leifson, E. 1962. Ibid.

P. indigofera (Voges 1893) Migula 1900
Voges, O. 1893. Zentralblatt fur Bakteriologie, Parasitenkunde, Infektionskrankheiten und Hygiene. Abteilung I. **14**:301-314; Migula, W. 1900. System der Bakterien, Vol. 2. Gustav Fischer, Jena. **Type strain:** ATCC 19706
Description: Elazari-Volcani, B. 1939. Archiv fur Mikrobiologie **10**:343-358.

P. iners Iizuka and Komagata 1964
Iizuka, H. and K. Komagata. 1964. Journal of General and Applied Microbiology **10**:223-231. **Type strain:** IAM 1419
Description: Iizuka, H. and K. Komagata. 1964. Ibid.

P. lanceolata Leifson 1962
Leifson, E. 1962. International Bulletin of Bacteriological Nomenclature and Taxon-

omy **12**:161-170. **Type strain:** ATCC 14669
Description: Leifson, E. 1962. Ibid.

P. lemoignei Delafield, Doudoroff, Palleroni, Lusty and Contopoulou 1965
Delafield, F.P., M. Doudoroff, N.J. Palleroni, C.J. Lusty and C.R. Contopoulou. 1965. Journal of Bacteriology **90**:1455-1466. **Type strain:** ATCC 17989
Description: Delafield, F.P., M. Doudoroff, N.J. Palleroni, C.L. Lusty and C.R. Contopoulou. 1965. Ibid.

P. mallei (Zopf 1885) Redfearn, Palleroni and Stanier 1966
Zopf, W. 1885. Die Spaltpilze, 3rd ed. Edward Trewendt, Breslau. pp. 1-127; Redfearn, M.S., N.J. Palleroni and R.Y. Stanier. 1966. Journal of General Microbiology **43**:293-313. **Type strain:** ATCC 23344
Description: Bergey 8.

P. marginalis (Brown 1918) Stevens 1925
Brown, N.A. 1918. Journal of Agricultural Research **13**:367-388; Stevens, F.L. 1925. Plant Disease Fungi. MacMillan Co., New York. pp. 469. **Type strain:** ATCC 10844
Description: Burkholder, W.J. 1957 in Breed, R.S., E.G.D. Murray and N.R. Smith (eds). Bergey's Manual of Determinative Bacteriology, 7th ed. The Williams and Wilkins Co., Baltimore. p. 135.

P. marina (Cobet, Wirsen and Jones 1970) Baumann, Baumann, Mandel and Allen 1972
Cobet, A.B., C. Wirsen and G.E. Jones. 1970. Journal of General Microbiology **62**:159-169; Baumann, L., P. Baumann, M. Mandel and R.D. Allen. 1972. Journal of Bacteriology **110**:402-429. **Type strain:** ATCC 25374
Description: Baumann, L., P. Baumann, M. Mandel and R.D. Allen. 1972. Ibid.

P. mendocina Palleroni 1970
Palleroni, N.J. 1970 in Palleroni, N.J., M. Doudoroff, R.Y. Stanier, R.E. Solanes and M. Mandel. Journal of General Microbiology **60**:215-231. **Type strain:** ATCC 25411
Description: Bergey 8.

P. mephitica Claydon and Hammer 1939
Claydon, T.J. and B.W. Hammer. 1939. Journal of Bacteriology **37**:252-258. **Type strain:** NCIB 9672
Description: Haynes, W.C. 1957 in Breed, R.S., E.G.D. Murray and N.R. Smith (eds). Bergey's Manual of Determinative Bacteriology, 7th ed. The Williams and Wilkins Co., Baltimore. p. 111.

P. mesophilica Austin and Goodfellow 1979
Austin, B. and M. Goodfellow. 1979. International Journal of Systematic Bacteriology **29**:373-378. **Type strain:** ATCC 29983
Description: Austin, B. and M. Goodfellow. 1979. Ibid.

P. mucidolens Levine and Anderson 1932
Levine, M. and D.Q. Anderson. 1932. Journal of Bacteriology **23**:337-347. **Type strain:** ATCC 4685
Description: Levine, M. and D.Q. Anderson. 1932. Ibid.

P. nautica Baumann, Baumann, Mandel and Allen 1972
Baumann, L., P. Baumann, M. Mandel and R.D. Allen. 1972. Journal of Bacteriology **110**:402-429. **Type strain:** ATCC 27132
Description: Baumann, L., P. Baumann, M. Mandel and R.D. Allen. 1972. Ibid.

P. nitroreducens Iizuka and Komagata 1964
Iizuka, H. and K. Komagata. 1964. Journal of General and Applied Microbiology **10**:207-221. **Type strain:** IAM 1439
Description: Iizuka, H. and K. Komagata. 1964. Ibid.

P. oleovorans Lee and Chandler 1941
Lee, M. and A.C. Chandler. 1941. Journal of Bacteriology **41**:373-386. **Type strain:** ATCC 8062
Description: Lee, M. and A.C. Chandler. 1941. Ibid.

P. palleronii Davis 1970
Davis, D.H. 1970 in Davis, D.H., R.Y. Stanier, M. Doudoroff and M. Mandel. Archiv fur Mikrobiologie **70**:1-13. **Type strain:** ATCC 17724
Description: Davis, D.H., R.Y. Stanier, M. Doudoroff and M. Mandel. 1970. Ibid.

P. paucimobilis Holmes, Owen, Evans, Malnick and Willcox 1977
Holmes, B., R.J. Owen, A. Evans, H. Malnick and W.R. Willcox. 1977. International Journal of Systematic Bacteriology **27**:133-146. **Type strain:** ATCC 29837
Description: Holmes, B., R.J. Owen, A.

Evans, H. Malnick and W.R. Willcox. 1977. Ibid.

P. pertucinogena Kawai and Yabuuchi 1975
Kawai, Y. and E. Yabuuchi. 1975. International Journal of Systematic Bacteriology **25**:317-323. **Type strain:** ATCC 190
Description: Kawai, Y. and E. Yabuuchi. 1975. Ibid.

P. phenazinium Bell and Turner 1973
Bell, S.C. and J.M. Turner. 1973. Biochemical Society Transactions **1**:751-753. **Type strain:** NCIB 11027
Description: Bell, S.C. and J.M. Turner. 1973. Ibid.

P. pickettii Ralston, Palleroni and Doudoroff 1973
Ralston, E., N.J. Palleroni and M. Doudoroff. 1973. International Journal of Systematic Bacteriology **23**:15-19. **Type strain:** ATCC 27511
Description: Ralston, E., N.J. Palleroni and M. Doudoroff. 1973. Ibid.

P. pictorum Gray and Thornton 1928
Gray, P.H.H. and H.G. Thornton. 1928. Zentralblatt fur Bakteriologie, Parasitenkunde, Infektionskrankheiten und Hygiene. Abteilung II. **73**:74-96. **Type strain:** ATCC 23328
Description: Gray, P.H.H. and H.G. Thornton. 1928. Ibid.

P. pseudoalcaligenes Stanier 1966
Stanier, R.Y. 1966 in Stanier, R.Y., N.J. Palleroni and M. Doudoroff. Journal of General Microbiology **43**:159-271. **Type strain:** ATCC 17440
Description: Bergey 8.

P. pseudoalcaligenes subsp. *citrulli* Schaad, Sowell, Goth, Colwell and Webb 1978
Schaad, N.W., G. Sowell, R.W. Goth, R.R. Colwell and R.E. Webb. 1978. International Journal of Systematic Bacteriology **28**:117-125. **Type strain:** ATCC 29625
Description: Schaad, N.W., G. Sowell, R.W. Goth, R.R. Colwell and R.E. Webb. 1978. Ibid.

P. pseudoalcaligenes subsp. *pseudoalcaligenes* Stanier 1966
Stanier, R.Y. 1966 in Stanier, R.Y., N.J. Palleroni and M. Doudoroff. Journal of General Microbiology **43**:159-271. **Type strain:** ATCC 17440
Description: Bergey 8.

P. pseudoflava Auling, Reh, Lee and Schlegel 1978
Auling, G., M. Reh, C.M. Lee and H.G. Schlegel. 1978. International Journal of Systematic Bacteriology **28**:82-95. **Type strain:** DSM 1034
Description: Auling, G., M. Reh, C.M. Lee and H.G. Schlegel. 1978. Ibid.

P. pseudomallei (Whitmore 1913) Haynes 1957
Whitmore, A. 1913. Journal of Hygiene **13**:1-34; Haynes, W.C. 1957 in Breed, R.S., E.G.D. Murray and N.R. Smith (eds). Bergey's Manual of Determinative Bacteriology, 7th ed. The Williams and Wilkins Co., Baltimore. pp. 89-152. **Type strain:** ATCC 23343
Description: Brindle, C.S. and S.T. Cowan. 1951. Journal of Pathology and Bacteriology **63**:571-575.

P. putida (Trevisan 1889) Migula 1895
Trevisan, V. 1889. I Generi e le Specie delle Battieriacee. Zanaboni and Gabuzzi, Milano; Migula, W. 1895 in Engler, A. and K. Prantl (eds). Die Naturlichen Pflanzenfamilien, W. Engelmann, Leipzig, Teil I, Abteilung Ia. pp. 1-44. **Type strain:** ATCC 12633
Description: Bergey 8.

P. pyrrocinia Imanaka, Kousaka, Tamura and Arima 1965
Imanaka, H., M. Kousaka, G. Tamura and K. Arima. 1965. Journal of Antibiotics (Tokyo) Series A **18**:205-206. **Type strain:** ATCC 15958
Description: Imanaka, H., M. Kousaka, G. Tamura and K. Arima. 1965. Ibid.

P. radiora Ito and Iizuka 1971
Ito, H. and H. Iizuka. 1971. Agricultural and Biological Chemistry **35**:1566-1571. **Type strain:** ATCC 27329
Description: Ito, H. and H. Iizuka. 1971. Ibid.

P. resinovorans Delaporte, Raymond and Daste 1961
Delaporte, B., M. Raymond and P. Daste. 1961. Comptes Rendus des Seances de l'Academie des Sciences Paris **252**: 1073-1075. **Type strain:** ATCC 14235
Description: Delaporte, B., M. Raymond and P. Daste. 1961. Ibid.

P. rhodos Heumann 1962
Heumann, W. 1962. Biologisches Zentralblatt **81**:341-354. **Type strain:** ATCC

14821
Description: Heumann, W. 1962. Ibid.

P. rubrilineans (Lee, Purdy, Barnum and Martin 1925) Stapp 1928
Lee, H.A., H.A. Purdy, C.C. Barnum and J.P. Martin. 1925. Bulletin, Experiment Station of the Hawaiian Sugar Planters' Association pp. 64-74; Stapp, C. 1928 in Sorauer, Handbuch der Pflanzenkrankheiten, 5th ed. Paul Parey, Berlin. pp. 758. **Type strain:** ATCC 19307
Description: Hayward, A.C. 1962. Mauritius Sugar Industry Research Institute Occasional Paper No. **13**:13-27.

P. rubrisubalbicans (Christopher and Edgerton 1930) Krassilnikov 1949
Christopher, W.N. and C.W. Edgerton. 1930. Journal of Agricultural Research **41**:259-267; Krassilnikov, N.A. 1949. Moskva - Leningrad, Izdatel'stvo Akademii Nauk SSSR. pp.1-830. **Type strain:** ATCC 19308
Description: Hayward, A.C. 1962. Mauritius Sugar Industry Research Institute Occasional Paper No. **13**:pp. 27.

P. saccharophila Doudoroff 1940
Doudoroff, M. 1940. Enzymologia **9**:59-72. **Type strain:** ATCC 15946
Description: Bergey 8.

P. solanacearum (Smith 1896) Smith 1914
Smith, E.F. 1896. Bulletin, Division of Vegetable Physiology and Pathology, United States Department of Agriculture **12**:1-28; Smith, E.F. 1914. Carnegie Institute, Washington **3**:1-309. **Type strain:** ATCC 11696
Description: Bergey 8.

P. spinosa Leifson 1962
Leifson, E. 1962. International Bulletin of Bacteriological Nomenclature and Taxonomy **12**:89-92. **Type strain:** ATCC 14606
Description: Leifson, E. 1962. Ibid.

P. straminae Iizuka and Komagata 1963
Iizuka, H. and K. Komagata. 1963. Journal of the Agricultural Chemical Society of Japan **37**:137-141. **Type strain:** IAM 1598
Description: Iizuka, H. and K. Komagata. 1963. Ibid.

P. stutzeri (Lehmann and Neumann 1896) Sijderius 1946
Lehmann, K.B. and R. Neumann. 1896. Atlas und Grundriss der Bakteriologie und Lehrbuch der speziellen bakteriologischen Diagnostik. 1st ed. J.F. Lehmann, Munchen; Sijderius, R. 1946. Heterotrophe Bacterien, die thiosulfaat oxydeeren. Thesis, University Amsterdam, pp. 1-146. **Type strain:** ATCC 17588
Description: Bergey 8.

P. synxantha (Ehrenberg 1840) Holland 1920
Ehrenberg, C.G. 1840. Bericht Bekannt Verhandlungen der Koniglisch-Preussichen Akademie der Wissenschaftlichen, Berlin **1840**:197-219; Holland, D.F. 1920 in Winslow, C.E.A., J. Broadhurst, R.E. Buchanan, C. Krumwiede Jr., L.A. Rogers and G.H. Smith. Journal of Bacteriology **5**:191-229. **Type strain:** ATCC 9890
Description: Haynes, W.C. 1957 in Breed, R.S., E.G.D. Murray and N.R. Smith (eds). Bergey's Manual of Determinative Bacteriology, 7th ed. The Williams and Wilkins Co., Baltimore. p. 104.

P. syringae van Hall 1902
van Hall, C.J.J. 1902. Inaugural Dissertation. Amsterdam. pp. 198. **Type strain:** ATCC 19310
Description: Bergey 8.

P. taetrolens Haynes 1957
Haynes, W.C. 1957 in Breed, R.S., E.G.D. Murray and N.R. Smith (eds). Bergey's Manual of Determinative Bacteriology, 7th ed. The Williams and Wilkins Co., Baltimore. pp. 89-152. **Type strain:** ATCC 4683
Description: Haynes, W.C. 1957. Ibid.

P. testosteroni Marcus and Talalay 1956
Marcus, P.I. and P. Talalay. 1956. Journal of Biological Chemistry **218**:661-674. **Type strain:** ATCC 11996
Description: Bergey 8.

P. tolaasii Paine 1919
Paine, S.G. 1919. Annals of Applied Biology **5**:206-219. **Type strain:** NCPPB 2192
Description: Burkholder, W.H. 1957 in Breed, R.S., E.G.D. Murray and N.R. Smith (eds). Bergey's Manual of Determinative Bacteriology, 7th ed. The Williams and Wilkins Co., Baltimore. p. 136.

P. vesicularis (Busing, Doll and Freytag 1953) Galarneault and Leifson 1964
Busing, K.H., W. Doll and K. Freytag. 1953. Archiv fur Mikrobiologie **19**:52-86; Galarneault, T.P. and E. Leifson. 1964. International Bulletin of Bacteriological

Nomenclature and Taxonomy **14**:165-168. **Type strain:** ATCC 11426
Description: Bergey 8.

P. viridiflava (Burkholder 1930) Dowson 1939
Burkholder, W.H. 1930. Memoirs, Cornell University Agricultural Experiment Station **127**:1-88; Dowson, W.J. 1939. Zentralblatt fur Bakteriologie, Parasitenkunde, Infektionskrankheiten und Hygiene. Abteilung II. **100**:177-193. **Type strain:** ATCC 13223
Description: Billing, E. 1970. Journal of Applied Bacteriology **33**:492-500.

P. woodsii (Smith 1911) Stevens 1925
Smith, E.F. 1911. Carnegie Institute, Washington **2**:1-368; Stevens, F.L. 1925. Plant disease fungi. MacMillan. New York. **Type strain:** ATCC 19311
Description: Burkholder, W.H. and C.E.F. Guterman. 1935. Phytopathology **25**:114-120.

Pseudonocardia Henssen 1957
Henssen, A. 1957. Archiv fur Mikrobiologie **26**:373-414. **Type species:** *P. thermophila* Henssen 1957
Description: Bergey 8.

P. spinosa Schafer 1971
Schafer, D. 1971 in Henssen, A. and D. Schafer. International Journal of Systematic Bacteriology **21**:29-34. **Type strain:** ATCC 25924
Description: Bergey 8.

P. thermophila Henssen 1957
Henssen, A. 1957. Archiv fur Mikrobiologie **26**:373-414. **Type strain:** ATCC 19285
Description: Bergey 8.

Rhizobium Frank 1889
Frank, B. 1889. Berichte der Deutschen Botanischen Gesellschaft **7**:332-346. **Type species:** *R. leguminosarum* (Frank 1879) Frank 1889
Description: Bergey 8.

R. japonicum (Kirchner 1896) Buchanan 1926
Kirchner, O. 1896. Beitrage zur Biologie der Pflanzen **7**:213-224; Buchanan, R.E. 1926. Proceedings of the Iowa Academy of Science **33**:81-90. **Type strain:** ATCC 10324
Description: Bergey 8.

R. leguminosarum (Frank 1879) Frank 1889
Frank, B. 1879. Bericht der Deutschen Botanischen Gesellschaft **37**:394-399; Frank, B. 1889. Berichte der Deutschen Botanischen Gesellschaft **7**:332-346. **Type strain:** ATCC 10004
Description: Bergey 8.

R. lupini (Schroeter 1886) Eckhardt, Baldwin and Fred 1931
Schroeter, J. 1885-1889 in Cohn, F. Kryptogamenflora von Schlesien. Band 3, Heft 3, Pilze J.U. Kern's Verlag, Breslau. pp. 1-814; Eckhardt, M.M., I.R. Baldwin and E.B. Fred. 1931. Journal of Bacteriology **21**:273-285. **Type strain:** ATCC 10319
Description: Bergey 8.

R. meliloti Dangeard 1926
Dangeard, P.A. 1926. Botaniste (Paris) **16**:1-275. **Type strain:** ATCC 9930
Description: Bergey 8.

R. phaseoli Dangeard 1926
Dangeard, P.A. 1926. Botaniste (Paris) **16**:1-275. **Type strain:** ATCC 14482
Description: Bergey 8.

R. trifolii Dangeard 1926
Dangeard, P.A. 1926. Botaniste (Paris) **16**:1-275. **Type strain:** ATCC 14480
Description: Bergey 8.

Rhodococcus Zopf 1891
Zopf, W. 1891. Berichte der Deutschen Botanischen Gesellschaft **9**:22-28. **Type species:** *R. rhodochrous* (Zopf 1891) Tsukamura 1974
Description: Goodfellow, M. and G. Alderson. 1977. Journal of General Microbiology **100**:99-122.

R. bronchialis (Tsukamura 1971) Tsukamura 1974
Tsukamura, M. 1971. Journal of General Microbiology **68**:15-26; Tsukamura, M. 1974. Japanese Journal of Microbiology **18**:37-44. **Type strain:** ATCC 25592
Description: Goodfellow, M. and G. Alderson. 1977. Journal of General Microbiology **100**:99-122.

R. coprophilus Rowbotham and Cross 1979
Rowbotham, T.J. and T. Cross. 1979. International Journal of Systematic Bacteriology **29**:79-80; Effective publication Rowbotham, T.J. and T. Cross. 1977. Journal of General Microbiology **100**:123-138. **Type strain:** ATCC 29080
Description: Rowbotham, T.J. and T. Cross. 1977. Ibid.

R. corallinus (Bergey, Harrison, Breed, Hammer and Huntoon 1923) Goodfellow and Alderson 1977
Bergey, D.H., F.C. Harrison, R.S. Breed, B.W. Hammer and F.M. Huntoon. 1923. Bergey's Manual of Determinative Bacteriology, 1st ed. The Williams and Wilkins Co., Baltimore. pp. 1-442; Effective publication Goodfellow, M. and G. Alderson. 1977. Journal of General Microbiology **100**:99-122. **Type strain:** ATCC 25593
Description: Goodfellow, M. and G. Alderson. 1977. Ibid.

R. equi (Magnusson 1923) Goodfellow and Alderson 1979
Magnusson, H. 1923. Archiv fur Wissenschaftliche und Praktische Tierheilkunde **50**:22-38; Goodfellow, M. and G. Alderson. 1979. International Journal of Systematic Bacteriology **29**:79-80; Effective publication Goodfellow, M. and G. Alderson. 1977. Journal of General Microbiology **100**:99-122. **Type strain:** ATCC 25729
Description: Goodfellow, M. and G. Alderson. 1977. Ibid.

R. erythropolis (Gray and Thornton 1928) Goodfellow and Alderson 1979
Gray, P.H.H. and H.G. Thornton. 1928. Zentralblatt fur Bakteriologie, Parasitenkunde, Infektionskrankheiten und Hygiene. Abteilung II. **73**:74-96; Goodfellow, M. and G. Alderson. 1979. International Journal of Systematic Bacteriology **29**:79-80; Effective publication Goodfellow, M. and G. Alderson. 1977. Journal of General Microbiology **100**:99-122. **Type strain:** ATCC 4277
Description: Goodfellow, M. and G. Alderson. 1977. Ibid.

R. rhodnii Goodfellow and Alderson 1979
Goodfellow, M. and G. Alderson. 1979. International Journal of Systematic Bacteriology **29**:79-80; Effective publication Goodfellow, M. and G. Alderson. 1977. Journal of General Microbiology **100**:99-122. **Type strain:** KCC A-0203
Description: Goodfellow, M. and G. Alderson. 1977. Ibid.

R. rhodochrous (Zopf 1891) Tsukamura 1974
Zopf, W. 1891. Berichte der Deutschen Botanischen Gesellschaft **9**:22-28; Tsukamura, M. 1974. Japanese Journal of Microbiology **18**:37-44. **Type strain:** ATCC 13808
Description: Goodfellow, M. and G. Alderson. 1977. Journal of General Microbiology **100**:99-122.

R. ruber (Kruse 1896) Goodfellow and Alderson 1977
Kruse, W. 1896 in Flugge, C. Die Mikroorganismen. Vol. 2, pp. 48-66; Effective publication Goodfellow, M. and G. Alderson. 1977. Journal of General Microbiology **100**:99-122. **Type strain:** KCC A-0205
Description: Goodfellow, M. and G. Alderson. 1977. Ibid.

R. rubropertinctus (Hefferan 1904) Tsukamura 1974
Hefferan, M. 1904. Zentralblatt fur Bakteriologie, Parasitenkunde, Infektionskrankheiten und Hygiene. Abteilung II. **11**:397-404; Tsukamura, M. 1974. Japanese Journal of Microbiology **18**:37-44. **Type strain:** ATCC 14352
Description: Goodfellow, M. and G. Alderson. 1977. Ibid.

R. terrae (Tsukamura 1971) Tsukamura 1974
Tsukamura, M. 1971. Journal of General Microbiology **68**:15-26; Tsukamura, M. 1974. Japanese Journal of Microbiology **18**:37-44. **Type strain:** ATCC 25594
Description: Goodfellow, M. and G. Alderson. 1977. Journal of General Microbiology **100**:99-122.

Rhodocyclus Pfennig 1978
Pfennig N. 1978. International Journal of Systematic Bacteriology **28**:283-288. **Type species:** *R. purpureus* Pfennig 1978
Description: Pfennig, N. 1978. Ibid.

R. purpureus Pfennig 1978
Pfennig, N. 1978. International Journal of Systematic Bacteriology **28**:283-288. **Type strain:** DSM 168
Description: Pfennig, N. 1978. Ibid.

Rhodomicrobium Duchow and Douglas 1949
Duchow, E. and H.C. Douglas. 1949. Journal of Bacteriology **58**:409-416. **Type species:** *R. vannielii* Duchow and Douglas 1949
Description: Bergey 8.

R. vannielii Duchow and Douglas 1949
Duchow, E. and H.C. Douglas. 1949.

Journal of Bacteriology **58**:409-416. **Type strain:** ATCC 17100
Description: Bergey 8.

Rhodopseudomonas Czurda and Maresch 1937
Czurda, V. and E. Maresch. 1937. Archiv fur Mikrobiologie **8**:99-124. **Type species:** *R. palustris* (Molisch 1907) van Niel 1944
Description: Bergey 8.

R. acidophila Pfennig 1969
Pfennig, N. 1969. Journal of Bacteriology **99**:597-602. **Type strain:** ATCC 25092
Description: Bergey 8.

R. capsulata (Molisch 1907) van Niel 1944
Molisch, H. 1907. Die Purpurbakterien nach neuen Untersuchungen. G. Fischer, Jena I-VII, pp. 1-95; van Niel, C.B. 1944. Bacteriological Reviews **8**:1-118. **Type strain:** ATCC 11166
Description: Weaver, P.F., J.D. Wall and H. Gest. 1975. Archives of Microbiology **105**:207-216.

R. gelatinosa (Molisch 1907) van Niel 1944
Molisch, H. 1907. Die Purpurbakterien nach neuen Untersuchungen. G. Fischer, Jena I-VII, pp. 1-95; van Niel, C.B. 1944. Bacteriological Reviews **8**:1-118. **Type strain:** ATCC 17011
Description: Bergey 8.

R. globiformis Pfennig 1974
Pfennig, N. 1974. Archives of Microbiology **100**:197-206. **Type strain:** DSM 161
Description: Pfennig, N. 1974. Ibid.

R. palustris (Molisch 1907) van Niel 1944
Molisch, H. 1907. Die Purpurbakterien nach neuen Untersuchungen. G. Fischer, Jena I-VII, pp. 1-95; van Niel, C.B. 1944. Bacteriological Reviews **8**:1-118. **Type strain:** ATCC 17001
Description: Bergey 8.

R. sphaeroides van Niel 1944
van Niel, C.B. 1944. Bacteriological Reviews **8**:1-118. **Type strain:** ATCC 17023
Description: Bergey 8.

R. sulfidophila Hansen and Veldkamp 1973
Hansen, T.A. and H. Veldkamp. 1973. Archiv fur Mikrobiologie **92**:45-58. **Type strain:** DSM 1374
Description: Hansen, T.A. and H. Veldkamp. 1973. Ibid.

R. sulfoviridis Keppen and Gorlenko 1975
Keppen, O.I. and V.M. Gorlenko. 1975. Mikrobiologiya **44**:258-264. **Type strain:** DSM 729
Description: Keppen, O.I. and V.M. Gorlenko. 1975. Ibid.

R. viridis Drews and Giesbrecht 1966
Drews, G. and P. Giesbrecht. 1966. Archiv fur Mikrobiologie **53**:255-262. **Type strain:** ATCC 19567
Description: Bergey 8.

Rhodospirillum Molisch 1907
Molisch, H. 1907. Die Purpurbakterien nach neuen Untersuchungen. G. Fischer, Jena I-VII, pp. 1-95. **Type species:** *R. rubrum* (Esmarch 1887) Molisch 1907
Description: Bergey 8.

R. fulvum van Niel 1944
van Niel, C.B. 1944. Bacteriological Reviews **8**:1-118. **Type strain:** ATCC 157980
Description: Bergey 8.

R. molischianum Giesberger 1947
Giesberger, G. 1947. Antonie van Leeuwenhoek Journal of Microbiology and Serology **13**:135-148. **Type strain:** ATCC 14031
Description: Bergey 8.

R. photometricum Molisch 1907
Molisch, H. 1907. Die Purpurbakterien nach neuen Untersuchungen. G. Fischer, Jena I-VII, pp. 1-95. **Type strain:** DSM 122
Description: Bergey 8.

R. rubrum (Esmarch 1887) Molisch 1907
Esmarch, E. 1887. Zentralblatt fur Bakteriologie, Parasitenkunde, Infektionskrankheiten und Hygiene. Abteilung I. **1**: 225-230; Molisch, H. 1907. Die Purpurbakterien nach neuen Untersuchungen. G. Fisher, Jena I-VII, pp. 1-95. **Type strain:** ATCC 11170
Description: Bergey 8.

R. tenue Pfennig 1969
Pfennig, N. 1969. Journal of Bacteriology **99**:619-920. **Type strain:** ATCC 25093
Description: Bergey 8.

Rickettsia da Rocha-Lima 1916
da Rocha-Lima, H. 1916. Berliner Klinische Wochenschrift **53**:567-569. **Type species:** *R. prowazekii* da Rocha-Lima 1916
Description: Bergey 8.

R. akari Huebner, Jellison and Pomerantz 1946
Huebner, R.J., W.L. Jellison and C. Po-

merantz. 1946. Public Health Report **61**: 1677-1682. **Type strain:** ATCC VR 148
Description: Bergey 8.

R. australis Philip 1950
Philip, C.B. 1950 in Pullen, R.L. Communicable Diseases. Lea and Febiger Co., Philadelphia. pp. 781-788. **Type strain:** NIAID Phillips 32
Description: Bergey 8.

R. canada McKiel, Bell and Lackmann 1967
McKiel, J.A., E.J. Bell and D.B. Lackmann. 1967. Canadian Journal of Microbiology **13**:503-510. **Type strain:** ATCC VR 610
Description: Bergey 8.

R. conorii Brumpt 1932
Brumpt, E. 1932. Comptes Rendus des Seances de la Societe de Biologie Filiales **110**:1199-1202. **Type strain:** NIAID Malish 7
Description: Bergey 8.

R. parkeri Lackman, Bell, Stoenner and Pickens 1965
Lackman, D.B., E.J. Bell, H.G. Stoenner and E.G. Pickens. 1965. Health Laboratory Science **2**:135-141. **Type strain:** NIAID maculatum 20
Description: Bergey 8.

R. prowazekii da Rocha-Lima 1916
da Rocha-Lima, H. 1916. Berliner Klinische Wochenschrift **53**:567-569. **Type strain:** ATCC VR 142
Description: Bergey 8.

R. rickettsii (Wolbach 1919) Brumpt 1922
Wolbach, S. 1919. Journal of Medical Research **41**:1-197; Brumpt, E. 1922. Precis de Parasitologie. 3rd ed. Masson et Cie, Paris. **Type strain:** ATCC VR 149
Description: Bergey 8.

R. sennetsu Misao and Kobayashi 1956
Misao, R. and Y. Kobayashi. 1956. Journal of the Japanese Association for Infectious Diseases **30**:453-465. **Type strain:** ATCC VR 367
Description: Bergey 8.

R. sibirica Zdrodovskii 1948
Zdrodovskii, P.F. 1948. Academy of Medical Science Moscow. **Type strain:** ATCC VR 151
Description: Bergey 8.

R. tsutsugamushi (Hayashi 1920) Ogata 1931
Hayashi, N. 1920. Journal of Parasitology **7**:53-68; Ogata, N. 1931. Zentralblatt fur Bakteriologie, Parasitenkunde, Infektionskrankheiten und Hygiene. Abteilung I. **122**:249-253. **Type strain:** ATCC VR 150
Description: Bergey 8.

R. typhi (Wolbach and Todd 1920) Philip 1943
Wolbach, S.B. and J.L. Todd. 1920. Annales de l'Institut Pasteur (Paris) **34**:153-158; Philip, C.B. 1943. American Journal of Hygiene **37**:301-309. **Type strain:** ATCC VR 144
Description: Bergey 8.

Rickettsiella Philip 1956
Philip, C.B. 1956. Canadian Journal of Microbiology **2**:261-270. **Type species:** *R. popilliae* (Dutky and Gooden 1952) Philip 1956
Description: Bergey 8.

R. popilliae (Dutky and Gooden 1952) Philip 1956
Dutky, S.R. and E.L. Gooden. 1952. Journal of Bacteriology **63**:743-750; Philip, C.B. 1956. Canadian Journal of Microbiology **2**:261-270. **Type strain:** no culture isolated.
Description: Bergey 8.

R. stethorae Hall and Badgley 1957
Hall, I.M. and M.E. Badgley. 1957. Journal of Bacteriology **74**:452-455. **Type strain:** no culture isolated.
Description: Bergey 8.

Rochalimaea (Macchiavello 1947) Krieg 1961
Macchiavello, A. 1947. Prima Reunione Interamericana del Tifo, Mexico. pp. 405-426; Krieg, A. 1961. Grundlagen der Insekten-pathologie. Dr. Dietrich Steinkofpp. Verlag, Darmstadt. **Type species:** *R. quintana* (Schmincke 1917) Krieg 1961
Description: Bergey 8.

R. quintana (Schmincke 1917) Krieg 1961
Schmincke, A. 1917. Munchener Medizinische Wochenschrift **64**:961; Krieg, A. 1961. Grundlagen der Insektenpathologie. Dr. Dietrich Steinkopff. Verlag, Darmstadt. **Type strain:** ATCC VR 358
Description: Bergey 8.

Rothia Georg and Brown 1967
Georg, L.K. and J.M. Brown. 1967. International Journal of Systematic Bacteriology **17**:79-88. **Type species:** *R.*

dentocariosa (Onishi 1949) Georg and Brown 1967
Description: Bergey 8.

R. dentocariosa (Onishi 1949) Georg and Brown 1967
Onishi, M. 1949. Journal of Dental Research **6**:273-282; Georg, L.K. and J.M. Brown. 1967. International Journal of Systematic Bacteriology **17**:79-88. **Type strain:** ATCC 17931
Description: Bergey 8.

Ruminococcus Sijpesteijn 1948
Sijpesteijn, A.K. 1948. Cellulose-decomposing bacteria from the rumen of cattle. Thesis. University of Leiden, The Netherlands. p. 152. **Type species:** *R. flavefaciens* Sijpesteijn 1948
Description: Bergey 8.

R. albus Hungate 1957
Hungate, R.E. 1957. Canadian Journal of Microbiology **3**:289-311. **Type strain:** ATCC 27210
Description: Holdeman, L.V., E.P. Cato and W.E.C. Moore (eds). 1977. Anaerobe Laboratory Manual, 4th ed. Virginia Polytechnic Institute and State University, Blacksburg, Virginia. pp. 1-156.

R. bromii Moore, Cato and Holdeman 1972
Moore, W.E.C., E.P. Cato and L.V. Holdeman. 1972. International Journal of Systematic Bacteriology **22**:78-80. **Type strain:** ATCC 27255
Description: Holdeman, L.V., E.P. Cato and W.E.C. Moore (eds). 1977. Anaerobe Laboratory Manual, 4th ed. Virginia Polytechnic Institute and State University, Blacksburg, Virginia. pp. 1-156.

R. callidus Holdeman and Moore 1974
Holdeman, L.V. and W.E.C. Moore. 1974. International Journal of Systematic Bacteriology **24**:260-277. **Type strain:** VPI 57-31
Description: Holdeman, L.V. and W.E.C. Moore. 1974. Ibid.

R. flavefaciens Sijpesteijn 1948
Sijpesteijn, A.K. 1948. Cellulose-decomposing bacteria from the rumen of cattle. Thesis, University of Leiden, The Netherlands. **Type strain:** ATCC 19208
Description: Holdeman, L.V., E.P. Cato and W.E.C. Moore (eds). 1977. Anaerobe Laboratory Manual, 4th ed. Virginia Polytechnic Institute and State University, Blacksburg, Virginia. pp. 1-156.

R. gnavus Moore, Johnson and Holdeman 1976
Moore, W.E.C., J.L. Johnson and L.V. Holdeman. 1976. International Journal of Systematic Bacteriology **26**:238-252. **Type strain:** ATCC 29149
Description: Moore, W.E.C., J.L. Johnson and L.V. Holdeman. 1976. Ibid.

R. lactaris Moore, Johnson and Holdeman 1976
Moore, W.E.C., J.L. Johnson and L.V. Holdeman. 1976. International Journal of Systematic Bacteriology **26**:238-252. **Type strain:** ATCC 29176
Description: Moore, W.E.C., J.L. Johnson and L.V. Holdeman. 1976. Ibid.

R. obeum Moore, Johnson and Holdeman 1976
Moore, W.E.C., J.L. Johnson and L.V. Holdeman. 1976. International Journal of Systematic Bacteriology **26**:238-252. **Type strain:** ATCC 29174
Description: Moore, W.E.C., J.L. Johnson and L.V. Holdeman. 1976. Ibid.

R. torques Holdeman and Moore 1974
Holdeman, L.V. and W.E.C. Moore. 1974. International Journal of Systematic Bacteriology **24**:260-277. **Type strain:** ATCC 27756
Description: Holdeman, L.V. and W.E.C. Moore. 1976. Ibid.

Runella Larkin and Williams 1978
Larkin, J.M. and P.M. Williams. 1978. International Journal of Systematic Bacteriology **28**:32-36. **Type species:** *R. slithyformis* Larkin and Williams 1978
Description: Larkin, J.M and P.M. Williams. 1978. Ibid.

R. slithyformis Larkin and Williams 1978
Larkin, J.M. and P.M. Williams. 1978. International Journal of Systematic Bacteriology **28**:32-36. **Type strain:** ATCC 29530
Description: Larkin, J.M. and P.M. Williams. 1978. Ibid.

Saccharomonospora Nonomura and Ohara 1971
Nonomura, H. and Y. Ohara. 1971. Journal of Fermentation Technology **49**:895-903. **Type species:** *S. viridis* (Schuurmans, Olson and San Clemente 1956) Nonomura and Ohara 1971

Description: Nonomura, H. and Y. Ohara. 1971. Ibid.

S. viridis (Schuurmans, Olson and San Clemente 1956) Nonomura and Ohara 1971
Schuurmans, D.M., B.H. Olson and C.L. San Clemente. 1956. Applied Microbiology **4**:61-66; Nonomura, H. and Y. Ohara. 1971. Journal of Fermentation Technology **49**:895-903. **Type strain:** ATCC 15386
Description: Nonomura, H. and Y. Ohara. 1971. Ibid.

Saccharopolyspora Lacey and Goodfellow 1975
Lacey, J. and M. Goodfellow. 1975. Journal of General Microbiology **88**:75-85. **Type species:** *S. hirsuta* Lacey and Goodfellow 1975
Description: Lacey, J. and M. Goodfellow. 1975. Ibid.

S. hirsuta Lacey and Goodfellow 1975
Lacey, J. and M. Goodfellow. 1975. Journal of General Microbiology **88**:75-85. **Type strain:** ATCC 27875
Description: Lacey, J. and M. Goodfellow. 1975. Ibid.

Salmonella Lignieres 1900
Lignieres, J. 1900. Bulletin of the Society for Central Medical Veterinarians ns **18**: 389-431. **Type species:** *S. choleraesuis* (Smith 1894) Weldin 1927
Description: Bergey 8.

S. arizonae (Borman 1957) Kauffman 1964
Borman, E.K. 1957 in Breed, R.S., E.G.D. Murray and N.R. Smith (eds). Bergey's Manual of Determinative Bacteriology, 7th ed. The Williams and Wilkins Co., Baltimore. pp. 346-348; Kauffman, F. 1964 in van Oye, E. The world problem of slamonellosis. Junk Publications, The Hague. **Type strain:** ATCC 13314
Description: Kauffman, F. 1964. Ibid.

S. choleraesuis (Smith 1894) Weldin 1927
Smith, T. 1894. Bulletin of the United States Bureau of Animal Industry **6**:6-40; Weldin, J.C. 1927. Iowa State Journal of Science **1**:121-197. **Type strain:** ATCC 13312
Description: Bergey 8.

S. enteritidis (Gaertner 1888) Castellani and Chalmers 1919
Gaertner, E. 1888. Korrespondenzblatt des Allgemeinen arztilchen Vereins von Thuringen **17**:573-600; Castellani, A. and A.J. Chalmers. 1919. Manual of tropical medicine, 3rd ed. Williams, Wood and Co., New York. p. 939. **Type strain:** ATCC 13076
Description: Bergey 8.

S. typhi (Schroeter 1886) Warren and Scott 1930
Schroeter, J. 1885-1889 in Cohn, F. Kryptogamenflora von Schlesien. Band 3, Heft 3, Pilze. J.U. Kern's Verlag, Breslau. pp. 1-814; Warren S.H. and W.M. Scott. 1930. Journal of Hygiene **29**:415-417. **Type strain:** ATCC 19430
Description: Bergey 8.

S. typhimurium (Loeffler 1892) Castellani and Chalmers 1919
Loeffler, F. 1892. Zentralblatt fur Bakteriologie, Parasitenkunde, Infektionskrankheiten und Hygiene. Abteilung I. **11**:129-141; Castellani, A. and A.J. Chalmers. 1919. Manual of tropical medicine, 3rd ed. Williams, Wood and Co., New York. p. 939. **Type strain:** ATCC 13311
Description: Bergey 8.

Saprospira Gross 1911
Gross, J. 1911. Mitteilungen aus der Zoologischen Station zu Neapel **20**:188-203. **Type species:** *S. grandis* Gross 1911
Description: Bergey 8.

S. grandis Gross 1911
Gross, J. 1911. Mitteilungen aus der Zoologischen Station zu Neapel **20**:188-203. **Type strain:** ATCC 23119
Description: Bergey 8.

Sarcina Goodsir 1842
Goodsir, J. 1842. Edinburgh Medical Surgery Journal **57**:430-443. **Type species:** *S. ventriculi* Goodsir 1842
Description: Bergey 8.

S. maxima Lindner 1888
Lindner, P. 1888. Inaugural Dissertation, Friedrich-Wilhems Universitat, pp. 1-59. **Type strain:** DSM 316
Description: Bergey 8.

S. ventriculi Goodsir 1842
Goodsir, J. 1842. Edinburgh Medical Surgery Journal **57**:430-443. **Type strain:** ATCC 19633
Description: Holdeman, L.V., E.P. Cato and W.E.C. Moore (eds). 1977. Anaerobe

Laboratory Manual, 4th ed. Virginia Polytechnic Institute and State University, Blacksburg, Virginia. pp. 1-156.

Selenomonas von Prowazek 1913
von Prowazek, S. 1913. Zentralblatt fur Bakteriologie Parasitenkunde, Infektionskrankheiten und Hygiene. Abteilung I. **70**:32-36. **Type species:** *S. sputigena* (Flugge 1886) Boskamp 1922
Description: Bergey 8.

S. ruminantium (Certes 1889) Wenyon 1926
Certes, A. 1889. Bulletin de la Societe Zoologie Francaise **14**:70-73; Wenyon, C.M. 1926. Protozoology, Vol. 1. Bailliere, Tindall and Cox, London. pp. 1-778. **Type strain:** ATCC 12561
Description: Holdeman, L.V., E.P. Cato and W.E.C. Moore (eds). 1977. Anaerobe Laboratory Manual, 4th ed. Virginia Polytechnic Institute and State University, Blacksburg, Virginia. pp. 1-156.

S. ruminantium subsp. *lactilytica* Bryant 1956
Bryant, M.P. 1956. Journal of Bacteriology **72**:162-167. **Type strain:** ATCC 19205
Description: Bergey 8.

S. ruminantium subsp. *ruminantium* Bryant 1974
Bryant, M.P. 1974 in Bergey 8. **Type strain:** ATCC 12561
Description: Bergey 8.

S. sputigena (Flugge 1886) Boskamp 1922
Flugge, C. 1886. Die Mikroorganismen. F.C.W. Vogel, Leipzig; Boskamp, E. 1922. Zentralblatt fur Bakteriologie Parasitenkunde, Infektionskrankheiten und Hygiene. Abteilung I. **88**:58-73. **Type strain:** VPI 10068
Description: Holdeman, L.V., E.P. Cato and W.E.C. Moore (eds). 1977. Anaerobe Laboratory Manual, 4th ed. Virginia Polytechnic Institute and State University, Blacksburg, Virginia. pp. 1-156.

Seliberia Aristovskaya and Parinkina 1963
Aristovskaya, T.V. and O.M. Parinkina. 1963. Bulletin of the Academy of Sciences USSR (Series Biology) **218**:49-56. **Type species:** *S. stellata* Aristovskaya and Parinkina 1963
Description: Bergey 8.

S. stellata Aristovskaya and Parinkina 1963
Aristovskaya, T.V. and O.M. Parinkina. 1963. Bulletin of the Academy of Sciences USSR (Series Biology) **218**:49-56. **Type strain:** INMI N-9
Description: Bergey 8.

Serpens Hespell 1977
Hespell, R.B. 1977. International Journal of Systematic Bacteriology **27**:371-381. **Type species:** *S. flexibilis* Hespell 1977
Description: Hespell, R.B. 1977. Ibid.

S. flexibilis Hespell 1977
Hespell, R.B. 1977. International Journal of Systematic Bacteriology **27**:371-381. **Type strain:** ATCC 29606
Description: Hespell, R.B. 1977. Ibid.

Serratia Bizio 1823
Bizio, B. 1823. Biblioteca Italiana o sia Giornale di Letteratura, Scienze e Arti (Anno VIII) **30**:275-295. **Type species:** *S. marcescens* Bizio 1823
Description: Bergey 8.

S. fonticola Gavini, Ferragut, Izard, Trinel, Leclerc, Lefebvre and Mossel 1979
Gavini, F., C. Ferragut, D. Izard, P.A. Trinel, H. Leclerc, B. Lefebvre and D.A.A. Mossel. 1979. International Journal of Systematic Bacteriology **29**:92-101. **Type strain:** ATCC 29844
Description: Gavini, F., C. Ferragut, D. Izard, P.A. Trinel, H. Leclerc, B. Lefebvre and D.A.A. Mossel. 1979. Ibid.

S. liquefaciens (Grimes and Hennerty 1931) Bascomb, Lapage, Willcox and Curtis 1971
Grimes, M. and P.J. Hennerty. 1931. Scientific Proceedings of the Royal Society Dublin **20**:89-97; Bascomb, S., S.P. Lapage, W.R. Willcox and M.A. Curtis. 1971. Journal of General Microbiology **66**:279-295. **Type strain:** ATCC 27592
Description: Bascomb, S., S.P. Lapage, W.R. Willcox and M.A. Curtis. 1971. Ibid.

S. marcescens Bizio 1823
Bizio, B. 1923. Biblioteca Italiana o sia Giornale di Letteratura, Scienze e Arti (Anno VIII) **30**:275-295. **Type strain:** ATCC 13880
Description: Grimont, P.A.D., F. Grimont, H.L.C. Dulong de Rosnay and P.H.A. Sneath. 1977. Journal of General Microbiology **98**:39-66.

S. marinorubra ZoBell and Upham 1944
ZoBell, C.E. and H.C. Upham. 1944. Bulletin Scripps Institution of Oceanography University of California (Technical Series) **5**:239-292. **Type strain:** ATCC 27593
Description: Grimont, P.A.D., F. Grimont, H.L.C. Dulong de Rosnay and P.H.A. Sneath. 1977. Journal of General Microbiology **98**:39-66.

S. odorifera Grimont, Grimont, Richard, Davis, Steigerwalt and Brenner 1978
Grimont, P.A.D., F. Grimont, C. Richard, B.R. Davis, A.G. Steigerwalt and D.J. Brenner. 1978. International Journal of Systematic Bacteriology **28**:453-463. **Type strain:** CDC 1979-77
Description: Grimont, P.A.D., F. Grimont, C. Richard, B.R. Davis, A.G. Steigerwalt and D.J. Brenner. 1978. Ibid.

S. plymuthica (Lehmann and Neumann 1896) Breed, Murray and Hitchens 1948
Lehmann, K.B. and R. Neumann. 1896. Atlas und Grundriss der Bakteriologie und Lehrbuch der speciellen bakteriologischen Diagnostik. 1st ed. J.F. Lehmann, Munchen; Breed, R.S., E.G.D. Murray and A.P. Hitchens. 1948. Bergey's Manual of Determinative Bacteriology, 6th ed. The Williams and Wilkins Co., Baltimore. pp. 1-1529. **Type strain:** ATCC 183
Description: Grimont, P.A.D., F. Grimont, H.L.C. Dulong de Rosnay and P.H.A. Sneath. 1977. Journal of General Microbiology **98**:39-66.

S. proteamaculans (Paine and Stansfield 1919) Grimont, Grimont and Starr 1978
Paine, S.G. and H. Stansfield. 1919. Annals of Applied Biology **6**:27-39; Grimont, P.A.D., F. Grimont and M.P. Starr. 1978. International Journal of Systematic Bacteriology **28**:503-510. **Type strain:** ATCC 19323
Description: Grimont, P.A.D., F. Grimont and M.P. Starr. 1978. Ibid.

S. rubidaea (Stapp 1940) Ewing, Davis, Fife and Lessel 1973
Stapp, C. 1940. Zentralblatt fur Bakteriologie, Parasitenkunde, Infektionskrankheiten und Hygiene. Abteilung II. **102**: 251-266; Ewing, W.H., B.R. Davis, M.A. Fife and E.F. Lessel. 1973. International Journal of Systematic Bacteriology **23**: 217-225. **Type strain:** ATCC 27593
Description: Ewing, W.H., B.R. Davis, M.A. Fife and E.F. Lessel. 1973. Ibid.

Shigella Castellani and Chalmers 1919
Castellani, A. and A.J. Chalmers. 1919. Manual of tropical medicine, 3rd ed. Williams, Wood and Co., New York. **Type species:** *S. dysenteriae* (Shiga 1898) Castellani and Chalmers 1919
Description: Bergey 8.

S. boydii Ewing 1949
Ewing, W.H. 1949. Journal of Bacteriology **57**:633-638. **Type strain:** ATCC 8700
Description: Ewing, W.H. 1949. Ibid.

S. dysenteriae (Shiga 1898) Castellani and Chalmers 1919
Shiga, K. 1898. Zentralblatt fur Bakteriologie, Parasitenkunde, Infektionskrankheiten und Hygiene. Abteilung I. **24**:817-828; Castellani, A. and A.J. Chalmers. 1919. Manual of tropical medicine, 3rd ed. Williams, Wood and Co., New York. p. 935. **Type strain:** ATCC 13313
Description: Bergey 8.

S. flexneri Castellani and Chalmers 1919
Castellani, A. and A.J. Chalmers. 1919. Manual of tropical medicine, 3rd ed. Williams, Wood and Co., New York. p. 937. **Type strain:** ATCC 29903
Description: Bergey 8.

S. sonnei (Levine 1920) Weldin 1927
Levine, M. 1920. Journal of Infectious Diseases **27**:31-39; Weldin, J.C. 1927. Iowa State Journal of Science **1**:121-197. **Type strain:** ATCC 29930
Description: Bergey 8.

Simonsiella Schmid 1922
Schmid, G. 1922 in Simons, H. Zentralblatt fur Bakteriologie, Parasitenkunde, Infektions- krankheiten und Hygiene. Abteilung I. **88**:501-510. **Type species:** *S. muelleri* Schmid 1922
Description: Bergey 8.

S. crassa Schmid 1922
Schmid, G. 1922 in Simons, H. Zentralblatt fur Bakteriologie, Parasitenkunde, Infektionskrankheiten und Hygiene. Abteilung I. **88**:501-510. **Type strain:** ATCC 15533
Description: Bergey 8.

S. muelleri Schmid 1922

Schmid, G. 1922 in Simons, H. Zentralblatt fur Bakteriologie, Parasitenkunde, Infektionskrankheiten und Hygiene. Abteilung I. **88**:501-510. **Type strain:** ATCC 29453
Description: Kuhn, D.A. and D.A. Gregory. 1978. Current Microbiology **1**:11-14.

S. steedae Kuhn and Gregory 1979
Kuhn, D.A. and D.A. Gregory. 1979 in International Journal of Systematic Bacteriology **29**:79-80; Effective publication Kuhn, D.A. and D.A. Gregory. 1978. Current Microbiology **1**:11-14. **Type strain:** ATCC 27409
Description: Kuhn, D.A. and D.A. Gregory. 1978. Ibid.

Sphaerotilus Kutzing 1833
Kutzing, F.T. 1833. Linnaea **8**:335-387. **Type species:** *S. natans* Kutzing 1833
Description: Bergey 8.

S. natans Kutzing 1833
Kutzing, F.T. 1833. Linnaea **8**:335-387. **Type strain:** ATCC 13338
Description: Stokes, J.L. 1954. Journal of Bacteriology **67**:278-291.

Spirillospora Couch 1963
Couch, N.J. 1963. Journal of the Elisha Mitchell Scientific Society **79**:53-70. **Type species:** *S. albida* Couch 1963
Description: Bergey 8.

S. albida Couch 1963
Couch, J.N. 1963. Journal of the Elisha Mitchell Scientific Society **79**:53-70. **Type strain:** ATCC 15331
Description: Bergey 8.

Spirillum Ehrenberg 1832
Ehrenberg, C.G. 1832. Abhandlungen der Koniglichen Akademie der Wissenschaften zu Berlin 1830, pp. 1-88. **Type species:** *S. volutans* Ehrenberg 1832
Description: Bergey 8.

S. volutans Ehrenberg 1832
Ehrenberg, C.G. 1832. Abhandlungen der Koniglichen Akademie der Wissenschaften zu Berlin 1830, pp. 1-88. **Type strain:** ATCC 19554
Description: Hylemon, P.B., J.S. Wells Jr., J.H. Bowdre, T.O. MacAdoo and N.R. Krieg. 1973. International Journal of Systematic Bacteriology **23**:20-27.

Spirochaeta Ehrenberg 1835
Ehrenberg, C.G. 1835. Abhandlungen der Preussischen Akademie der Wissenschaften (Berlin) aus den Jahre 1833-1835, pp. 143-336. **Type species:** *S. plicatilis* Ehrenberg 1835
Description: Bergey 8.

S. halophila Greenberg and Canale-Parola 1977
Greenberg, E.P. and E. Canale-Parola. 1977 in International Journal of Systematic Bacteriology **27**:306; Effective publication Greenberg, E.P. and E. Canale-Parola. 1976. Archives of Microbiology **110**: 185-194. **Type strain:** ATCC 29478
Description: Greenberg, E.P. and E. Canale-Parola. 1976. Ibid.

S. plicatilis Ehrenberg 1835
Ehrenberg, C.G. 1835. Abhandlungen der Preussischen Akademie der Wissenschaften (Berlin) aus den Jahre 1833-1835, pp. 143-336. **Type strain:** no culture available.
Description: Bergey 8.

S. stenostrepta Zuelzer 1912
Zuelzer, M. 1912. Archiv fur Protistenkunde **24**:1-59. **Type strain:** ATCC 25083
Description: Bergey 8.

Spiroplasma Saglio, Lhospital, Lafleche, Dupont, Bove, Tully and Freundt 1973
Saglio, P., M. Lhospital, D. Lafleche, G. Dupont, J.M. Bove, J.G. Tully and E.A. Freundt. 1973. International Journal of Systematic Bacteriology **23**:191-204. **Type species:** *S. citri* Saglio, Lhospital, Lafleche, Dupont, Bove, Tully and Freundt 1973
Description: Bergey 8.

S. citri Saglio, Lhospital, Lafleche, Dupont, Bove, Tully and Freundt 1973
Saglio, P., M. Lhospital, D. Lafleche, G. Dupont, J.M. Bove, J.G. Tully and E.A. Freundt. 1973. International Journal of Systematic Bacteriology **23**:191-204. **Type strain:** ATCC 27556
Description: Saglio, P., M. Lhospital, D. Lafleche, G. Dupont, J.M. Bove, J.G. Tully and E.A. Freundt. 1973. Ibid.

Spirosoma Migula 1894
Migula, W. 1894. Arbeiten aus dem Bakteriologischen Institut der Technischen Hochschule zu Karlsruhe **1**:235-238. **Type species:** *S. linguale* Migula 1894

Description: Larkin, J.M., P.M. Williams and R. Taylor. 1977. International Journal of Systematic Bacteriology **27**:147-156.

S. linguale Migula 1894
Migula, W. 1894. Arbeiten aus dem Bakteriologischen Institut der Technischen Hochschule zu Karlsruhe **1**:235-238. **Type strain:** DSM 74
Description: Larkin, J.M., P.M. Williams and R. Taylor. 1977. International Journal of Systematic Bacteriology **27**:147-156.

Sporichthya Lechevalier, Lechevalier and Holbert 1968
Lechevalier, M.P., H.A. Lechevalier and P.E. Holbert. 1968. Annales de l'Institut Pasteur (Paris) **114**:277-285. **Type species:** *S. polymorpha* Lechevalier, Lechevalier and Holbert 1968
Description: Bergey 8.

S. polymorpha Lechevalier, Lechevalier and Holbert 1968
Lechevalier, H.A., M.P. Lechevalier and P.E. Holbert. 1968. Annales de l'Institut Pasteur (Paris) **114**:277-285. **Type strain:** ATCC 23823
Description: Lechevalier, M.P., H.A. Lechevalier and P.E. Holbert. 1968. Ibid.

Sporocytophaga Stanier 1940
Stanier, R.Y. 1940. Journal of Bacteriology **40**:619-636. **Type species:** *S. myxococcoides* (Krzemieniewska 1933) Stanier 1940
Description: Bergey 8.

S. myxococcoides (Krzemieniewska 1933) Stanier 1940
Krzemieniewska, H. 1933. Archiv fur Mikrobiologie **4**:394-408; Stanier, R.Y. 1940. Journal of Bacteriology **40**:619-636. **Type strain:** ATCC 10010
Description: Bergey 8.

Sporolactobacillus Kitahara and Suzuki 1963
Kitahara, K. and J. Suzuki. 1963. Journal of General and Applied Microbiology **9**: 59-71. **Type species:** *S. inulinus* (Kitahara and Suzuki 1963) Kitahara and Lai 1967
Description: Bergey 8.

S. inulinus (Kitahara and Suzuki 1963) Kitahara and Lai 1967
Kitahara, K. and J. Suzuki. 1963. Journal of General and Applied Microbiology **9**: 59-71; Kitahara, K. and C.L. Lai. 1967. Journal of General and Applied Microbiology **13**:197-203. **Type strain:** ATCC 15538
Description: Holdeman, L.V., E.P. Cato and W.E.C. Moore (eds). 1977. Anaerobe Laboratory Manual, 4th ed. Virginia Polytechnic Institute and State University, Blacksburg, Virginia. pp. 1-156.

Sporosarcina Kluyver and van Niel 1936
Kluyver, A.J. and C.B. van Niel. 1936. Zentralblatt fur Bakteriologie, Parasitenkunde, Infektionskrankheiten und Hygiene. Abteilung II. **94**:369-403. **Type species:** *S. ureae* (Beijerinck 1901) Kluyver and van Niel 1936
Description: Bergey 8.

S. ureae (Beijerinck 1901) Kluyver and van Niel 1936
Beijerinck, M.W. 1901. Zentralblatt fur Bakteriologie, Parasitenkunde, Infektionskrankheiten und Hygiene. Abteilung II. **7**:33-61; Kluyver, A.J. and C.B. van Niel. 1936. Zentralblatt fur Bakteriologie, Parasitenkunde, Infektionskrankheiten und Hygiene. Abteilung II. **94**:369-403. **Type strain:** ATCC 6473
Description: Bergey 8.

Staphylococcus Rosenbach 1884
Rosenbach, F.J. 1884. Microorganismen bei den Wund-Infections-Krankheiten des Menschen. J.F. Bergmann, Wiesbaden. pp. 1-122. **Type species:** *S. aureus* Rosenbach 1884
Description: Bergey 8.

S. aureus Rosenbach 1884
Rosenbach, F.J. 1884. Microorganismen bei den Wund-Infections-Krankheiten des Menschen. J.F. Bergmann, Wiesbaden. pp. 1-122. **Type strain:** ATCC 12600
Description: Bergey 8.

S. capitis Kloos and Schleifer 1975
Kloos, W.E. and K.H. Schleifer. 1975. International Journal of Systematic Bacteriology **25**:62-79. **Type strain:** ATCC 27840
Description: Kloos, W.E. and K.H. Schleifer. 1975. Ibid.

S. cohnii Schleifer and Kloos 1975
Schleifer, K.H. and W.E. Kloos. 1975. International Journal of Systematic Bac-

teriology **25**:50-61. **Type strain:** ATCC 29974
Description: Schleifer, K.H. and W.E. Kloos. 1975. Ibid.

S. epidermidis (Winslow and Winslow 1908) Evans 1916
Winslow, C.E.A. and A.R. Winslow. 1908. The systematic relationships of the Coccaceae. John Wiley, New York; Evans, A.C. 1916. Journal of Infectious Diseases **18**:437-476. **Type strain:** ATCC 14990
Description: Bergey 8.

S. haemolyticus Schleifer and Kloos 1975
Schleifer, K.H. and W.E. Kloos. 1975. International Journal of Systematic Bacteriology **25**:50-61. **Type strain:** ATCC 29970
Description: Schleifer, K.H. and W.E. Kloos. 1975. Ibid.

S. hominis Kloos and Schleifer 1975
Kloos, W.E. and K.H. Schleifer. 1975. International Journal of Systematic Bacteriology **25**:62-79. **Type strain:** ATCC 27844
Description: Kloos, W.E. and K.H. Schleifer. 1975. Ibid.

S. hyicus (Sompolinski 1953) Devriese, Hajek, Oeding, Meyer and Schleifer 1978
Sompolinski, D. 1953. Schweizerishces Archiv fur Tierheilkunde **95**:302-339; Devriese, L.A., V. Hajek, P. Oeding, S.A. Meyer and K.H. Schleifer. 1978. International Journal of Systematic Bacteriology **28**:482-490. **Type strain:** ATCC 11249
Description: Devriese, L.A., V. Hajek, P. Oeding, S.A. Meyer and K.H. Schleifer. 1978. Ibid.

S. hyicus subsp. *chromogenes* Devriese, Hajek, Oeding, Meyer and Schleifer 1978
Devriese, L.A., V. Hajek, P. Oeding, S.A. Meyer and K.H. Schleifer. 1978. International Journal of Systematic Bacteriology **28**:482-490. **Type strain:** NCTC 10530
Description: Devriese, L.A., V. Hajek, P. Oeding, S.A. Meyer and K.H. Schleifer. 1978. Ibid.

S. hyicus subsp. *hyicus* (Sompolinski 1953) Devriese, Hajek, Oeding, Meyer and Schleifer 1978
Sompolinski, D. 1953. Schweizerisches Archiv fur Tierheilkunde **95**:302-339; Devriese, L.A., V. Hajek, P. Oeding, S.A. Meyer and K.H. Schleifer. 1978. International Journal of Systematic Bacteriology **28**:482-490. **Type strain:** ATCC 11249
Description: Devriese, L.A., V. Hajek, P. Oeding, S.A. Meyer and K.H. Schleifer. 1978. Ibid.

S. intermedius Hajek 1976
Hajek, V. 1976. International Journal of Systematic Bacteriology **26**:401-408. **Type strain:** ATCC 29663
Description: Hajek, V. 1976. Ibid.

S. saprophyticus (Fairbrother 1940) Shaw, Stitt and Cowan 1951
Fairbrother, R.W. 1940. Journal of Pathology and Bacteriology **50**:83-88; Shaw, C., J.M. Stitt and S.T. Cowan. 1951. Journal of General Microbiology **5**:1010-1023. **Type strain:** ATCC 15305
Description: Bergey 8.

S. sciuri Kloos, Schleifer and Smith 1976
Kloos, W.E., K.H. Schleifer and R.F. Smith. 1976. International Journal of Systematic Bacteriology **26**:22-37. **Type strain:** ATCC 29062
Description: Kloos, W.E., K.H. Schleifer and R.F. Smith. 1976. Ibid.

S. sciuri subsp. *lentus* Kloos, Schleifer and Smith 1976
Kloos, W.E., K.H. Schleifer and R.F. Smith. 1976. International Journal of Systematic Bacteriology **26**:22-37. **Type strain:** ATCC 29070
Description: Kloos, W.E., K.H. Schleifer and R.F. Smith. 1976. Ibid.

S. sciuri subsp. *sciuri* Kloos, Schleifer and Smith 1976
Kloos, W.E., K.H. Schleifer and R.F. Smith. 1976. International Journal of Systematic Bacteriology **26**:22-37. **Type strain:** ATCC 29062
Description: Kloos, W.E., K.H. Schleifer and R.F. Smith. 1976. Ibid.

S. simulans Kloos and Schleifer 1975
Kloos, W.E. and K.H. Schleifer. 1975. International Journal of Systematic Bacteriology **25**:62-79. **Type strain:** ATCC 27848
Description: Kloos, W.E. and K.H. Schleifer. 1975. Ibid.

S. warneri Kloos and Schleifer 1975
Kloos, W.E. and K.H. Schleifer. 1975. International Journal of Systematic Bac-

teriology **25**:62-79. **Type strain:** ATCC 27836
Description: Kloos, W.E. and K.H. Schleifer. 1975. Ibid.

S. xylosus Schleifer and Kloos 1975
Schleifer, K.H. and W.E. Kloos. 1975. International Journal of Systematic Bacteriology **25**:50-61. **Type strain:** ATCC 29971
Description: Schleifer, K.H. and W.E. Kloos. 1975. Ibid.

Stibiobacter Lyalikova 1974
Lyalikova, N.N. 1974. Mikrobiologiya **43**: 799-805. **Type species:** *S. senarmontii* Lyalikova 1974
Description: Lyalikova, N.N. 1974. Ibid.

S. senarmontii Lyalikova 1974
Lyalikova, N.N. 1974. Mikrobiologiya **43**: 799-805. **Type strain:** VKM MV-1130
Description: Lyalikova, N.N. 1974. Ibid.

Stigmatella Berkeley and Curtis 1875
Berkeley, M.J. and Curtis. 1875 in Berkeley, M.J. Grevillea **3**:97-112. **Type species:** *S. aurantiaca* Berkeley and Curtis 1875
Description: Bergey 8.

S. aurantiaca Berkeley and Curtis 1875
Berkeley, M.J. and Curtis. 1875 in Berkeley, M.J. Grevillea **3**:97-112. **Type strain:** ATCC 25190
Description: Reichenbach, H. and M. Dworkin. 1969. Journal of General Microbiology **58**:3-14.

S. erecta (Schroeter 1886) McCurdy 1971
Schroeter, J. 1885-1889 in Cohn, F. Kryptogamenflora von Schlesien. Band 3, Heft 3, Pilze J.U. Kern's Verlag, Breslau. pp. 1-814; McCurdy, H.D. 1971. International Journal of Systematic Bacteriology **21**: 40-49. **Type strain:** ATCC 25191
Description: McCurdy, H.D. 1971. Ibid.

Streptobacillus Levaditi, Nicolau and Poincloux 1925
Levaditi, C., S. Nicolau and P. Pioncloux. 1925. Compte Rendu Hebdomadaire des Seances de l'Academie des Sciences (Paris) **180**:1188-1190. **Type species:** *S. moniliformis* Levaditi, Nicolau and Poincloux 1925
Description: Bergey 8.

S. moniliformis Levaditi, Nicolau and Poincloux 1925
Levaditi, C., S. Nicolau and P. Poincloux. 1925. Compte Rendu Hebdomadaire des Seances de l'Academie des Sciences (Paris) **180**:1188-1190. **Type strain:** ATCC 14647
Description: Bergey 8.

Streptococcus Rosenbach 1884
Rosenbach, F.J. 1884. Microorganismen bei den Wund-Infections- Krankheiten des Menschen. K.F. Bergmann, Wiesbaden, pp. 1-122. **Type species:** *S. pyogenes* Rosenbach 1884
Description: Bergey 8.

S. acidominimus Ayers and Mudge 1922
Ayers, S.H. and C.S. Mudge. 1922. Journal of Infectious Diseases **31**:40-50. **Type strain:** NCDO 2025
Description: Bergey 8.

S. agalactiae Lehmann and Neumann 1896
Lehmann, K.B. and R. Neumann. 1896. Atlas und Grundriss der Bakteriologie und Lehrbuch der speziellen bakteriologischen Diagnostik. 1st ed. J.F. Lehmann, Munchen. **Type strain:** NCTC 8181
Description: Bergey 8.

S. anginosus (Andrewes and Horder 1906) Smith and Sherman 1938
Andrewes, F.W. and T.J. Horder. 1906. Lancet **2**:708-713; Smith, F.R. and J.M. Sherman. 1939. Journal of Infectious Diseases **62**:186-189. **Type strain:** NCTC 10713
Description: Bergey 8.

S. bovis Orla-Jensen 1919
Orla-Jensen, S. 1919. The lactic acid bacteria. Host, Copenhagen. pp. 1-196. **Type strain:** NCDO 597
Description: Garvie, E.I. and A.J. Bramley. 1979. Journal of Applied Bacteriology **46**:557-566.

S. constellatus (Prevot 1924) Holdeman and Moore 1974
Prevot, A.R. 1924. Comptes Rendus des Seances de la Societe de Biologie (Paris) **91**:426-428; Holdeman, L.V. and W.E.C. Moore. 1974. International Journal of Systematic Bacteriology **24**:260-277. **Type strain:** ATCC 27823
Description: Holdeman, L.V. and W.E.C. Moore. 1974. Ibid.

S. cremoris Orla-Jensen 1919
Orla-Jensen, S. 1919. The lactic acid bac-

teria. Host, Copenhagen. pp. 1-196. **Type strain:** NCDO 607
Description: Whitehead, H.R., C.A.E. Briggs, E.I. Garvie and L.G.M. Newland. 1956. Journal of Dairy Research **23**:315-318.

S. cricetus Coykendall 1977
Coykendall, A.L. 1977. International Journal of Systematic Bacteriology **27**:26-30. **Type strain:** ATCC 19642
Description: Coykendall, A.L. 1977. Ibid.

S. equi Sand and Jensen 1888
Sand, G. and C.O. Jensen. 1888. Deutsche Zeitschrift fur Tiermedizin und Vergleichende Pathologie **13**:437-464. **Type strain:** NCTC 9682
Description: Bergey 8.

S. equinus Andrewes and Horder 1906
Andrewes, F.W. and T.J. Horder. 1906. Lancet **2**:708-713. **Type strain:** ATCC 9812
Description: Fuller, R. and L.G.M. Newland. 1963. Journal of General Microbiology **31**:431-434.

S. faecalis Andrewes and Horder 1906
Andrewes, F.W. and T.J. Horder. 1906. Lancet **2**:708-713. **Type strain:** NCTC 775
Description: Shattock, P.Y.F. and A.T.R. Mattick. 1943. Journal of Hygiene Cambridge **43**:173-188.

S. faecium Orla-Jensen 1919
Orla-Jensen, S. 1919. The lactic acid bacteria. Host, Copenhagen. pp. 1-196. **Type strain:** NCTC 7171
Description: Bergey 8.

S. hansenii Holdeman and Moore 1974
Holdeman, L.V. and W.E.C. Moore. 1974. International Journal of Systematic Bacteriology **24**:260-277. **Type strain:** ATCC 27752
Description: Holdeman, L.V. and W.E.C. Moore. 1974. Ibid.

S. iniae Pier and Madin 1976
Pier, G.B. and S.H. Madin. 1976. International Journal of Systematic Bacteriology **26**:545-553. **Type strain:** ATCC 29178
Description: Pier, G.B. and S.H. Madin. 1976. Ibid.

S. intermedius Prevot 1925
Prevot, A.R. 1925. Annales de l'Institut Pasteur (Paris) **39**:417-447. **Type strain:** ATCC 27335
Description: Holdeman, L.V., E.P. Cato and W.E.C. Moore (eds). 1977. Anaerobe Laboratory Manual, 4th ed. Virginia Polytechnic Institute and State University, Blacksburg, Virginia. pp. 1-156.

S. lactis (Lister 1873) Lohnis 1909
Lister, J. 1873. Quarterly Journal of Microbiological Science **13**:380-408; Lohnis, F. 1909. Zentralblatt fur Bakteriologie, Parasitenkunde, Infektionskrankheiten und Hygiene. Abteilung II. **22**:553-555. **Type strain:** NCDO 604
Description: Shattock, P.Y.F. and A.T.R. Mattick. 1943. Journal of Hygiene Cambridge **43**:173-188.

S. mitis Andrewes and Horder 1906
Andrewes, F.W. and T.J. Horder. 1906. Lancet **2**:708-713. **Type strain:** NCTC 3165
Description: Bergey 8.

S. morbillorum (Prevot 1933) Holdeman and Moore 1974
Prevot, A.R. 1933. Annales des sciences naturelles **15**:23-260; Holdeman, L.V. and W.E.C. Moore. 1974. International Journal of Systematic Bacteriology **24**:260-277. **Type strain:** ATCC 27824
Description: Holdeman, L.V. and W.E.C. Moore. 1974. Ibid.

S. mutans Clarke 1924
Clarke, J.K. 1924. British Journal of Experimental Pathology **5**:141-147. **Type strain:** ATCC 25175
Description: Coykendall, A.L. 1977. International Journal of Systematic Bacteriology **27**:26-30.

S. pleomorphus Barnes, Impey, Stevens and Peel 1979
Barnes, E.M., C.S. Impey, B.J.H. Stevens and J.L. Peel. 1977. International Journal of Systematic Bacteriology **29**:79-80; Effective publication Barnes, E.M., C.S. Impey, B.J.H. Stevens and J.L. Peel. 1977. Journal of General Microbiology **102**:45-53. **Type strain:** NCTC 11087
Description: Barnes, E.M., C.S. Impey, B.J.H. Stevens and J.L. Peel. 1977. Ibid.

S. pneumoniae (Klein 1884) Chester 1901
Klein, E. 1884. Practitioner **32**:321-352; Chester, F.D. 1901. A manual of determinative bacteriology. The MacMillan Co., New York. pp. 1-401. **Type strain:** NCTC 7465
Description: Bergey 8.

S. pyogenes Rosenbach 1884
Rosenbach, F.J. 1884. Microorganismen bei den Wund-Infections- Krankheiten des Menschen. J.F. Bergmann, Wiesbaden pp. 1-122. **Type strain:** ATCC 12344
Description: Bergey 8.

S. raffinolactis Orla-Jensen and Hansen 1932
Orla-Jensen, A.D. and P.A. Hansen. 1932. Zentralblatt fur Bakteriologie, Parasitenkunde, Infektionskrankheiten und Hygiene. Abteilung II. **86**:6-29. **Type strain:** NCDO 617
Description: Garvie, E.I. 1979. International Journal of Systematic Bacteriology **29**:152.

S. rattus Coykendall 1977
Coykendall, A.L. 1977. International Journal of Systematic Bacteriology **27**:26-30. **Type strain:** ATCC 19645
Description: Coykendall, A.L. 1977. Ibid.

S. salivarius Andrewes and Horder 1906
Andrewes, F.W. and T.J. Horder. 1906. Lancet **2**:708-713. **Type strain:** ATCC 7073
Description: Bergey 8.

S. sanguis White and Niven 1946
White, J.C. and C.F. Niven Jr. 1946. Journal of Bacteriology **51**:717-722. **Type strain:** ATCC 10556
Description: Bergey 8.

S. thermophilus Orla-Jensen 1919
Orla-Jensen, S. 1919. The lactic acid bacteria. Host, Copenhagen. pp. 1-196. **Type strain:** NCDO 573
Description: Bergey 8.

S. uberis Diernhofer 1932
Diernhofer, K. 1932. Milchwirtschaftliche Forschungen **13**:368-374. **Type strain:** NCTC 3858
Description: Garvie, E.I. and A.J. Bramley. 1979. Journal of Applied Bacteriology **46**:295-304.

Streptomyces Waksman and Henrici 1943
Waksman, S.A. and A.T. Henrici. 1943. Journal of Bacteriology **46**:337-341. **Type species:** *S. albus* (Rossi Doria 1891) Waksman and Henrici 1943
Description: Bergey 8.

S. aburaviensis Nishimura, Kimura, Tawara, Sasaki, Nakajima, Shimaoka, Okamoto, Shimohara and Isono 1957
Nishimura, H., T. Kimura, K. Tawara, K. Sasaki, K. Nakajima, N. Shimaoka, S. Okamoto, M. Shimohara and J. Isono. 1957. Journal of Antibiotics (Tokyo) Series A **10**:205-212. **Type strain:** ATCC 23869
Description: Shirling, E.B. and D. Gottlieb. 1968. International Journal of Systematic Bacteriology **18**:279-392.

S. achromogenes Okami and Umezawa 1953
Okami, Y. and H. Umezawa. 1953 in Umezawa, H., T. Takeuchi, Y. Okami and T. Tazaki. Japanese Journal of Medical Science and Biology **6**:261-268. **Type strain:** ATCC 12767
Description: Shirling, E.B. and D. Gottlieb. 1968. International Journal of Systematic Bacteriology **18**:69-189.

S. achromogenes subsp. *achromogenes* Okami and Umezawa 1953
Okami, Y. and H. Umezawa. 1953 in Umezawa, H., T. Takeuchi, Y. Okami and T. Tazaki. Japanese Journal of Medical Science and Biology **6**:261-268. **Type strain:** ATCC 12767
Description: Shirling, E.B. and D. Gottlieb. 1968. International Journal of Systematic Bacteriology **18**:69-189.

S. achromogenes subsp. *rubradiris* Bhuyan, Owen and Dietz 1965
Bhuyan, B.K., S.P. Owen and A. Dietz. 1965. Antimicrobial Agents and Chemotherapy **1964**:91-96. **Type strain:** CBS 566.-70
Description: Bhuyan, B.K., S.P. Owen and A. Dietz. 1965. Ibid.

S. acrimycini (Preobrazhenskaya, Blinov and Ryabova 1957) Pridham, Hesseltine and Benedict 1958
Preobrazhenskaya, T.P., N.O. Blinov and I.D. Ryabova. 1957 in Gauze, G.F., T.P. Preobrazhenskaya, E.S. Kudrina, N.O. Blinov, I.D. Ryabova and M.A. Sveshnikova. Problems of classification of actinomycetes-antagonists. Government Publishing House of Medical Literature, Medgiz, Moscow. 1-398; Pridham, T.G., C.W. Hesseltine and R.G. Benedict. 1958. Applied Microbiology **6**:52-79. **Type strain:** ATCC 19885
Description: Shirling, E.B. and D. Gottlieb. 1968. International Journal of Systematic Bacteriology **18**:69-189.

S. afghaniensis Shimo, Shiga, Tomosugi and

Kamoi 1959
Shimo, M., T. Shiga, T. Tomosugi and I. Kamoi. 1959. Journal of Antibiotics (Tokyo) Series A **12**:1-6. **Type strain:** ATCC 23871
Description: Shirling, E.B. and D. Gottlieb. 1968. International Journal of Systematic Bacteriology **18**:279-392.

S. alanosinicus Thiemann and Beretta 1966
Thiemann, J.E. and G. Beretta. 1966. Journal of Antibiotics (Tokyo) Series A **19**:155-160. **Type strain:** ATCC 15710
Description: Shirling, E.B. and D. Gottlieb. 1972. International Journal of Systematic Bacteriology **22**:265-394.

S. albaduncus Tsukiura, Okanishi, Ohmori, Koshiyama, Miyaki, Kitazima and Kawaguchi 1964
Tsukiura, H., M. Okanishi, T. Ohmor, H. Koshiyama, T. Miyaki, H. Kitazima and H. Kawaguchi. 1964. Journal of Antibiotics (Tokyo) Series A **17**:39-47. **Type strain:** ATCC 14698
Description: Shirling, E.B. and D. Gottlieb. 1972. International Journal of Systematic Bacteriology **22**:265-394.

S. albidoflavus (Rossi Doria 1891) Waksman and Henrici 1948
Rossi Doria, T. 1891. Annali dell' Istituto d'Igiene Sperimentale della Rendiconti Universita di Roma **1**:399-438; Waksman, S.A. and A.T. Henrici. 1948 in Breed, R.S., E.G.D. Murray and A.P. Hitchens (eds). Bergey's Manual of Determinative Bacteriology, 6th ed. The Williams and Wilkins Co., Baltimore. pp. 929-980. **Type strain:** ATCC 25422
Description: Shirling, E.B. and D. Gottlieb. 1969. International Journal of Systematic Bacteriology **19**:391-512.

S. albofaciens Thirumalachar and Bhatt 1960
Thirumalachar, M.J. and V.V. Bhatt. 1960. Hindustan Antibiotics Bulletin **3**:61-63. **Type strain:** ATCC 25184
Description: Shirling, E.B. and D. Gottlieb. 1968. International Journal of Systematic Bacteriology **18**:279-392.

S. alboflavus (Waksman and Curtis 1916) Waksman and Henrici 1948
Waksman, S.A. and R.E. Curtis. 1916. Soil Science **1**:99-134; Waksman, S.A. and A.T. Henrici. 1948 in Breed, R.S., E.G.D. Murray and A.P. Hitchens (eds). Bergey's Manual of Determinative Bacteriology, 6th ed. The Williams and Wilkins Co., Baltimore. pp. 929-980. **Type strain:** ATCC 12626
Description: Shirling, E.B. and D. Gottlieb. 1968. International Journal of Systematic Bacteriology **18**:279-392.

S. albogriseolus Benedict, Shotwell, Pridham, Lindenfelser and Haynes 1954
Benedict, R.G., O.L. Shotwell, T.G. Pridham, L.A. Lindenfelser and W.C. Haynes. 1954. Antibiotics and Chemotherapy **4**:653-656. **Type strain:** NRRL B-1305
Description: Shirling, E.B. and D. Gottlieb. 1968. International Journal of Systematic Bacteriology **18**:279-392.

S. albolongus Tsukiura, Okanishi, Koshiyama, Ohmori, Miyaki and Kawaguchi 1964
Tsukiura, H., M. Okanishi, H. Koshiyama, T. Ohmori, T. Miyaki and H. Kawaguchi. 1964. Journal of Antibiotics (Tokyo) Series A **17**:223-229. **Type strain:** ATCC 27414
Description: Shirling, E.B. and D. Gottlieb. 1972. International Journal of Systematic Bacteriology **22**:265-394.

S. alboniger Porter, Hewitt, Hesseltine, Krupka, Lowery, Wallace, Bohonos and Williams 1952
Porter, J.N., R.I. Hewitt, C.W. Hesseltine, G. Krupka, J.A. Lowery, W.S. Wallace, N. Bohonos and J.H. Williams. 1952. Antibiotics and Chemotherapy **2**:409-410. **Type strain:** ATCC 12461
Description: Shirling, E.B. and D. Gottlieb. 1968. International Journal of Systematic Bacteriology **18**:69-189.

S. albospinus Wang, Hamada, Okami and Umezawa 1966
Wang, E.L., M. Hamada, Y. Okami and H. Umezawa. 1966. Journal of Antibiotics (Tokyo) Series A **19**:216-221. **Type strain:** ATCC 29808
Description: Wang, E.L., M. Hamada, Y. Okami and H. Umezawa. 1966. Ibid.

S. albosporeus (Krainsky 1914) Waksman and Henrici 1948
Krainsky, A. 1914. Zentralblatt fur Bakteriologie, Parasitenkunde, Infektionskrankheiten und Hygiene. Abteilung II. **41**:649-688; Waksman, S.A. and A.T. Henrici. 1948 in Breed, R.S., E.G.D.

Murray and A.P. Hitchens (eds). Bergey's Manual of Determinative Bacteriology, 6th ed. The Williams and Wilkins Co., Baltimore. pp. 929-980. **Type strain:** ATCC 15394
Description: Pridham, T.G., A.J. Lyons Jr. and H.L. Seckinger. 1965. International Bulletin of Bacteriological Nomenclature and Taxonomy **15**:191-237.

S. albosporeus subsp. *albosporeus* (Krainsky 1914) Waksman and Henrici 1948
Krainsky, A. 1914. Zentralblatt fur Bakteriologie, Parasitenkunde, Infektionskrankheiten und Hygiene. Abteilung II. **41**:649-688; Waksman, S.A. and A.T. Henrici. 1948 in Breed, R.S., E.G.D. Murray and A.P. Hitchens (eds). Bergey's Manual of Determinative Bacteriology, 6th ed. The Williams and Wilkins Co., Baltimore. pp. 929-980. **Type strain:** ATCC 15394
Description: Pridham, T.G., A.J. Lyons Jr. and H.L. Seckinger. 1965. International Bulletin of Bacteriological Nomenclature and Taxonomy **15**:191-237.

S. albosporeus subsp. *labilomyceticus* Okami, Suzuki and Umezawa 1963
Okami, Y., M. Suzuki and H. Umezawa. 1963. Journal of Antibiotics (Tokyo) Series A **16**:152-154. **Type strain:** NIHJ 425
Description: Okami, Y., M. Suzuki and H. Umezawa. 1963. Ibid.

S. albovinaceus (Kudrina 1957) Pridham, Hesseltine and Benedict 1958
Kudrina, E.S. 1957 in Gauze, G.F., T.P. Preobrazhenskaya, E.S. Kudrina, N.O. Blinov, I.D. Ryabova and M.A. Sveshnikova. Problems in the classification of actinomycetes-antagonists. State Publishing House of Medical Literature, Medgiz, Moscow, USSR; Pridham, T.G., C.W. Hesseltine and R.G. Benedict. 1958. Applied Microbiology **6**:52-79. **Type strain:** ATCC 15823
Description: Shirling, E.B. and D. Gottlieb. 1968. International Journal of Systematic Bacteriology **18**:69-189.

S. alboviridis (Duche 1934) Pridham, Hesseltine and Benedict 1958
Duche, J. 1934. Encyclopedie Mycologique **6**:1-375; Pridham, T.G., C.W. Hesseltine and R.G. Benedict. 1958. Applied Microbiology **6**:52-79. **Type strain:** ATCC 25425
Description: Shirling, E.B. and D. Gottlieb. 1969. International Journal of Systematic Bacteriology **19**:391-512.

S. albulus Routien 1969
Routien, J.B. 1969 in Pridham, T.G. and A.J. Lyons, Jr. Developments in Industrial Microbiology **10**:183-221. **Type strain:** ATCC 12757
Description: Shirling, E.B. and D. Gottlieb. 1972. International Journal of Systematic Bacteriology **22**:265-394.

S. albus (Rossi Doria 1891) Waksman and Henrici 1943
Rossi Doria, T. 1891. Annali dell' Istituto d'Igiene Sperimentale della Rendiconti Universita di Roma **1**:399-438; Waksman, S.A. and A.T. Henrici. 1943. Journal of Bacteriology **46**:337-341. **Type strain:** ATCC 3004
Description: Shirling, E.B. and D. Gottlieb. 1969. International Journal of Systematic Bacteriology **19**:391-512.

S. albus subsp. *albus* (Rossi Doria 1891) Waksman and Henrici 1943
Rossi Doria, T. 1891. Annali dell' Istituto d'Igiene Sperimentale della Rendiconti Universita di Roma **1**:399-438; Waksman, S.A. and A.T. Henrici. 1943. Journal of Bacteriology **46**:337-341. **Type strain:** ATCC 3004
Description: Shirling, E.B. and D. Gottlieb. 1969. International Journal of Systematic Bacteriology **19**:391-512.

S. albus subsp. *pathocidicus* Nagatsu, Anzai and Suzuki 1962
Nagatsu, J., K. Anzai and S. Suzuki. 1962. Journal of Antibiotics (Tokyo) Series A **15**:103-106. **Type strain:** ATCC 14510
Description: Bergey 8.

S. almquistii (Duche 1934) Pridham, Hesseltine and Benedict 1958
Duche, J. 1934. Encyclopedie Mycologique **6**:1-375; Pridham, T.G., C.W. Hesseltine and R.G. Benedict. 1958. Applied Microbiology **6**:52-79. **Type strain:** ATCC 618
Description: Shirling, E.B. and D. Gottlieb. 1969. International Journal of Systematic Bacteriology **19**:391-512.

S. althioticus Yamaguchi, Nakayama, Takeda, Tawara, Maeda, Takeuchi and Umezawa

1957
Yamaguchi, H., Y. Nakayama, K. Takeda, K. Tawara, K. Maeda, T. Takeuchi and H. Umezawa. 1957. Journal of Antibiotics (Tokyo) Series A **10**:195-200. **Type strain:** ATCC 19724
Description: Shirling, E.B. and D. Gottlieb. 1968. International Journal of Systematic Bacteriology **18**:69-189.

S. amakusaensis Nagatsu, Anzai, Ohkuma and Suzuki 1963
Nagatsu, J., K. Anzai, K. Ohkuma and S. Suzuki. 1963. Journal of Antibiotics (Tokyo) Series A **16**:207-210. **Type strain:** ATCC 23876
Description: Shirling, E.B. and D. Gottlieb. 1968. International Journal of Systematic Bacteriology **18**:279-392.

S. ambofaciens Pinnert-Sindico 1954
Pinnert-Sindico, S. 1954. Annales de l'Institut Pasteur (Paris) **87**:702-707. **Type strain:** ATCC 23877
Description: Shirling, E.B. and D. Gottlieb. 1968. International Journal of Systematic Bacteriology **18**:279-392.

S. aminophilus Foster 1961
Foster, W.J. 1961 in Oswald, E.J., R.J. Reedy and W.A. Randall. 1956 in Hutter, R. 1961. Archiv fur Mikrobiologie **38**:367-383. **Type strain:** ATCC 14961
Description: Shirling, E.B. and D. Gottlieb. 1968. International Journal of Systematic Bacteriology **18**:279-392.

S. anandii Batra and Bajaj 1965
Batra, S.K. and B.S. Bajaj. 1965. Indian Journal of Experimental Biology **3**:240-242. **Type strain:** ATCC 19388
Description: Shirling, E.B. and D. Gottlieb. 1972. International Journal of Systematic Bacteriology **22**:265-394.

S. anthocyanicus (Krassilnikov, Sorokina, Alferova and Bezzubenkova 1965) Pridham 1970
Krassilnikov, N.A., E.J. Sorokina, V.A. Alferova and A.P. Bezzubenkova. 1965 in Krassilnikov, N.A. (ed). Biology of selected groups of Actinomycetes (in Russian). Institute of Microbiology, Academy of Science, Publishing Firm Nauka, Moscow, USSR. pp. 74-123; Pridham, T.G. 1970. Bulletin of the United States Department of Agriculture **1424**:1-55. **Type strain:** ATCC 19821
Description: Bergey 8.

S. antibioticus (Waksman and Woodruff 1941) Waksman and Henrici 1948
Waksman, S.A. and H.B. Woodruff. 1941. Journal of Bacteriology **42**:231-249; Waksman, S.A. and A.T. Henrici. 1948 in Breed, R.S., E.G.D. Murray and A.P. Hitchens (eds). Bergey's Manual of Determinative Bacteriology, 6th ed. The Williams and Wilkins Co., Baltimore. pp. 929-980. **Type strain:** ATCC 8663
Description: Shirling, E.B. and D. Gottlieb. 1968. International Journal of Systematic Bacteriology **18**:279-392.

S. antimycoticus Waksman 1957
Waksman, S.A. 1957 in Breed, R.S., E.G.D. Murray and N.R. Smith (eds). Bergey's Manual of Determinative Bacteriology, 7th ed. The Williams and Wilkins Co., Baltimore. pp. 744-825. **Type strain:** NRRL 2421
Description: Shirling, E.B. and D. Gottlieb. 1968. International Journal of Systematic Bacteriology **18**:279-392.

S. anulatus (Beijerinck 1912) Waksman 1953
Beijerinck, M.W. 1912. Folia Mikrobiologiya (Delft) **1**:4-100; Waksman, S.A. 1953 in Waksman, S.A. and H.A. Lechevalier. A Guide to the Classification of the Actinomycetes and their Antibiotics. The Williams and Wilkins Co., Baltimore. pp. 1-246. **Type strain:** CBS 100.18
Description: Shirling, E.B. and D. Gottlieb. 1972. International Journal of Systematic Bacteriology **22**:265-394.

S. arabicus Shibata, Nakazawa, Miyake, Inoue and Okabori 1957
Shibata, M., K. Nakazawa, A. Miyake, M. Inoue and A. Okabori. 1957. Annual Report of the Takeda Research Laboratory **16**:32-37. **Type strain:** ATCC 23881
Description: Shirling, E.B. and D. Gottlieb. 1968. International Journal of Systematic Bacteriology **18**:279-392.

S. arenae Pridham, Hesseltine and Benedict 1958
Pridham, T.G., C.W. Hesseltine and R.G. Benedict. 1958. Applied Microbiology **6**:52-79. **Type strain:** NRRL 2377
Description: Shirling, E.B. and D. Gottlieb. 1969. International Journal of Systematic Bacteriology **19**:391-512.

S. argenteolus Tresner, Davies and Backus 1961

Tresner, H.D., M.C. Davies and E.J. Backus. 1961. Journal of Bacteriology **81**: 70-80. **Type strain:** ATCC 11009
Description: Shirling, E.B. and D. Gottlieb. 1968. International Journal of Systematic Bacteriology **18**:279-392.

S. atratus Shibata, Higashide, Yamamoto and Nakazawa 1962
Shibata, M., E. Higashide, H. Yamamoto and K. Nakazawa. 1962. Agricultural and Biological Chemistry **26**:228-233. **Type strain:** IFO 3897
Description: Shibata, M., E. Higashide, H. Yamamoto and K. Nakazawa. 1962. Ibid.

S. atroolivaceus (Preobrazhenskaya, Blinov and Ryabova 1957) Pridham, Hesseltine and Benedict 1958
Preobrazhenskaya, T.P., N.O. Blinov and I.D. Ryabova. 1957 in Gauze, G.F., T.P. Preobrazhenskova, E.S. Kudrina, N.O. Blinov, I.D. Ryabova and M.A. Sveshnikova. Problems in the classification of actinomycetes-antagonists. Government Publishing House of Medical Literature, Medgiz, Moscow. pp. 1-398; Pridham, T.G., C.W. Hesseltine and R.G. Benedict. 1958. Applied Microbiology **6**:52-79. **Type strain:** ATCC 19725
Description: Shirling, E.B. and D. Gottlieb. 1968. International Journal of Systematic Bacteriology **18**:69-189.

S. aurantiacus (Rossi Doria 1891) Waksman 1953
Rossi Doria, T. 1891. Annali dell' Istituto d'Igiene Sperimentale della Rendiconti Universita di Roma **1**:399-438; Waksman, S.A. 1953 in Waksman, S.A. and H.A. Lechevalier. Guide to the classification and identification of the actinomycetes and their antibiotics. The Williams and Wilkins Co., Baltimore. pp. 1-246. **Type strain:** ATCC 19822
Description: Shirling, E.B. and D. Gottlieb. 1969. International Journal of Systematic Bacteriology **19**:391-512.

S. aurantiogriseus (Preobrazhenskaya 1957) Pridham, Hesseltine and Benedict 1958
Preobrazhenskaya, T.P. 1957 in Gauze, G.F., T.P. Preobrazhenskaya, E.S. Kudrina, N.O. Blinov, I.D. Ryabova and M.A. Sveshnikova. Problems of classification of actinomycetes-antagonists. Government Publishing House of Medical Literature, Medgiz, Moscow. pp. 1-398; Pridham, T.G., C.W. Hesseltine and R.G. Benedict. 1958. Applied Microbiology **6**:52-79. **Type strain:** ATCC 19887
Description: Shirling, E.B. and D. Gottlieb. 1968. International Journal of Systematic Bacteriology **18**:279-392.

S. aureocirculatus (Krassilnikov and Yuan 1965) Pridham 1970
Krassilnikov, N.A. and T. Yuan. 1965 in Krassilnikov, N.A. (ed). Biology of individual groups of Actinomycetes (in Russian). pp. 28-57; Pridham, T.G. 1970. Bulletin of the United States Department of Agriculture **1424**:1-55. **Type strain:** ATCC 19823
Description: Shirling, E.B. and D. Gottlieb. 1969. International Journal of Systematic Bacteriology **19**:391-512.

S. aureofaciens Duggar 1948
Duggar, B. 1948. Annals of the New York Academy of Science **51**:177-181. **Type strain:** NRRL 2209
Description: Shirling, E.B. and D. Gottlieb. 1968. International Journal of Systematic Bacteriology **18**:279-392.

S. aureoverticillatus (Krassilnikov and Yuan 1960) Pridham 1970
Krassilnikov, N.A. and C.-S. Yuan. 1960. Mikrobiologiya **29**:482-489 (English Translation 354-358); Pridham, T.G. 1970. Bulletin of the United States Department of Agriculture **1424**:1-55. **Type strain:** ATCC 15854
Description: Shirling, E.B. and D. Gottlieb. 1968. International Journal of Systematic Bacteriology **18**:69-189.

S. avellaneus Baldacci and Grein 1966
Baldacci, E. and A. Grein. 1966. Giornale di Microbiologia **14**:185-198. **Type strain:** ATCC 23730
Description: Shirling, E.B. and D. Gottlieb. 1972. International Journal of Systematic Bacteriology **22**:265-394.

S. avidinii Stapley, Mata, Miller, Demny and Woodruff 1964
Stapley, E.O., J.M. Mata, I.M. Miller, T.C. Demny and H.B. Woodruff. 1964. Antimicrobial Agents and Chemotherapy **1963**:20-27. **Type strain:** NRRL 3077
Description: Shirling, E.B. and D. Gottlieb. 1972. International Journal of

Systematic Bacteriology **22**:265-394.

S. azureus Kelly, Kutscher and Tuoti 1959
Kelly, J., A.H. Kutscher and F. Tuoti. 1959. Oral Surgery, Oral Medicine and Oral Pathology **12**:1334-1339. **Type strain:** ATCC 14921
Description: Shirling, E.B. and D. Gottlieb. 1968. International Journal of Systematic Bacteriology **18**:69-189.

S. baarnensis Pridham, Hesseltine and Benedict 1958
Pridham, T.G., C.W. Hesseltine and R.G. Benedict. 1958. Applied Microbiology **6**:52-79. **Type strain:** CBS 306.55
Description: Shirling, E.B. and D. Gottlieb. 1968. International Journal of Systematic Bacteriology **18**:279-392.

S. bacillaris (Krassilnikov 1958) Pridham 1970
Krassilnikov, N.A. 1958. Folia Biologica (Praha) **4**:257-265; Pridham, T.G. 1970. Bulletin of the United States Department of Agriculture **1424**:1-55. **Type strain:** ATCC 15855
Description: Shirling, E.B. and D. Gottlieb. 1972. International Journal of Systematic Bacteriology **22**:265-394.

S. badius (Kudrina 1957) Pridham, Hesseltine and Benedict 1958
Kudrina, E.S. 1957 in Gauze, G.F., T.P. Preobrazhenskaya, E.S. Kudrina, N.O. Blinov, I.D. Ryabova and M.A. Sveshnikova. Problems of classification of actinomycetes-antagonists. Government Publishing House of Medical Literature, Medgiz, Moscow. pp. 1-398; Pridham, T.G., C.W. Hesseltine and R.G. Benedict. 1958. Applied Microbiology **6**:52-79. **Type strain:** ATCC 19888
Description: Shirling, E.B. and D. Gottlieb. 1968. International Journal of Systematic Bacteriology **18**:69-189.

S. bambergiensis Wallhausser, Nesemann, Prave and Steigler 1966
Wallhausser, K., G. Nesemann, P. Prave and A. Steigler. 1966. Antimicrobial Agents and Chemotherapy **1965**:734-736. **Type strain:** ATCC 13879
Description: Shirling, E.B. and D. Gottlieb. 1972. International Journal of Systematic Bacteriology **22**:265-394.

S. bellus Margalith and Beretta 1960
Margalith, P. and G. Beretta. 1960. Mycopathologia et Mycologia Applicata **12**: 189-195. **Type strain:** ATCC 14925
Description: Shirling, E.B. and D. Gottlieb. 1968. International Journal of Systematic Bacteriology **18**:279-392.

S. bikiniensis Johnstone and Waksman 1947
Johnstone, D.B. and S.A. Waksman. 1947. Proceedings of the Society for Experimental Biology and Medicine **65**:294-295. **Type strain:** ATCC 11062
Description: Shirling, E.B. and D. Gottlieb. 1968. International Journal of Systematic Bacteriology **18**:279-392.

S. bluensis Mason, Dietz and Hanka 1963
Mason, D.J., A. Dietz and L.J. Hanka. 1963. Antimicrobial Agents and Chemotherapy **1962**:607-613. **Type strain:** NRRL 2876
Description: Shirling, E.B. and D. Gottlieb. 1972. International Journal of Systematic Bacteriology **22**:265-394.

S. bobili (Waksman and Curtis 1916) Waksman and Henrici 1948
Waksman, S.A. and R.E. Curtis. 1916. Soil Science **1**:99-134; Waksman, S.A. and A.T. Henrici. 1948 in Breed, R.S., E.G.D. Murray and A.P. Hitchens (eds). Bergey's Manual of Determinative Bacteriology, 6th ed. The Williams and Wilkins Co., Baltimore. pp. 929-980. **Type strain:** ATCC 3310
Description: Shirling, E.B. and D. Gottlieb. 1968. International Journal of Systematic Bacteriology **18**:279-392.

S. bottropensis Waksman 1961
Waksman, S.A. 1961. The Actinomycetes. Vol. 2. Classification, identification and descriptions of genera and species. The Williams and Wilkins Co., Baltimore. pp. 1-363. **Type strain:** ATCC 25435
Description: Shirling, E.B. and D. Gottlieb. 1969. International Journal of Systematic Bacteriology **19**:391-512.

S. cacaoi (Waksman 1932) Waksman and Henrici 1948
Waksman, S.A. 1932 in Bunting, R.H. Annals of Applied Biology **19**:515-517; Waksman, S.A. and A.T. Henrici. 1948 in Breed, R.S., E.G.D. Murray and A.P. Hitchens (eds). Bergey's Manual of Determinative Bacteriology, 6th ed. The Williams and Wilkins Co., Baltimore. pp. 929-980. **Type strain:** ATCC 3082
Description: Shirling, E.B. and D.

Gottlieb. 1968. International Journal of Systematic Bacteriology **18**:69-189.

S. cacaoi subsp. *asoensis* Isono, Nagatsu, Kawashima and Suzuki 1965
Isono, K., J. Nagatsu, Y. Kawashima and S. Suzuki. 1965. Agricultural and Biological Chemistry **29**:848-854. **Type strain:** IFO 13813
Description: Isono, K., J. Nagatsu, Y. Kawashima and S. Suzuki. 1965. Ibid.

S. cacaoi subsp. *cacaoi* (Waksman 1932) Waksman and Henrici 1948
Waksman, S.A. 1932 in Bunting, R.H. Annals of Applied Biology **19**:515-517; Waksman, S.A. and A.T. Henrici. 1948 in Breed, R.S., E.G.D. Murray and A.P. Hitchens (eds). Bergey's Manual of Determinative Bacteriology, 6th ed. The Williams and Wilkins Co., Baltimore. pp. 929-980. **Type strain:** ATCC 3082
Description: Shirling, E.B. and D. Gottlieb. 1968. International Journal of Systematic Bacteriology **18**:69-189.

S. caelestis De Boer, Dietz, Wilkins, Lewis and Savage 1955
De Boer, C., A. Dietz, J.R. Wilkins, C.N. Lewis and G.M. Savage. 1955. Antibiotics Annual **1954**:831-841. **Type strain:** NRRL 2418
Description: Shirling, E.B. and D. Gottlieb. 1968. International Journal of Systematic Bacteriology **18**:69-189.

S. caeruleus (Baldacci 1944) Pridham, Hesseltine and Benedict 1958
Baldacci, E. 1944. Atti dell'Istituto Botanico della Universita Laboratorio Crittogamico di Pavia Series 5, **3**:139-193; Pridham, T.G., C.W. Hesseltine and R.G. Benedict. 1958. Applied Microbiology **6**:52-79. **Type strain:** ATCC 27421
Description: Shirling, E.B. and D. Gottlieb. 1972. International Journal of Systematic Bacteriology **22**:265-394.

S. californicus (Waksman and Curtis 1916) Waksman and Henrici 1948
Waksman, S.A. and R.E. Curtis. 1916. Soil Science **1**:99-134; Waksman, S.A. and A.T. Henrici. 1948 in Breed, R.S., E.G.D. Murray and A.P. Hitchens (eds). Bergey's Manual of Determinative Bacteriology, 6th ed. The Williams and Wilkins Co., Baltimore. pp. 929-980. **Type strain:** ATCC 3312
Description: Shirling, E.B. and D. Gottlieb. 1968. International Journal of Systematic Bacteriology **18**:69-189.

S. calvus Backus, Tresner and Campbell 1957
Backus, E.J., H.D. Tresner and T.H. Campbell. 1957. Antibiotics and Chemotherapy **7**:532-541. **Type strain:** ATCC 13382
Description: Shirling, E.B. and D. Gottlieb. 1968. International Journal of Systematic Bacteriology **18**:279-392.

S. canarius Vavra and Dietz 1965
Vavra, J. and A. Dietz. 1965. Antimicrobial Agents and Chemotherapy **1964**:75-79. **Type strain:** NRRL 2976
Description: Shirling, E.B. and D. Gottlieb. 1972. International Journal of Systematic Bacteriology **22**:265-394.

S. canescens Waksman 1957
Waksman, S.A. 1957 in Breed, R.S., E.G.D. Murray and N.R. Smith (eds). Bergey's Manual of Determinative Bacteriology, 7th ed. The Williams and Wilkins Co., Baltimore. **Type strain:** ATCC 19736
Description: Shirling, E.B. and D. Gottlieb. 1968. International Journal of Systematic Bacteriology **18**:69-189.

S. canus Heinemann, Kaplan, Muir and Hooper 1953
Heinemann, B., M.A. Kaplan, R.D. Muir and I.R. Hooper. 1953. Antibiotics and Chemotherapy **3**:1239-1242. **Type strain:** ATCC 12237
Description: Shirling, E.B. and D. Gottlieb. 1968. International Journal of Systematic Bacteriology **18**:69-189.

S. capoamus Goncalves de Lima, Albert and Goncalves de Lima 1964
Goncalves de Lima, V.O., C.A. Albert and O. Goncalves de Lima. 1964. Anais da Academia Brazileira de Ciencias **36**: 317-322. **Type strain:** ATCC 19006
Description: Shirling, E.B. and D. Gottlieb. 1972. International Journal of Systematic Bacteriology **22**:265-394.

S. catenulae Davisson and Finlay 1961
Davisson, J.W. and A.C. Finlay. 1961 in Waksman, S.A. The Actinomycetes. Vol. 2. Classification, identification and descriptions of genera and species. The Williams and Wilkins Co., Baltimore. pp. 1-363. **Type strain:** ATCC 12476
Description: Shirling, E.B. and D.

Gottlieb. 1968. International Journal of Systematic Bacteriology **18**:279-392.

S. cavourensis Skarbek and Brady 1978
Skarbek, J.D. and L.R. Brady. 1978. International Journal of Systematic Bacteriology **28**:45-53. **Type strain:** ATCC 14889
Description: Skarbek, J.D. and L.R. Brady. 1978. Ibid.

S. cavourensis subsp. *cavourensis* Skarbek and Brady 1978
Skarbek, J.D. and L.R. Brady. 1978. International Journal of Systematic Bacteriology **28**:45-53. **Type strain:** ATCC 14889
Description: Skarbek, J.D. and L.R. Brady. 1978. Ibid.

S. cavourensis subsp. *washingtonensis* Skarbek and Brady 1978
Skarbek, J.D. and L.R. Brady. 1978. International Journal of Systematic Bacteriology **28**:45-53. **Type strain:** ATCC 27732
Description: Skarbek, J.D. and L.R. Brady. 1978. Ibid.

S. cellostaticus Hamada 1958
Hamada, S. 1958. Tohoku Journal of Experimental Medicine **67**:173-179. **Type strain:** ATCC 23894
Description: Shirling, E.B. and D. Gottlieb. 1968. International Journal of Systematic Bacteriology **18**:279-392.

S. celluloflavus Nishimura, Kimura and Kuroya 1953
Nishimura, H., T. Kimura and M. Kuroya. 1953. Journal of Antibiotics (Tokyo) Series A **6**:57-65. **Type strain:** IFO 13780
Description: Bergey 8.

S. cellulosae (Krainsky 1914) Waksman and Henrici 1948
Krainsky, A. 1914. Zentralblatt fur Bakteriologie, Parasitenkunde, Infektionskrankheiten und Hygiene. Abteilung II. **41**:649-688; Waksman, S.A. and A.T. Henrici. 1948 in Breed, R.S., E.G.D. Murray and A.P. Hitchens (eds). Bergey's Manual of Determinative Bacteriology, 6th ed. The Williams and Wilkins Co., Baltimore. pp. 929-980. **Type strain:** CBS 122.18
Description: Shirling, E.B. and D. Gottlieb. 1969. International Journal of Systematic Bacteriology **19**:391-512.

S. champavatii Uma and Narasimha Rao 1959
Uma, B.N. and P.L. Narasimha Rao. 1959. Indian Institute of Sciences Golden Jubilee Research Vol. 1909-1959. Bangalore, India. pp. 130-141. **Type strain:** NRRL B-5682
Description: Bergey 8.

S. chartreusis Leach, Calhoun, Johnson, Teeters and Jackson 1953
Leach, B.E., K.M. Calhoun, L.E. Johnson, C.M. Teeters and W.G. Jackson. 1953. Journal of the American Chemical Society **75**:4011-4012. **Type strain:** NRRL 2287
Description: Shirling, E.B. and D. Gottlieb. 1968. International Journal of Systematic Bacteriology **18**:69-189.

S. chattanoogensis Burns and Holtman 1959
Burns, J. and D.F. Holtman. 1959. Antibiotics and Chemotherapy **9**:398-405. **Type strain:** ATCC 13358
Description: Shirling, E.B. and D. Gottlieb. 1968. International Journal of Systematic Bacteriology **18**:69-189.

S. chibaensis Suzuki, Nakamura, Okuma and Tomiyama 1958
Suzuki, S., G. Nakamura, K. Okuma and Y. Tomiyama. 1958. Journal of Antibiotics (Tokyo) Series A **11**:81-83. **Type strain:** ATCC 23895
Description: Shirling, E.B. and D. Gottlieb. 1968. International Journal of Systematic Bacteriology **18**:279-392.

S. chrestomyceticus Canevazzi and Scotti 1959
Canevazzi, G. and T. Scotti. 1959. Giornale di Microbiologia **7**:242-250. **Type strain:** ATCC 14947
Description: Shirling, E.B. and D. Gottlieb. 1972. International Journal of Systematic Bacteriology **22**:265-394.

S. chromofuscus (Preobrazhenskaya, Blinov and Ryabova 1957) Pridham, Hesseltine and Benedict 1958
Preobrazhenskaya, E.S., N.O. Blinov and I.D. Ryabova. 1957 in Gauze, G.F., T.P. Preobrazhenskaya, E.S. Kudrina, N.O. Blinov, I.D. Ryabova and M.A. Sveshnikova. Problems of classification of actinomycetes-antagonists. Government Publishing House of Medical Literature, Medgiz, Moscow. pp. 1-398; Pridham, T.G., C.W. Hesseltine and R.G. Benedict. 1958. Applied Microbiology **6**:52-79. **Type strain:** ATCC 23896

Description: Shirling, E.B. and D. Gottlieb. 1968. International Journal of Systematic Bacteriology **18**:69-189.

S. chryseus (Krassilnikov, Korenyako and Nikitina 1965) Pridham 1970
Krassilnikov, N.A., A.I. Korenyako and N.I. Nikitina. 1965 in Krassilnikov, N.A. (ed). Biology of selected groups of actinomycetes (in Russian). Institute of Microbiology, Academy of Science, Publishing Firm Nauka, Moscow, USSR. pp. 1-372; Pridham, T.G. 1970. Bulletin of the United States Department of Agriculture **1424**: 1-55. **Type strain:** ATCC 19829
Description: Shirling, E.B. and D. Gottlieb. 1972. International Journal of Systematic Bacteriology **22**:265-394.

S. chrysomallus Lindenbein 1952
Lindenbein, W. 1952. Archiv fur Mikrobiologie **17**:361-383. **Type strain:** ATCC 11523
Description: Shirling, E.B. and D. Gottlieb. 1968. International Journal of Systematic Bacteriology **18**:69-189.

S. chrysomallus subsp. *chrysomallus* Lindenbein 1952
Lindenbein, W. 1952. Archiv fur Mikrobiologie **17**:361-383. **Type strain:** ATCC 11523
Description: Shirling, E.B. and D. Gottlieb. 1968. International Journal of Systematic Bacteriology **18**:69-189.

S. chrysomallus subsp. *fumigatus* Frommer 1959
Frommer, W. 1959. Archiv fur Mikrobiologie **32**:187-206. **Type strain:** NRRL B-2289
Description: Hutter, R. 1963. Giornale di Microbiologia **11**:191-246.

S. cinereoruber Corbaz, Ettlinger, Keller-Schierlein and Zahner 1957
Corbaz, R., L. Ettlinger, W. Keller-Schierlein and H. Zahner. 1957. Archiv fur Mikrobiologie **25**:325-332. **Type strain:** ATCC 19740
Description: Shirling, E.B. and D. Gottlieb. 1968. International Journal of Systematic Bacteriology **18**:69-189.

S. cinereoruber subsp. *cinereoruber* Corbaz, Ettlinger, Keller-Schierlein and Zahner 1957
Corbaz, R., L. Ettlinger, W. Keller-Schierlein and H. Zahner. 1957. Archiv fur Mikrobiologie **25**:325-332. **Type strain:** ATCC 19740
Description: Shirling, E.B. and D. Gottlieb. 1968. International Journal of Systematic Bacteriology **18**:69-189.

S. cinereoruber subsp. *fructofermentans* Corbaz, Ettlinger, Keller-Schierlein and Zahner 1957
Corbaz, R., L. Ettlinger, W. Keller-Schierlein and H. Zahner. 1957. Archiv fur Mikrobiologie **25**:325-332. **Type strain:** NRRL 2588
Description: Ettlinger, L., R. Corbaz and R. Hutter. 1958. Archiv fur Mikrobiologie **31**:326-358.

S. cinerochromogenes Miyairi, Takashima, Shimizu and Sakai 1966
Miyairi, N., M. Takashima, K. Shimizu and H. Sakai. 1966. Journal of Antibiotics (Tokyo) Series A **19**:56-62. **Type strain:** IFO 13922
Description: Miyairi, N., M. Takashima, K. Shimizu and H. Sakai. 1966. Ibid.

S. cinnabarinus (Ryabova and Preobrazhenskaya 1957) Pridham, Hesseltine and Benedict 1958
Ryabova, I.D. and T.P. Preobrazhenskaya. 1957 in Gauze, G.F., T.P. Preobrazhenskaya, E.S. Kudrina, N.O. Blinov, I.D. Ryabova and M.A. Sveshnikova. Problems of classification of actinomycetes-antagonists. Government Publishing House of Medical Literature, Medgiz, Moscow. pp. 1-398; Pridham, T.G., C.W. Hesseltine and R.G. Benedict. 1958. Applied Microbiology **6**:52-79. **Type strain:** ATCC 23617
Description: Shirling, E.B. and D. Gottlieb. 1969. International Journal of Systematic Bacteriology **19**:391-512.

S. cinnamonensis Okami 1952
Okami, Y. 1952 in Maeda, K., Y. Okami, H. Kosaka, O. Taya and H. Umezawa. Journal of Antibiotics (Tokyo) **5**:572-573. **Type strain:** ATCC 12308
Description: Ettlinger, L., R. Corbaz and R. Hutter. 1958. Archiv fur Mikrobiologie **31**:326-358.

S. cirratus Koshiyama, Okanishi, Ohmori, Miyaki, Tsukiura, Matsuzaki and Kawaguchi 1963
Koshiyama, H., M. Okanishi, T. Ohmori, T. Miyaki, H. Tsukiura, M. Matsuzaki

and H. Kawaguchi. 1963. Journal of Antibiotics (Tokyo) Series A **16**:59-66. **Type strain:** ATCC 14699
Description: Shirling, E.B. and D. Gottlieb. 1972. International Journal of Systematic Bacteriology **22**:265-394.

S. citreofluorescens (Korenyako, Krassilnikov, Nikitina and Sokolova 1960) Pridham 1970
Korenyako, A.I., N.A. Krasillnilov, N.I. Nikitina and A.I. Sokolova. 1960 in Rautenshtein, Ya. I. Transactions of the Institute of Microbiology, Academy of Science USSR **8**:1-344; Pridham, T.G. 1970. Bulletin of the United States Department of Agriculture **1424**:1-55. **Type strain:** ATCC 15858
Description: Shirling, E.B. and D. Gottlieb. 1968. International Journal of Systematic Bacteriology **18**:279-392.

S. clavifer (Millard and Burr 1926) Waksman 1953
Millard, W.A. and S. Burr. 1926. Annals of Applied Biology **13**:580-644; Waksman, S.A. 1953 in Waksman, S.A. and H.A. Lechevalier. Guide to the classification and identification of the actinomycetes and their antibiotics. The Williams and Wilkins Co., Baltimore. pp. 1-246. **Type strain:** CBS 101.27
Description: Bergey 8.

S. clavuligerus Higgens and Kastner 1971
Higgens, C.E. and R.E. Kastner. 1971. International Journal of Systematic Bacteriology **21**:326-331. **Type strain:** NRRL 3585
Description: Higgens, C.E. and R.E. Kastner. 1971. Ibid.

S. coelescens (Krassilnikov, Sorokina, Alferova and Bezzubenkova 1965) Pridham 1970
Krassilnikov, N.A., E.J. Sorokina, V.A. Alferova and A.P. Bezzubenkova. 1965 in Krassilnikov, N.A. (ed). Biology of selected groups of actinomycetes (in Russian). Institute of Microbiology, Academy of Science, Publishing Firm Nauka, Moscow, USSR. pp. 1-372; Pridham, T.G. 1970. Bulletin of the United States Department of Agriculture **1424**:1-55. **Type strain:** ATCC 19830
Description: Shirling, E.B. and D. Gottlieb. 1972. International Journal of Systematic Bacteriology **22**:265-394.

S. coelicolor (Muller 1908) Waksman and Henrici 1948
Muller, R. 1908. Zentralblatt fur Bakteriologie, Parasitenkunde, Infektionskrankheiten und Hygiene. Abteilung I. **46**:195-222; Waksman, S.A. and A.T. Henrici. 1948 in Breed, R.S., E.G.D. Murray and A.P. Hitchens (eds). Bergey's Manual of Determinative Bacteriology, 6th ed. The Williams and Wilkins Co., Baltimore. pp. 929-980. **Type strain:** CBS 210.27
Description: Shirling, E.B. and D. Gottlieb. 1968. International Journal of Systematic Bacteriology **18**:279-392.

S. coeruleofuscus (Preobrazhenskaya 1957) Pridham, Hesseltine and Benedict 1958
Preobrazhenskaya, T.P. 1957 in Gauze, G.F., T.P. Preobrazhenskaya, E.S. Kudrina, N.O. Blinov, I.D. Ryabova and M.A. Sveshnikova. Problems of classification of actinomycetes-antagonists. Government Publishing House of Medical Literature, Medgiz, Moscow. pp. 1-398; Pridham, T.G., C.W. Hesseltine and R.G. Benedict. 1958. Applied Microbiology **6**:52-79. **Type strain:** ATCC 23618
Description: Shirling, E.B. and D. Gottlieb. 1968. International Journal of Systematic Bacteriology **18**:69-189.

S. coeruleorubidus (Preobrazhenskaya 1957) Pridham, Hesseltine and Benedict 1958
Preobrazhenskaya, T.P. 1957 in Gauze, G.F., T.P. Preobrazhenskaya, E.S. Kudrina, N.O. Blinov, I.D. Ryabova and M.A. Sveshnikova. Problems of classification of actinomycetes-antagonists. Government Publishing House of Medical Literature, Medgiz, Moscow. pp. 1-398; Pridham, T.G., C.W. Hesseltine and R.G. Benedict. 1958. Applied Microbiology **6**:52-79. **Type strain:** ATCC 13740
Description: Shirling, E.B. and D. Gottlieb. 1968. International Journal of Systematic Bacteriology **18**:279-392.

S. coerulescens (Preobrazhenskaya 1957) Pridham, Hesseltine and Benedict 1958
Preobrazhenskaya, T.P. 1957 in Gauze, G.F., T.P. Preobrazhenskaya, E.S. Kudrina, N.O. Blinov, I.D. Ryabova and M.A. Sveshnikova. Problems of classification of actinomycetes-antagonists. Government Publishing House of Medical Literature, Medgiz, Moscow. pp. 1-398;

Pridham, T.G., C.W. Hesseltine and R.G. Benedict. 1958. Applied Microbiology **6**:52-79. **Type strain:** ATCC 19896
Description: Shirling, E.B. and D. Gottlieb. 1968. International Journal of Systematic Bacteriology **18**:69-189.

S. collinus Lindenbein 1952
Lindenbein, W. 1952. Archiv fur Mikrobiologie **17**:361-383. **Type strain:** ATCC 19743
Description: Shirling, E.B. and D. Gottlieb. 1968. International Journal of Systematic Bacteriology **18**:69-189.

S. colombiensis Pridham, Hesseltine and Benedict 1958
Pridham, T.G., C.W. Hesseltine and R.G. Benedict. 1958. Applied Microbiology **6**:52-79. **Type strain:** ATCC 27425
Description: Shirling, E.B. and D. Gottlieb. 1972. International Journal of Systematic Bacteriology **22**:265-394.

S. corchorusii Ahmad and Bhuiyan 1958
Ahmad, K. and J.A.M. Bhuiyan. 1958. Pakistan Journal of Biological and Agricultural Sciences **1**:137-143. **Type strain:** ATCC 25444
Description: Shirling, E.B. and D. Gottlieb. 1969. International Journal of Systematic Bacteriology **19**:391-512.

S. cremeus (Kudrina 1957) Pridham, Hesseltine and Benedict 1958
Kudrina, E.S. 1957 in Gauze, G.F., T.P. Preobrazhenskaya, E.S. Kudrina, N.O. Blinov, I.D. Ryabova and M.A. Sveshnikova. Problems in the classification of actinomycetes-antagonists. State Publishing House of Medical Literature, Medgiz, Moscow. USSR; Pridham, T.G., C.W. Hesseltine and R.G. Benedict. 1958. Applied Microbiology **6**:52-79. **Type strain:** ATCC 19897
Description: Shirling, E.B. and D. Gottlieb. 1968. International Journal of Systematic Bacteriology **18**:69-189.

S. crystallinus Tresner, Davies and Backus 1961
Tresner, H.D., M.C. Davies and E.J. Backus. 1961. Journal of Bacteriology **81**: 70-80. **Type strain:** NRRL B-3629
Description: Bergey 8.

S. curacoi Cataldi 1963
Cataldi, M. 1963 in Trejo, W. and R.E. Bennett. Journal of Bacteriology **85**:676-690. **Type strain:** ATCC 13385
Description: Shirling, E.B. and D. Gottlieb. 1968. International Journal of Systematic Bacteriology **18**:69-189.

S. cuspidosporus Higashide, Hasegawa, Shibata, Mizuno and Akaike 1966
Higashide, E., T. Hasegawa, M. Shibata, K. Mizuno and H. Akaike. 1966. Annual Report of the Takeda Research Laboratory **25**:1-14. **Type strain:** IFO 12378
Description: Higashide, E., T. Hasegawa, M. Shibata, K. Mizuno and H. Akaike. 1966. Ibid.

S. cyaneofuscatus (Kudrina 1957) Pridham, Hesseltine and Benedict 1958
Kudrina, E.S. 1957 in Gauze, G.F., T.P. Preobrazhenskaya, E.S. Kudrina, N.O. Blinov, I.D. Ryabova and M.A. Sveshnikova. Problems of classification of actinomycetes-antagonists. Government Publishing House of Medical Literature, Medgiz, Moscow. pp. 1-398; Pridham, T.G., C.W. Hesseltine and R.G. Benedict. 1958. Applied Microbiology **6**:52-79. **Type strain:** ATCC 23619
Description: Shirling, E.B. and D. Gottlieb. 1968. International Journal of Systematic Bacteriology **18**:69-189.

S. cyaneus (Krassilnikov 1941) Waksman 1953
Krassilnikov, N.A. 1941. Guide to the bacteria and Actinomycetes (in Russian). Institute of Microbiology, Academy of Science, Publishing Firm Nauka, Moscow, USSR. pp. 1-830; Waksman, S.A. 1953 in Waksman, S.A. and H.A. Lechevalier. Guide to the classification and identification of the Actinomycetes and their antibiotics. The Williams and Wilkins Co., Baltimore. pp. 1-245. **Type strain:** ATCC 14923
Description: Shirling, E.B. and D. Gottlieb. 1972. International Journal of Systematic Bacteriology **22**:265-394.

S. cyanoalbus (Krassilnikov and Agre 1960) Pridham 1970
Krassilnikov, N.A. and N.S. Agre. 1960 in Rautenshtein, Ya. I. Transactions of the Institute of Microbiology, Academy of Science USSR **8**:1-344; Pridham, T.G. 1970. Bulletin of the United States Department of Agriculture **1424**:1-55. **Type strain:** ATCC 15859
Description: Shirling, E.B. and D. Gottlieb. 1968. International Journal of

Systematic Bacteriology **18**:279-392.

S. daghestanicus (Sveshnikova 1957) Pridham, Hesseltine and Benedict 1958
Sveshnikova, M.A. 1957 in Gauze, G.F., T.P. Preobrazhenskaya, E.S. Kudrina, N.O. Blinov, I.D. Ryabova and M.A. Sveshnikova. Problems of classification of actinomycetes-antagonists. Government Publishing House of Medical Literature, Medgiz, Moscow. pp. 1-398; Pridham, T.G., C.W. Hesseltine and R.G. Benedict. 1958. Applied Microbiology **6**:52-79. **Type strain:** ATCC 23620
Description: Shirling, E.B. and D. Gottlieb. 1968. International Journal of Systematic Bacteriology **18**:69-189.

S. diastaticus (Krainsky 1914) Waksman and Henrici 1948
Krainsky, A. 1914. Zentralblatt fur Bakteriologie, Parasitenkunde, Infektionskrankheiten und Hygiene. Abteilung II. **41**:649-688; Waksman, S.A. and A.T. Henrici. 1948 in Breed, R.S., E.G.D. Murray and A.P. Hitchens (eds). Bergey's Manual of Determinative Bacteriology, 6th ed. The Williams and Wilkins Co., Baltimore. pp. 929-980. **Type strain:** ATCC 3315
Description: Shirling, E.B. and D. Gottlieb. 1972. International Journal of Systematic Bacteriology **22**:265-394.

S. diastaticus subsp. *ardesiacus* (Baldacci, Grein and Spalla 1955) Pridham, Hesseltine and Benedict 1958
Baldacci, E., A. Grein and C. Spalla. 1955. Giornale di Microbiologia **1**:127-143; Pridham, T.G., C.W. Hesseltine and R.G. Benedict. 1958. Applied Microbiology **6**:52-79. **Type strain:** CBS 100.56
Description: Hutter, R. 1964. Zentralblatt fur Bakteriologie, Parasitenkunde, Infektionskrankheiten und Hygiene. Abteilung II. **117**:603-661.

S. diastaticus subsp. *diastaticus* (Krainsky 1914) Waksman and Henrici 1948 Krainsky, A. 1914. Zentralblatt fur Bakteriologie, Parasitenkunde, Infektionskrankheiten und Hygiene. Abteilung II. **41**:649-688; Waksman, S.A. and A.T. Henrici. 1948 in Breed, R.S., E.G.D. Murray and A.P. Hitchens (eds). Bergey's Manual of Determinative Bacteriology, 6th ed. The Williams and Wilkins Co., Baltimore. pp. 929-980. **Type strain:** ATCC 3315
Description: Shirling, E.B. and D. Gottlieb. 1972. International Journal of Systematic Bacteriology **22**:265-394.

S. diastatochromogenes (Krainsky 1914) Waksman and Henrici 1948
Krainsky, A. 1914. Zentralblatt fur Bakteriologie, Parasitenkunde, Infektionskrankheiten und Hygiene. Abteilung II. **41**:649-688; Waksman, S.A. and A.T. Henrici. 1948 in Breed, R.S., E.G.D. Murray and A.P. Hitchens (eds). Bergey's Manual of Determinative Bacteriology, 6th ed. The Williams and Wilkins Co., Baltimore. pp. 929-980. **Type strain:** ATCC 12309
Description: Shirling, E.B. and D. Gottlieb. 1972. International Journal of Systematic Bacteriology **22**:265-394.

S. djakartensis Huber, Wallhausser, Fries, Steigler and Weidenmuller 1962
Huber, G., K.H. Wallhausser, L. Fries, A. Steigler and H.-L. Weidenmuller. 1962. Arzneimittel Forschung **12**:1191-1195. **Type strain:** DSM 40743
Description: Bergey 8.

S. durhamensis Gordon and Lapa 1966
Gordon, M.A. and E.W. Lapa. 1966. Applied Microbiology **14**:754-760. **Type strain:** ATCC 23194
Description: Shirling, E.B. and D. Gottlieb. 1972. International Journal of Systematic Bacteriology **22**:265-394.

S. echinatus Corbaz, Ettlinger, Gaumann, Keller-Schierlein, Kradolfer, Neipp, Prelog, Reusser and Zahner 1957
Corbaz, R., L. Ettlinger, E. Gaumann, W. Keller-Schierlein, F. Kradolfer, L. Neipp, V. Prelog, P. Reusser and H. Zahner. 1957. Helvetica Chimica Acta **40**:199-204. **Type strain:** ATCC 19748
Description: Shirling, E.B. and D. Gottlieb. 1968. International Journal of Systematic Bacteriology **18**:69-189.

S. ederensis Wallhausser, Nesemann, Prave and Steigler 1966
Wallhausser, K.H., G. Nesemann, P. Prave and A. Steigler. 1966. Antimicrobial Agents and Chemotherapy **1965**:734-736. **Type strain:** ATCC 15304
Description: Bergey 8.

S. endus Anderson and Gottlieb 1952
Anderson, H.W. and D. Gottlieb. 1952.

Economic Botany **6**:294-308. **Type strain:** NRRL 2339
Description: Shirling, E.B. and D. Gottlieb. 1968. International Journal of Systematic Bacteriology **18**:279-392.

S. erumpens Calot and Cercos 1963
Calot, L. and A.P. Cercos. 1963. Annales de l'Institut Pasteur (Paris) **105**:159-161. **Type strain:** ATCC 23266
Description: Bergey 8.

S. erythraeus (Waksman 1923) Waksman and Henrici 1948
Waksman, S.A. 1923 in Bergey, D.H., F.C. Harrison, R.S. Breed, B.W. Hammer and F.M. Huntoon. Bergey's Manual of Determinative Bacteriology, 1st ed. The Williams and Wilkins Co., Baltimore. pp. 339-371; Waksman, S.A. and A.T. Henrici. 1948 in Breed, R.S., E.G.D. Murray and A.P. Hitchens (eds). Bergey's Manual of Determinative Bacteriology, 6th ed. The Williams and Wilkins Co., Baltimore. pp. 929-980. **Type strain:** ATCC 11635
Description: Shirling, E.B. and D. Gottlieb. 1972. International Journal of Systematic Bacteriology **22**:265-394.

S. erythrogriseus Falcao de Morais and Dalia Maia 1959
Falcao de Morais, J.O. and M.H. Dalia Maia. 1959. Revista do Instituto de Antibioticos, Universidade do Recife **2**:63-67. **Type strain:** CBS 485.74
Description: Shirling, E.B. and D. Gottlieb. 1972. International Journal of Systematic Bacteriology **22**:265-394.

S. eurythermus Corbaz, Ettlinger, Gaumann, Keller-Schierlein, Kradolfer, Neipp, Prelog, Reusser and Zahner 1957
Corbaz, R., L. Ettlinger, E. Gaumann, W. Keller-Schierlein, F. Kradolfer, L. Neipp, V. Prelog, P. Reusser and H. Zahner. 1957. Helvetica Chimica Acta **40**:199-203. **Type strain:** ATCC 14975
Description: Shirling, E.B. and D. Gottlieb. 1968. International Journal of Systematic Bacteriology **18**:69-189.

S. exfoliatus (Waksman and Curtis 1916) Waksman and Henrici 1948
Waksman, S.A. and R.E. Curtis. 1916. Soil Science **1**:99-134; Waksman, S.A. and A.T. Henrici. 1948 in Breed, R.S., E.G.D. Murray and A.P. Hitchens (eds). Bergey's Manual of Determinative Bacteriology, 6th ed. The Williams and Wilkins Co., Baltimore. pp. 929-980. **Type strain:** ATCC 12627
Description: Shirling, E.B. and D. Gottlieb. 1968. International Journal of Systematic Bacteriology **18**:69-189.

S. felleus Lindenbein 1952
Lindenbein, W. 1952. Archiv fur Mikrobiologie **17**:361-383. **Type strain:** ATCC 19752
Description: Shirling, E.B. and D. Gottlieb. 1968. International Journal of Systematic Bacteriology **18**:69-189.

S. filamentosus Okami and Umezawa 1953
Okami, Y. and H. Umezawa. 1953 in Okami, Y., T. Okuda, T. Takeuchi, K. Nitta and H. Umezawa. Journal of Antibiotics (Tokyo) Series A **6**:153-157. **Type strain:** ATCC 19753
Description: Shirling, E.B. and D. Gottlieb. 1968. International Journal of Systematic Bacteriology **18**:69-189.

S. filipinensis Ammann, Gottlieb, Brock, Carter and Whitfield 1955
Ammann, A., D. Gottlieb, T.D. Brock, H.E. Carter and G.B. Whitfield. 1955. Phytopathology **45**:559-563. **Type strain:** NRRL 2437
Description: Shirling, E.B. and D. Gottlieb. 1968. International Journal of Systematic Bacteriology **18**:279-392.

S. fimbriatus (Millard and Burr 1926) Waksman and Lechevalier 1953
Millard, W.A. and S. Burr. 1926. Annals of Applied Biology **13**:580-644; Waksman, S.A. and H.A. Lechevalier. 1953. Guide to the classification and identification of the actinomycetes and their antibiotics. The Williams and Wilkins Co., Baltimore. pp. 1-246. **Type strain:** ATCC 15051
Description: Bergey 8.

S. fimicarius (Duche 1934) Waksman and Henrici 1948
Duche, J. 1934. Encyclopedie Mycologique **6**:1-375; Waksman, S.A. and A.T. Henrici. 1948 in Breed, R.S., E.G.D. Murray and A.P. Hitchens (eds). Bergey's Manual of Determinative Bacteriology, 6th ed. The Williams and Wilkins Co., Baltimore. pp. 929-980. **Type strain:** CBS 420.34

Description: Shirling, E.B. and D. Gottlieb. 1969. International Journal of Systematic Bacteriology **19**:391-512.

S. finlayi (Szabo, Martin, Buti and Partai 1963) Pridham 1970
Szabo, I., M. Marton, I. Buti and G. Partai. 1963. Acta Microbiologica Academiae Scientiarum Hungaricae **10**:207-214; Pridham, T.G. 1970. Bulletin of the United States Department of Agriculture **1424**: 1-55. **Type strain:** ATCC 23340
Description: Shirling, E.B. and D. Gottlieb. 1968. International Journal of Systematic Bacteriology **18**:279-392.

S. flaveolus (Waksman 1923) Waksman and Henrici 1948
Waksman, S.A. 1923 in Bergey, D.H., F.C. Harrison, R.S. Breed, B.W. Hammer and F.M. Huntoon. Bergey's Manual of Determinative Bacteriology, 1st ed. The Williams and Wilkins Co., Baltimore. pp. 339-371; Waksman, S.A. and A.T. Henrici. 1948 in Breed, R.S., E.G.D. Murray and A.P. Hitchens (eds). Bergey's Manual of Determinative Bacteriology, 6th ed. The Williams and Wilkins Co., Baltimore. pp. 929-980. **Type strain:** ATCC 3319
Description: Shirling, E.B. and D. Gottlieb. 1968. International Journal of Systematic Bacteriology **18**:69-189.

S. flavidovirens (Kudrina 1957) Pridham, Hesseltine and Benedict 1958
Kudrina, E.S. 1957 in Gauze, G.F., T.P. Preobrazhenskaya, E.S. Kudrina, N.O. Blinov, I.D. Ryabova and M.A. Sveshnikova. Problems of classification of actinomycetes-antagonists. Government Publishing House of Medical Literature, Medgiz, Moscow. pp. 1-398; Pridham, T.G., C.W. Hesseltine and R.G. Benedict. 1958. Applied Microbiology **6**:52-79. **Type strain:** ATCC 19900
Description: Shirling, E.B. and D. Gottlieb. 1969. International Journal of Systematic Bacteriology **19**:391-512.

S. flavogriseus (Duche 1934) Waksman and Lechevalier 1953
Duche, J. 1934. Encyclopedie Mycologique **6**:1-375; Waksman, S.A. and H.A. Lechevalier. 1953. Guide to the classification and identification of the actinomycetes and their antibiotics. The Williams and Wilkins Co., Baltimore. pp. 1-246. **Type strain:** CBS 101.34
Description: Shirling, E.B. and D. Gottlieb. 1969. International Journal of Systematic Bacteriology **19**:391-512.

S. flavotricini (Preobrazhenskaya and Sveshnikova 1957) Pridham, Hesseltine and Benedict 1958
Preobrazhenskaya, T.P. and M.A. Sveshnikova. 1957 in Gauze, G.F., T.P. Preobrazhenskaya, E.S. Kudrina, N.O. Blinov, I.D. Ryabova and M.A. Sveshnikova. Problems of classification of actinomycetes-antagonists. Government Publishing House of Medical Literature, Medgiz, Moscow. pp. 1-398; Pridham, T.G., C.W. Hesseltine and R.G. Benedict. 1958. Applied Microbiology **6**:52-79. **Type strain:** ATCC 23621
Description: Shirling, E.B. and D. Gottlieb. 1968. International Journal of Systematic Bacteriology **18**:69-189.

S. flavovirens (Waksman 1923) Waksman and Henrici 1948
Waksman, S.A. 1923 in Bergey, D.H., F.C. Harrison, R.S. Breed, B.W. Hammer and F.M. Huntoon. Bergey's Manual of Determinative Bacteriology, 1st ed. The Williams and Wilkins Co., Baltimore. pp. 339-371; Waksman, S.A. and A.T. Henrici. 1948 in Breed, R.S., E.G.D. Murray and A.P. Hitchens (eds). Bergey's Manual of Determinative Bacteriology, 6th ed. The Williams and Wilkins Co., Baltimore. pp. 929-980. **Type strain:** ATCC 3320
Description: Shirling, E.B. and D. Gottlieb. 1968. International Journal of Systematic Bacteriology **18**:69-189.

S. flocculus (Duche 1934) Waksman and Henrici 1948
Duche, J. 1934. Encyclopedie Mycologique **6**:1-375; Waksman, S.A. and A.T. Henrici. 1948 in Breed, R.S., E.G.D. Murray and A.P. Hitchens (eds). Bergey's Manual of Determinative Bacteriology, 6th ed. The Williams and Wilkins Co., Baltimore. pp. 929-980. **Type strain:** ATCC 25453
Description: Shirling, E.B. and D. Gottlieb. 1969. International Journal of Systematic Bacteriology **19**:391-512.

S. floridae Bartz, Ehrlich, Mold, Penner and

Smith 1951
Bartz, Q., J. Ehrlich, J.D. Mold, M.A. Penner and R.M. Smith. 1951. American Review of Tuberculosis **63**:4-6. **Type strain:** NRRL 2423
Description: Ettlinger, L., R. Corbaz and R. Hutter. 1958. Archiv fur Mikrobiologie **31**:326-358.

S. fluorescens (Krassilnikov 1958) Pridham 1970
Krassilnikov, N.A. 1958. Folia Biologica (Praha) **4**:257-265; Pridham, T.G. 1970. Bulletin of the United States Department of Agriculture **1424**:1-55. **Type strain:** ATCC 15860
Description: Shirling, E.B. and D. Gottlieb. 1968. International Journal of Systematic Bacteriology **18**:279-392.

S. fradiae (Waksman and Curtis 1916) Waksman and Henrici 1948
Waksman, S.A. and R.E. Curtis. 1916. Soil Science **1**:99-134; Waksman, S.A. and A.T. Henrici. 1948 in Breed, R.S., E.G.D. Murray and A.P. Hitchens (eds). Bergey's Manual of Determinative Bacteriology, 6th ed. The Williams and Wilkins Co., Baltimore. pp. 929-980. **Type strain:** ATCC 10745
Description: Shirling, E.B. and D. Gottlieb. 1968. International Journal of Systematic Bacteriology **18**:69-189.

S. fragilis Anderson, Ehrlich, Sun and Burkholder 1956
Anderson, L.E., J. Ehrlich, S.H. Sun and P.R. Burkholder. 1956. Antibiotics and Chemotherapy **6**:100-115. **Type strain:** NRRL 2424
Description: Shirling, E.B. and D. Gottlieb. 1968. International Journal of Systematic Bacteriology **18**:279-392.

S. fulvissimus (Jensen 1930) Waksman and Henrici 1948
Jensen, H.L. 1930. Proceedings of the Linnean Society of New South Wales **55**: 231-249; Waksman, S.A. and A.T. Henrici. 1948 in Breed, R.S., E.G.D. Murray and A.P. Hitchens (eds). Bergey's Manual of Determinative Bacteriology, 6th ed. The Williams and Wilkins Co., Baltimore. pp. 929-980. **Type strain:** NRRL B-1453
Description: Shirling, E.B. and D. Gottlieb. 1972. International Journal of Systematic Bacteriology **22**:265-394.

S. fumanus (Sveshnikova 1957) Pridham, Hesseltine and Benedict 1958
Sveshnikova, M.A. 1957 in Gauze, G.F., T.P. Preobrazhenskaya, E.S. Kudrina, N.O. Blinov, I.D. Ryabova and M.A. Sveshnikova. Problems of classification of actinomycetes-antagonists. Government Publishing House of Medical Literature, Medgiz, Moscow. pp. 1-398; Pridham, T.G., C.W. Hesseltine and R.G. Benedict. 1958. Applied Microbiology **6**:52-79. **Type strain:** ATCC 19904
Description: Shirling, E.B. and D. Gottlieb. 1969. International Journal of Systematic Bacteriology **19**:391-512.

S. galbus Frommer 1959
Frommer, W. 1959. Archiv fur Mikrobiologie **32**:187-206. **Type strain:** ATCC 23910
Description: Shirling, E.B. and D. Gottlieb. 1968. International Journal of Systematic Bacteriology **18**:279-392.

S. galilaeus Ettlinger, Corbaz and Hutter 1958
Ettlinger, L., R. Corbaz and R. Hutter. 1958. Archiv fur Mikrobiologie **31**:326-358. **Type strain:** NRRL 2722
Description: Shirling, E.B. and D. Gottlieb. 1972. International Journal of Systematic Bacteriology **22**:265-394.

S. gancidicus Suzuki 1957
Suzuki, M. 1957. Journal Chiba Medical Society **33**:535-542. **Type strain:** NRRL B-1872
Description: Bergey 8.

S. gardneri (Waksman 1942) Waksman 1961
Waksman, S.A. 1942 in Waksman, S.A., E.S. Horning, M. Welsch and H.B. Woodruff. Soil Science **54**:281-296; Waksman, S.A. 1961. The Actinomycetes. Vol. 2. Classification, identification and descriptions of genera and species. The Williams and Wilkins Co., Baltimore. pp. 1-363. **Type strain:** ATCC 9604
Description: Shirling, E.B. and D. Gottlieb. 1968. International Journal of Systematic Bacteriology **18**:279-392.

S. gelaticus (Waksman 1923) Waksman and Henrici 1948
Waksman, S.A. 1923 in Bergey, D.H., F.C. Harrison, R.S. Breed, B.W. Hammer and F.M. Huntoon. Bergey's Manual of Determinative Bacteriology, 1st ed. The Williams and Wilkins Co., Baltimore.

pp. 339-371; Waksman, S.A. and A.T. Henrici. 1948 in Breed, R.S., E.G.D. Murray and A.P. Hitchens (eds). Bergey's Manual of Determinative Bacteriology, 6th ed. The Williams and Wilkins Co., Baltimore. pp. 929-980. **Type strain:** ATCC 3323

Description: Shirling, E.B. and D. Gottlieb. 1968. International Journal of Systematic Bacteriology **18**:279-392.

S. geysiriensis Wallhausser, Nesemann, Prave and Steigler 1966

Wallhausser, K.H., G. Nesemann, P. Prave and A. Steigler. 1966. Antimicrobial Agents and Chemotherapy **1965**:734-736. **Type strain:** ATCC 15303

Description: Bergey 8.

S. ghanaensis Wallhausser, Nesemann, Prave and Steigler 1966

Wallhausser, K.H., G. Nesemann, P. Prave and A. Steigler. 1966. Antimicrobial Agents and Chemotherapy **1965**:734-736. **Type strain:** ATCC 14672

Description: Bergey 8.

S. gibsonii (Erikson 1935) Waksman and Henrici 1948

Erikson, D. 1935. Medical Research Council (Great Britain) Special Report Series **203**:5-61; Waksman, S.A. and A.T. Henrici. 1948 in Breed, R.S., E.G.D. Murray and A.P. Hitchens (eds). Bergey's Manual of Determinative Bacteriology, 6th ed. The Williams and Wilkins Co., Baltimore. pp. 929-980. **Type strain:** ATCC 6852

Description: Bergey 8.

S. glaucescens (Preobrazhenskaya 1957) Pridham, Hesseltine and Benedict 1958

Preobrazhenskaya, T.P. 1957 in Gauze, G.F., T.P. Preobrazhenskaya, E.S. Kudrina, N.O. Blinov, I.D. Ryabova and M.A. Sveshnikova. Problems of classification of actinomycetes-antagonists. Government Publishing House of Medical Literature, Medgiz, Moscow. pp. 1-398; Pridham, T.G., C.W. Hesseltine and R.G. Benedict. 1958. Applied Microbiology **6**:52-79. **Type strain:** ATCC 23622

Description: Shirling, E.B. and D. Gottlieb. 1968. International Journal of Systematic Bacteriology **18**:69-189.

S. globisporus (Krassilnikov 1941) Waksman 1953

Krassilnikov, N.A. 1941. Guide to the Bacteria and Actinomycetes (in Russian). Akademiya Nauk SSSR, Moscow. pp. 1-830; Waksman, S.A. 1953 in Waksman, S.A. and H.A. Lechevalier. Guide to the classification and identification of the Actinomycetes and their antibiotics. The Williams and Wilkins Co., Baltimore. pp. 1-246. **Type strain:** ATCC 15864

Description: Shirling, E.B. and D. Gottlieb. 1968. International Journal of Systematic Bacteriology **18**:279-392.

S. globisporus subsp. *caucasicus* (Kudrina 1957) Pridham, Hesseltine and Benedict 1958

Kudrina, E.S. 1957 in Gauze, G.F., T.P. Preobrazhenskaya, E.S. Kudrina, N.O. Blinov, I.D. Ryabova and M.A. Sveshnikova. Problems of classification of actinomycetes-antagonists. Government Publishing House of Medical Literature, Medgiz, Moscow. pp. 1-398; Pridham, T.G., C.W. Hesseltine and R.G. Benedict. 1958. Applied Microbiology **6**:52-79. **Type strain:** ATCC 19907

Description: Bergey 8.

S. globisporus subsp. *flavofuscus* (Kudrina 1957) Pridham, Hesseltine and Benedict 1958

Kudrina, E.S. 1957 in Gauze, G.F., T.P. Preobrazhenskaya, E.S. Kudrina, N.O. Blinov, I.D. Ryabova and M.A. Sveshnikova. Problems of classification of actinomycetes-antagonists. Government Publishing House of Medical Literature, Medgiz, Moscow. pp. 1-398; Pridham, T.G., C.W. Hesseltine and R.G. Benedict. 1958. Applied Microbiology **6**:52-79. **Type strain:** ATCC 19908

Description: Bergey 8.

S. globisporus subsp. *globisporus* (Krassilnikov 1941) Waksman 1953

Krassilnikov, N.A. 1941. Guide to the Bacteria and Actinomycetes (in Russian). Akademiya Nauk SSSR, Moscow. pp. 1-830; Waksman, S.A. 1953 in Waksman, S.A. and H.A. Lechevalier. Guide to the classification and identification of the Actinomycetes and their antibiotics. The Williams and Wilkins Co., Baltimore. pp. 1-246. **Type strain:** ATCC 15864

Description: Shirling, E.B. and D. Gottlieb. 1968. International Journal of Systematic Bacteriology **18**:279-392.

S. globosus (Krassilnikov 1941) Waksman 1953
Krassilnikov, N.A. 1941. Guide to the Bacteria and Actinomycetes (in Russian). Akademiya Nauk SSSR, Moscow. pp. 1-830; Waksman, S.A. 1953 in Waksman, S.A. and H.A. Lechevalier. **Type strain:** NRRL B-2292
Description: Bergey 8.

S. glomeroaurantiacus (Krassilnikov and Yuan 1965) Pridham 1970
Krassilnikov, N.A. and T. Yuan. 1965 in Krassilnikov, N.A. (ed). Biology of individual groups of Actinomycetes (in Russian). pp. 28-57; Pridham, T.G. 1970. Bulletin of the United States Department of Agriculture **1424**:1-55. **Type strain:** ATCC 15866
Description: Shirling, E.B. and D. Gottlieb. 1972. International Journal of Systematic Bacteriology **22**:265-394.

S. gobitricini (Preobrazhenskaya and Sveshnikova 1957) Pridham, Hesseltine and Benedict 1958
Preobrazhenskaya, T.P. and M.A. Sveshnikova. 1957 in Gauze, G.F., T.P. Preobrazhenskaya, E.S. Kudrina, N.O. Blinov, I.D. Ryabova and M.A. Sveshnikova. Problems of classification of actinomycetes-antagonists. Government Publishing House of Medical Literature, Medgiz, Moscow. pp. 1-398; Pridham, T.G., C.W. Hesseltine and R.G. Benedict. 1958. Applied Microbiology **6**:52-79. **Type strain:** CBS 123.60
Description: Bergey 8.

S. goshikiensis Niida 1966
Niida, T. 1966 in Shirling, E.B. and D. Gottlieb. International Journal of Systematic Bacteriology **16**:313-340. **Type strain:** ATCC 23914
Description: Shirling, E.B. and D. Gottlieb. 1968. International Journal of Systematic Bacteriology **18**:279-392.

S. gougerotii (Duche 1934) Waksman and Henrici 1948
Duche, J. 1934. Encyclopedie Mycologique **6**:1-375; Waksman, S.A. and A.T. Henrici. 1948 in Breed, R.S., E.G.D. Murray and A.P. Hitchens (eds). Bergey's Manual of Determinative Bacteriology, 6th ed. The Williams and Wilkins Co., Baltimore. pp. 929-980. **Type strain:** CBS 422.34
Description: Shirling, E.B. and D. Gottlieb. 1969. International Journal of Systematic Bacteriology **19**:391-512.

S. graminofaciens Charney, Fisher, Curran, Machlowitz and Tytell 1953
Charney, J., W.P. Fisher, C. Curran, R.A. Machlowitz and A.A. Tytell. 1953. Antibiotics and Chemotherapy **3**:1283-1286. **Type strain:** ATCC 12705
Description: Shirling, E.B. and D. Gottlieb. 1972. International Journal of Systematic Bacteriology **22**:265-394.

S. griseinus Waksman 1959
Waksman, S.A. 1959. Proceedings of the National Academy of Science, U.S.A. **45**: 1043-1047. **Type strain:** ATCC 23915
Description: Shirling, E.B. and D. Gottlieb. 1968. International Journal of Systematic Bacteriology **18**:279-392.

S. griseoaurantiacus (Krassilnikov and Yuan 1965) Pridham 1970
Krassilnikov, N.A. and T. Yuan. 1965. in Krassilnikov, N.A. (ed). Biology of individual groups of Actinomycetes (in Russian). pp. 28-57; Pridham, T.G. 1970. Bulletin of the United States Department of Agriculture **1424**:1-55. **Type strain:** ATCC 19840
Description: Shirling, E.B. and D. Gottlieb. 1972. International Journal of Systematic Bacteriology **22**:265-394.

S. griseobrunneus Waksman 1961
Waksman, S.A. 1961. The Actinomycetes. Vol. 2. Classification, identification and descriptions of genera and species. The Williams and Wilkins Co., Baltimore. pp. 1-363. **Type strain:** ATCC 19762
Description: Shirling, E.B. and D. Gottlieb. 1968. International Journal of Systematic Bacteriology **18**:69-189.

S. griseochromogenes Fukunaga 1955
Fukunaga, K. 1955 in Fukunaga, K., T. Misato, I. Ishii and M. Asakawa. Bulletin of the Agricultural Chemical Society of Japan **19**:181-188. **Type strain:** ATCC 14511
Description: Shirling, E.B. and D. Gottlieb. 1972. International Journal of Systematic Bacteriology **22**:265-394.

S. griseoflavus (Krainsky 1914) Waksman and Henrici 1948
Krainsky, A. 1914. Zentralblatt fur Bakteriologie, Parasitenkunde, Infektion-

skrankheiten und Hygiene. Abteilung II. **41**:649-688; Waksman, S.A. and A.T. Henrici. 1948 in Breed, R.S., E.G.D. Murray and A.P. Hitchens (eds). Bergey's Manual of Determinative Bacteriology, 6th ed. The Williams and Wilkins Co., Baltimore. pp. 929-980. **Type strain:** ATCC 25456
Description: Shirling, E.B. and D. Gottlieb. 1969. International Journal of Systematic Bacteriology **19**:391-512.

S. griseofuscus Sakamoto, Kondo, Yumoto and Arishima 1962
Sakamoto, J.M., S.-i. Kondo, H. Yumoto and M. Arishima. 1962. Journal of Antibiotics (Tokyo) Series A **15**:98-102. **Type strain:** ATCC 23916
Description: Shirling, E.B. and D. Gottlieb. 1968. International Journal of Systematic Bacteriology **18**:279-392.

S. griseoincarnatus (Preobrazhenskaya, Ryabova and Blinov 1957) Pridham, Hesseltine and Benedict 1958
Preobrazhenskaya, T.P., I.D. Ryabova and N.O. Blinov. 1957 in Gauze, G.F., T.P. Preobrazhenskaya, E.S. Kudrina, N.O. Blinov, I.D. Ryabova and M.A. Sveshnikova. Problems of classification of actinomycetes-antagonists. Government Publishing House of Medical Literature, Medgiz, Moscow. pp. 1-398; Pridham, T.G., C.W. Hesseltine and R.G. Benedict. 1958. Applied Microbiology **6**:52-79. **Type strain:** ATCC 23623
Description: Shirling, E.B. and D. Gottlieb. 1968. International Journal of Systematic Bacteriology **18**:279-392.

S. griseoloalbus (Kudrina 1957) Pridham, Hesseltine and Benedict 1958
Kudrina, E.S. 1957 in Gauze, G.F., T.P. Preobrazhenskaya, E.S. Kudrina, N.O. Blinov, I.D. Ryabova and M.A. Sveshnikova. Problems of classification of actinomycetes-antagonists. Government Publishing House of Medical Literature, Medgiz, Moscow. pp. 1-398; Pridham, T.G., C.W. Hesseltine and R.G. Benedict. 1958. Applied Microbiology **6**:52-79. **Type strain:** ATCC 23624
Description: Shirling, E.B. and D. Gottlieb. 1969. International Journal of Systematic Bacteriology **19**:391-512.

S. griseolus (Waksman 1923) Waksman and Henrici 1948
Waksman, S.A. 1923 in Bergey, D.H., F.C. Harrison, R.S. Breed, B.W. Hammer and F.M. Huntoon. Bergey's Manual of Determinative Bacteriology, 1st ed. The Williams and Wilkins Co., Baltimore. pp. 339-371; Waksman, S.A. and A.T. Henrici. 1948 in Breed, R.S., E.G.D. Murray and A.P. Hitchens (eds). Bergey's Manual of Determinative Bacteriology, 6th ed. The Williams and Wilkins Co., Baltimore. pp. 929-980. **Type strain:** ATCC 3325
Description: Shirling, E.B. and D. Gottlieb. 1968. International Journal of Systematic Bacteriology **18**:69-189.

S. griseoluteus Umezawa, Hayano, Maeda, Ogata and Okami 1950
Umezawa, H., S. Hayano, K. Maeda, Y. Ogata and Y. Okami. 1950. Japanese Medical Journal **3**:111-117. **Type strain:** ATCC 12768
Description: Shirling, E.B. and D. Gottlieb. 1972. International Journal of Systematic Bacteriology **22**:265-394.

S. griseomycini (Preobrazhenskaya, Blinov and Ryabova 1957) Pridham, Hesseltine and Benedict 1958
Preobrazhenskaya, T.P., N.O. Blinov and I.D. Ryabova. 1957 in Gauze, G.F., T.P. Preobrazhenskaya, E.S. Kudrina, N.O. Blinov, I.D. Ryabova and M.A. Sveshnikova. Problems of classification of antinomycetes-antagonists. Government Publishing House of Medical Literature, Medgiz, Moscow. pp. 1-398; Pridham, T.G., C.W. Hesseltine and R.G. Benedict. 1958. Applied Microbiology **6**:52-79. **Type strain:** ATCC 23625
Description: Shirling, E.B. and D. Gottlieb. 1968. International Journal of Systematic Bacteriology **18**:69-189.

S. griseoplanus Backus, Tresner and Campbell 1957
Backus, E.J., H.D. Tresner and T.H. Campbell. 1957. Antibiotics and Chemotherapy **7**:532-541. **Type strain:** ATCC 19766
Description: Shirling, E.B. and D. Gottlieb. 1968. International Journal of Systematic Bacteriology **18**:69-189.

S. griseorubens (Preobrazhenskaya, Blinov and Ryabova 1957) Pridham, Hesseltine and

Benedict 1958
Preobrazhenskaya, T.P., N.O. Blinov and I.D. Ryabova. 1957 in Gauze, G.F., T.P. Preobrazhenskaya, E.S. Kudrina, N.O. Blinov, I.D. Ryabova and M.A. Sveshnikova. Problems of classification of actinomycetes-antagonists. Government Publishing House of Medical Literature, Medgiz, Moscow. pp. 1-398; Pridham, T.G., C.W. Hesseltine and R.G. Benedict. 1958. Applied Microbiology **6**:52-79. **Type strain:** ATCC 19909
Description: Shirling, E.B. and D. Gottlieb. 1968. International Journal of Systematic Bacteriology **18**:69-189.

S. griseoruber Yamaguchi and Saburi 1955
Yamaguchi, T. and Y. Saburi. 1955. Journal of General and Applied Microbiology **1**:201-235. **Type strain:** ATCC 23919
Description: Shirling, E.B. and D. Gottlieb. 1968. International Journal of Systematic Bacteriology **18**:279-392.

S. griseorubiginosus (Ryabova and Preobrazhenskaya 1957) Pridham, Hesseltine and Benedict 1958
Ryabova, I.D. and T.P. Preobrazhenskaya. 1957 in Gauze, G.F., T.P. Preobrazhenskaya, E.S. Kudrina, N.O. Blinov, I.D. Ryabova and M.A. Sveshnikova. Problems of classification of actinomycetes-antagonists. Government Publishing House of Medical Literature, Medgiz, Moscow. pp. 1-398; Pridham, T.G., C.W. Hesseltine and R.G. Benedict. 1958. Applied Microbiology **6**:52-79. **Type strain:** ATCC 23627
Description: Shirling, E.B. and D. Gottlieb. 1969. International Journal of Systematic Bacteriology **19**:391-512.

S. griseosporeus Niida and Ogasawara 1960
Niida, T. and M. Ogasawara. 1960. Scientific Reports of Meiji Seika Kaisha **3**:23-26. **Type strain:** ATCC 27435
Description: Shirling, E.B. and D. Gottlieb. 1972. International Journal of Systematic Bacteriology **22**:265-394.

S. griseostramineus (Preobrazhenskaya, Kudrina, Blinov and Ryabova 1957) Pridham, Hesseltine and Benedict 1958
Preobrazhenskaya, T.P., E.S. Kudrina, N.O. Blinov and I.D. Ryabova. 1957 in Gauze, G.F., T.P. Preobrazhenskaya, E.S. Kudrina, N.O. Blinov, I.D. Ryabova and M.A. Sveshnikova. Problems of classification of actinomycetes-antagonists. Government Publishing House of Medical Literature, Medgiz, Moscow. pp. 1-398; Pridham, T.G., C.W. Hesseltine and R.G. Benedict. 1958. Applied Microbiology **6**:52-79. **Type strain:** ATCC 23628
Description: Shirling, E.B. and D. Gottlieb. 1968. International Journal of Systematic Bacteriology **18**:69-189.

S. griseoviridis Anderson, Ehrlich, Sun and Burkholder 1956
Anderson, L.E., J. Ehrlich, S.H. Sun and P.R. Burkholder. 1956. Antibiotics and Chemotherapy **6**:100-115. **Type strain:** NRRL 2427
Description: Shirling, E.B. and D. Gottlieb. 1968. International Journal of Systematic Bacteriology **18**:279-392.

S. griseus (Krainsky 1914) Waksman and Henrici 1948
Krainsky, A. 1914. Zentralblatt fur Bakteriologie, Parasitenkunde, Infektionskrankheiten und Hygiene. Abteilung II. **41**:649-688; Waksman, S.A. and A.T. Henrici. 1948 in Breed, R.S., E.G.D. Murray and A.P. Hitchens (eds). Bergey's Manual of Determinative Bacteriology, 6th ed. The Williams and Wilkins Co., Baltimore. pp. 929-980. **Type strain:** ATCC 23345
Description: Shirling, E.B. and D. Gottlieb. 1968. International Journal of Systematic Bacteriology **18**:279-392.

S. griseus subsp. *alpha* (Ciferri 1927) Pridham 1970
Ciferri, R. 1927. Zentralblatt fur Bakteriologie, Parasitenkunde, Infektionskrankheiten und Hygiene. Abteilung II. **71**:80-93; Pridham, T.G. 1970. Bulletin of the United States Department of Agriculture **1424**:1-55. **Type strain:** CBS 219.25
Description: Bergey 8.

S. griseus subsp. *cretosus* Pridham 1970
Pridham, T.G. 1970. Bulletin of the United States Department of Agriculture **1424**: 1-55. **Type strain:** CBS 137.21
Description: Bergey 8.

S. griseus subsp. *griseus* (Krainsky 1914) Waksman and Henrici 1948
Krainsky, A. 1914. Zentralblatt fur Bakteriologie, Parasitenkunde,

Infektionskrankheiten und Hygiene. Abteilung II. **41**:649-688; Waksman, S.A. and A.T. Henrici. 1948 in Breed, R.S., E.G.D. Murray and A.P. Hitchens (eds). Bergey's Manual of Determinative Bacteriology, 6th ed. The Williams and Wilkins Co., Baltimore. pp. 929-980. **Type strain:** ATCC 23345
Description: Shirling, E.B. and D. Gottlieb. 1968. International Journal of Systematic Bacteriology **18**:279-392.

S. griseus subsp. *solvifaciens* Pridham 1970
Pridham, T.G. 1970. Bulletin of the United States Department of Agriculture **1424**: 1-55. **Type strain:** NRRL B-1561
Description: Bergey 8.

S. halstedii (Waksman and Curtis 1916) Waksman and Henrici 1948
Waksman, S.A. and R.E. Curtis. 1916. Soil Science **1**:99-134; Waksman, S.A. and A.T. Henrici. 1948 in Breed, R.S., E.G.D. Murray and A.P. Hitchens (eds). Bergey's Manual of Determinative Bacteriology, 6th ed. The Williams and Wilkins Co., Baltimore. pp. 929-980. **Type strain:** ATCC 10897
Description: Shirling, E.B. and D. Gottlieb. 1968. International Journal of Systematic Bacteriology **18**:69-189.

S. hawaiiensis Cron, Whitehead, Hooper, Heinemann and Lein 1956
Cron, M.J., D.F. Whitehead, I.R. Hooper, B. Heinemann and J. Lein. 1956. Antibiotics and Chemotherapy **6**:63-67. **Type strain:** ATCC 12236
Description: Shirling, E.B. and D. Gottlieb. 1968. International Journal of Systematic Bacteriology **18**:69-189.

S. helvaticus (Krassilnikov, Korenyako and Nikitina 1965) Pridham 1970
Krassilnikov, N.A., A.I. Korenyako and N.I. Nikitina. 1965 in Krassilnikov N.A. Biology of selected groups of actinomycetes (in Russian). Institute of Microbiology, Academy of Science, Publishing Firm Nauka, Moscow, USSR. pp. 1-372; Pridham, T.G. 1970. Bulletin of the United States Department of Agriculture **1424**:1-55. **Type strain:** ATCC 19841
Description: Shirling, E.B. and D. Gottlieb. 1972. International Journal of Systematic Bacteriology **22**:265-394.

S. herbaricolor Kawato and Shinobu 1959
Kawato, N. and R. Shinobu. 1959. Memoirs of the Osaka University of the Liberal Arts and Education B. Natural Science **8**:114-119. **Type strain:** ATCC 23922
Description: Shirling, E.B. and D. Gottlieb. 1968. International Journal of Systematic Bacteriology **18**:279-392.

S. hirsutus Ettlinger, Corbaz and Hutter 1958
Ettlinger, L., R. Corbaz and R. Hutter. 1958. Archiv fur Mikrobiologie **31**:326-358. **Type strain:** ATCC 19773
Description: Shirling, E.B. and D. Gottlieb. 1968. International Journal of Systematic Bacteriology **18**:69-189.

S. humidus Nakazawa and Shibata 1956
Nakazawa, K. and M. Shibata. 1956 in Imamura, A., M. Hori, K. Nakazawa, M. Shibata, S. Tatsuoka and A. Miyake. Proceedings of the Japanese Academy of Science **32**:648-653. **Type strain:** ATCC 12760
Description: Shirling, E.B. and D. Gottlieb. 1968. International Journal of Systematic Bacteriology **18**:279-392.

S. hydrogenans Lindner, Junk, Nesemann and Schmidt-Thome 1958
Lindner, F., R. Junk, G. Nesemann and J. Schmidt-Thome. 1958. Hoppe-Seyler's Zeitschrift fur Physiologische Chemie **313**: 117-123. **Type strain:** ATCC 19631
Description: Shirling, E.B. and D. Gottlieb. 1972. International Journal of Systematic Bacteriology **22**:265-394.

S. hygroscopicus (Jensen 1931) Waksman and Henrici 1948
Jensen, H.L. 1931. Proceedings of the Linnean Society of New South Wales **56**: 345-370; Waksman, S.A. and A.T. Henrici. 1948 in Breed, R.S., E.G.D. Murray and A.P. Hitchens (eds). Bergey's Manual of Determinative Bacteriology, 6th ed. The Williams and Wilkins Co., Baltimore. pp. 929-980. **Type strain:** ATCC 27438
Description: Shirling, E.B. and D. Gottlieb. 1972. International Journal of Systematic Bacteriology **22**:265-394.

S. hygroscopicus subsp. *angustmyceticus* Yuntsen, Ohkuma, Ishii and Yonehara 1956
Yuntsen, H., K. Ohkuma, Y. Ishii and H. Yonehara. 1956. Journal of Antibiotics (Tokyo) Series A **9**:195-201. **Type strain:** ATCC 15484
Description: Bergey 8.

S. hygroscopicus subsp. *decoyicus* Vavra, Dietz, Churchill, Siminoff and Koepsell 1959
Vavra, J.J., A. Dietz, B.W. Churchill, P. Siminoff and H.J. Koepsell. 1959. Antibiotics and Chemotherapy **9**:427-431.
Type strain: NRRL 2666
Description: Bergey 8.

S. hygroscopicus subsp. *glebosus* Ohmori, Okanishi and Kawaguchi 1962
Ohmori, T., M. Okanishi and H. Kawaguchi. 1962. Journal of Antibiotics (Tokyo) Series A **15**:21-27. **Type strain:** ATCC 14607
Description: Ohmori, T., M. Okanishi and H. Kawaguchi. 1962. Ibid.

S. hygroscopicus subsp. *hygroscopicus* (Jensen 1931) Waksman and Henrici 1948
Jensen, H.L. 1931. Proceedings of the Linnean Society of New South Wales **56**: 345-370; Waksman, S.A. and A.T. Henrici. 1948 in Breed, R.S., E.G.D. Murray and A.P. Hitchens (eds). Bergey's Manual of Determinative Bacteriology, 6th ed. The Williams and Wilkins Co., Baltimore. pp. 929-980. **Type strain:** ATCC 27438
Description: Shirling, E.B. and D. Gottlieb. 1972. International Journal of Systematic Bacteriology **22**:265-394.

S. hygroscopicus subsp. *ossamyceticus* Schmitz, Jubinski, Hooper, Crook, Price and Lein 1965
Schmitz, H., S.D. Jubinski, I.R. Hooper, K.E. Crook Jr., K.E. Price and J. Lein. 1965. Journal of Antibiotics (Tokyo) Series A **18**:82-88. **Type strain:** ATCC 15420
Description: Schmitz, H., S.D. Jubinski, I.R. Hooper, K.E. Crook Jr., K.E. Price and J. Lein. 1965. Ibid.

S. iakyrus deQuerioz and Albert 1962
deQuerioz, V.M. and C.A. Albert. 1962. Revista do Instituto de Antibioticos, Universidade do Recife **4**:33-46. **Type strain:** ATCC 15375
Description: Shirling, E.B. and D. Gottlieb. 1972. International Journal of Systematic Bacteriology **22**:265-394.

S. indigoferus Shinobu and Kawato 1960
Shinobu, R. and M. Kawato. 1960. Memoirs of the Osaka University of the Liberal Arts and Education B Natural Science **9**:49-53. **Type strain:** ATCC 23924
Description: Shirling, E.B. and D. Gottlieb. 1968. International Journal of Systematic Bacteriology **18**:279-392.

S. intermedius (Kruger 1904) Waksman 1953
Kruger, F. 1904. Arbeiten aus der Biologischen Abteilung fur Land- und Forstwirtschaft am Kaiserlichen Gesundheitsamte Band IV, Heft 3, Verlagsbuchhandlung Paul Parey, Verlagsbuchhandlung Julius Springer, Berlin. pp. 275-318; Waksman, S.A. 1953 in Waksman, S.A. and H.A. Lechevalier. Guide to the classification and identification of the Actinomycetes and their antibiotics. The Williams and Wilkins Co., Baltimore. pp. 1-246. **Type strain:** ATCC 3329
Description: Shirling, E.B. and D. Gottlieb. 1969. International Journal of Systematic Bacteriology **19**:391-512.

S. inusitatus Hasegawa, Yamano and Masahiko 1978
Hasegawa, T., T. Yamano and Y. Masahiko. 1978. International Journal of Systematic Bacteriology **28**:407-410. **Type strain:** IFO 13601
Description: Hasegawa, T., T. Yamano and Y. Masahiko. 1978. Ibid.

S. ipomoeae (Person and Martin 1940) Waksman and Henrici 1948
Person, L.H. and W.J. Martin. 1940. Phytopathology **30**:913-926; Waksman, S.A. and A.T. Henrici. 1948 in Breed, R.S., E.G.D. Murray and A.P. Hitchens (eds). Bergey's Manual of Determinative Bacteriology, 6th ed. The Williams and Wilkins Co., Baltimore. pp. 929-980. **Type strain:** ATCC 25462
Description: Shirling, E.B. and D. Gottlieb. 1969. International Journal of Systematic Bacteriology **19**:391-512.

S. janthinus (Artamonova and Krassilnikov 1960) Pridham 1970
Artamonova, O.I. and N.A. Krassilnikov. 1960 in Rautenshtein, Ya. I. Transactions of the Institute of Microbiology, Academy of Sciences USSR **8**:1-344; Pridham, T.G. 1970. Bulletin of the United States Department of Agriculture **1424**:1-55. **Type strain:** ATCC 15870
Description: Shirling, E.B. and D. Gottlieb. 1968. International Journal of Systematic Bacteriology **18**:279-392.

S. kanamyceticus Okami and Umezawa 1957
Okami, Y. and H. Umezawa 1957 in

Umezawa, H., M. Ueda, K. Maeda, K. Yagashita, S. Kondo, Y. Okami, R. Utahara, Y. Osato, K. Nitta and T. Takeuchi. Journal of Antibiotics (Tokyo) Series A **10**:181-188. **Type strain:** ATCC 12853

Description: Shirling, E.B. and D. Gottlieb. 1972. International Journal of Systematic Bacteriology **22**:265-394.

S. katrae Gupta and Chopra 1963

Gupta, K.C. and I.C. Chopra. 1963. Indian Journal of Microbiology **3**:1-4. **Type strain:** ATCC 27440

Description: Shirling, E.B. and D. Gottlieb. 1972. International Journal of Systematic Bacteriology **22**:265-394.

S. kurssanovii (Preobrazhenskaya, Kudrina, Ryabova and Blinov 1957) Pridham, Hesseltine and Benedict 1958

Preobrazhenskaya, T.P., E.S. Kudrina, I.D. Ryabova and N.O. Blinov. 1957 in Gauze, G.F., T.P. Preobrazhenskaya, E.S. Kudrina, N.O. Blinov, I.D. Ryabova and M.A. Sveshnikova. Problems of classification of actinomycetes-antagonists. Government Publishing House of Medical Literature, Medgiz, Moscow. pp. 1-398; Pridham, T.G., C.W. Hesseltine and R.G. Benedict. 1958. Applied Microbiology **6**:52-79. **Type strain:** ATCC 15824

Description: Shirling, E.B. and D. Gottlieb. 1968. International Journal of Systematic Bacteriology **18**:69-189.

S. lanatus Frommer 1959

Frommer, W. 1959. Archiv fur Mikrobiologie **32**:187-206. **Type strain:** ATCC 19775

Description: Shirling, E.B. and D. Gottlieb. 1968. International Journal of Systematic Bacteriology **18**:69-189.

S. lateritius (Sveshnikova 1957) Pridham, Hesseltine and Benedict 1958

Sveshnikova, M.A. 1957 in Gauze, G.F., T.P. Preobrazhenskaya, E.S. Kudrina, N.O. Blinov, I.D. Ryabova and M.A. Sveshnikova. Problems of classification of actinomycetes-antagonists. Government Publishing House of Medical Literature, Medgiz, Moscow. pp. 1-398; Pridham, T.G., C.W. Hesseltine and R.G. Benedict. 1958. Applied Microbiology **6**:52-79. **Type strain:** ATCC 19913

Description: Shirling, E.B. and D. Gottlieb. 1968. International Journal of Systematic Bacteriology **18**:69-189.

S. laurentii Trejo, Dean, Pluscec, Meyers and Brown 1979

Trejo, W.H., L.D. Dean, J. Pluscec, E. Meyers and W.E. Brown. 1979. International Journal of Systematic Bacteriology **29**:79-80; Effective publication Trejo, W.H., L.D. Dean, J. Pluscec, E. Meyers and W.E. Brown. 1977. Journal of Antibiotics **30**:639-643. **Type strain:** ATCC 31255

Description: Trejo, W.H., L.D. Dean, J. Pluscec, E. Meyers and W.E. Brown. 1977. Ibid.

S. lavendofoliae (Kuchaeva, Krassilnikov, Taptykova and Gesheva 1961) Pridham 1970

Kuchaeva, A.G., N.A. Krassilnikov, S.D. Taptykova and R.L. Gesheva. 1961. Izvestiya Na Mikrobiologicheskiya Institut, Bulgarian Academy of sciences, Class of Biological sciences, Sofia **13**:103-124; Pridham, T.G. 1970. Bulletin of the United States Department of Agriculture **1424**: 1-55. **Type strain:** ATCC 15872

Description: Shirling, E.B. and D. Gottlieb. 1968. International Journal of Systematic Bacteriology **18**:279-392.

S. lavendulae (Waksman and Curtis 1916) Waksman and Henrici 1948

Waksman, S.A. and R.E. Curtis. 1916. Soil Science **1**:99-134; Waksman, S.A. and A.T. Henrici. 1948 in Breed, R.S., E.G.D. Murray and A.P. Hitchens (eds). Bergey's Manual of Determinative Bacteriology, 6th ed. The Williams and Wilkins Co., Baltimore. pp. 929-980. **Type strain:** ATCC 8664

Description: Shirling, E.B. and D. Gottlieb. 1968. International Journal of Systematic Bacteriology **18**:69-189.

S. lavendulae subsp. *grasserius* (Kuchaeva, Krassilnikov, Taptykova and Gesheva 1961) Pridham 1970

Kuchaeva, A.G., N.A. Krassilnikov, S.D. Taptykova and R.L. Gesheva. 1961. Izvestiya Na Mikrobiologicheskiya Institut, Bulgarian Academy of Sciences, Class of Biological Sciences, Sofia **13**:103-124; Pridham, T.G. 1970. Bulletin of the United States Department of Agriculture **1424**: 1-55. **Type strain:** ATCC 15875

Description: Bergey 8.

S. lavendulae subsp. *lavendulae* (Waksman and Curtis 1916) Waksman and Henrici 1948
Waksman, S.A. and R.E. Curtis. 1916. Soil Science **1**:99-134; Waksman, S.A. and A.T. Henrici. 1948 in Breed, R.S., E.G.D. Murray and A.P. Hitchens (eds). Bergey's Manual of Determinative Bacteriology, 6th ed. The Williams and Wilkins Co., Baltimore. pp. 929-980. **Type strain:** ATCC 8664
Description: Shirling, E.B. and D. Gottlieb. 1968. International Journal of Systematic Bacteriology **18**:69-189.

S. lavendulocolor (Kuchaeva, Krassilnikov, Taptykova and Gesheva 1961) Pridham 1970
Kuchaeva, A.G., N.A. Krassilnikov, S.D. Taptykova and R.L. Gesheva. 1961. Izvestiya Na Mikrobiologicheskiya Institut, Bulgarian Academy of sciences, Class of Biological sciences, Sofia **13**:103-124; Pridham, T.G. 1970. Bulletin of the United States Department of Agriculture **1424**: 1-55. **Type strain:** ATCC 15871
Description: Bergey 8.

S. libani Baldacci and Grein 1966
Baldacci, E. and A. Grein. 1966. Giornale di Microbiologia **14**:185-198. **Type strain:** CBS 753.72
Description: Shirling, E.B. and D. Gottlieb. 1972. International Journal of Systematic Bacteriology **22**:265-394.

S. libani subsp. *libani* Baldacci and Grein 1966
Baldacci, E. and A. Grein. 1966. Giornale di Microbiologia **14**:185-198. **Type strain:** CBS 753.72
Description: Shirling, E.B. and D. Gottlieb. 1972. International Journal of Systematic Bacteriology **22**:265-394.

S. libani subsp. *rufus* Baldacci and Grein 1966
Baldacci, E. and A. Grein. 1966. Giornale di Microbiologia **14**:185-198. **Type strain:** ATCC 23731
Description: Bergey 8.

S. limosus Lindenbein 1952
Lindenbein, W. 1952. Archiv fur Mikrobiologie **17**:361-383. **Type strain:** ATCC 19778
Description: Shirling, E.B. and D. Gottlieb. 1968. International Journal of Systematic Bacteriology **18**:69-189.

S. lincolnensis Mason, Dietz and De Boer 1963
Mason, D.J., A. Dietz and C. DeBoer. 1963. Antimicrobial Agents and Chemotherapy **1962**:554-559. **Type strain:** NRRL 2936
Description: Shirling, E.B. and D. Gottlieb. 1969. International Journal of Systematic Bacteriology **19**:391-512.

S. lipmanii (Waksman and Curtis 1916) Waksman and Henrici 1948
Waksman, S.A. and R.E. Curtis. 1916. Soil Science **1**:99-134; Waksman, S.A. and A.T. Henrici. 1948 in Breed, R.S., E.G.D. Murray and A.P. Hitchens (eds). Bergey's Manual of Determinative Bacteriology, 6th ed. The Williams and Wilkins Co., Baltimore. pp. 929-980. **Type strain:** ATCC 3331
Description: Shirling, E.B. and D. Gottlieb. 1968. International Journal of Systematic Bacteriology **18**:69-189.

S. litmocidini (Ryabova and Preobrazhenskaya 1957) Pridham, Hesseltine and Benedict 1958
Ryabova, I.D. and T.P. Preobrazhenskaya. 1957 in Gauze, G.F., T.P. Preobrazhenskaya, E.S. Kudrina, N.O. Blinov, I.D. Ryabova and M.A. Sveshnikova. Problems of classification of actinomycetes-antagonists. Government Publishing House of Medical Literature, Medgiz, Moscow. pp. 1-398; Pridham, T.G., C.W. Hesseltine and R.G. Benedict. 1958. Applied Microbiology **6**:52-79. **Type strain:** ATCC 19914
Description: Shirling, E.B. and D. Gottlieb. 1968. International Journal of Systematic Bacteriology **18**:69-189.

S. lomondensis Johnson and Dietz 1969
Johnson, L.E. and A. Dietz. 1969. Applied Microbiology **17**:755-759. **Type strain:** NRRL 3252
Description: Johnson, L.E. and A. Dietz. 1969. Ibid.

S. longisporoflavus Waksman 1953
Waksman, S.A. 1953 in Waksman, S.A. and H.A. Lechevalier. Guide to the classification and identification of the Actinomycetes and their antibiotics. The Williams and Wilkins Co., Baltimore. pp. 1-246. **Type strain:** ATCC 19915
Description: Shirling, E.B. and D. Gottlieb. 1968. International Journal of Systematic Bacteriology **18**:279-392.

S. longispororuber Waksman 1953

Waksman, S.A. 1953 in Waksman, S.A. and H.A. Lechevalier. Guide to the classification and identification of the Actinomycetes and their antibiotics. The Williams and Wilkins Co., Baltimore. pp. 1-246. **Type strain:** ATCC 27443
Description: Shirling, E.B. and D. Gottlieb. 1972. International Journal of Systematic Bacteriology **22**:265-394.

S. longisporus (Krassilnikov 1941) Waksman 1953
Krassilnikov, N.A. 1941. Guide to the bacteria and Actinomycetes (in Russian). Akademiya Nauk SSSR, Moscow. pp. 1-830; Waksman, S.A. 1953 in Waksman, S.A. and H.A. Lechevalier. Guide to the classification and identification of the Actinomycetes and their antibiotics. The Williams and Wilkins Co., Baltimore. pp. 1-246. **Type strain:** ATCC 23931
Description: Shirling, E.B. and D. Gottlieb. 1968. International Journal of Systematic Bacteriology **18**:279-392.

S. lucensis Arcamone, Bertazzoli, Canevazzi, diMarco, Ghione and Grein 1957
Arcamone, F., C. Bertazzoli, G. Canevazzi, A. diMarco, M. Ghione and A. Grein. 1957. Giornale di Microbiologia **4**:119-128. **Type strain:** ATCC 17804
Description: Shirling, E.B. and D. Gottlieb. 1969. International Journal of Systematic Bacteriology **19**:391-512.

S. luridus (Krassilnikov, Korenyako, Meksina, Valedinskaya and Veselov 1957) Waksman 1961
Krassilnikov, N.A., A.I. Korenyako, M.M. Meksina, L.K. Valedinskaya and N.M. Veselov. 1957. Mikrobiologiya **26**: 558-564; Waksman, S.A. 1961. The Actinomycetes. Vol. 2. Classification, identification and descriptions of genera and species. The Williams and Wilkins Co., Baltimore. pp. 1-363. **Type strain:** ATCC 19782
Description: Shirling, E.B. and D. Gottlieb. 1968. International Journal of Systematic Bacteriology **18**:69-189.

S. lusitanus Villax 1963
Villax, I. 1963. Antimicrobial Agents and Chemotherapy **1962**:661-668. **Type strain:** NCIB 9585
Description: Shirling, E.B. and D. Gottlieb. 1972. International Journal of Systematic Bacteriology **22**:265-394.

S. luteogriseus Schmitz, Deak, Crook and Hooper 1964
Schmitz, H., S.B. Deak, K.E. Crook Jr. and I.R. Hooper. 1964. Antimicrobial Agents and Chemotherapy **1963**:89-94. **Type strain:** ATCC 15072
Description: Shirling, E.B. and D. Gottlieb. 1972. International Journal of Systematic Bacteriology **22**:265-394.

S. lydicus De Boer, Dietz, Silver and Savage 1956
De Boer, C., A. Dietz, W.S. Silver and G.M. Savage. 1956. Antibiotics Annual **1955-1956**:886-892. **Type strain:** NRRL 2433
Description: Shirling, E.B. and D. Gottlieb. 1969. International Journal of Systematic Bacteriology **19**:391-512.

S. massasporeus Shinobu and Kawato 1959
Shinobu, R. and M. Kawato. 1959. Botanical Magazine (Tokyo) **72**:283-288. **Type strain:** ATCC 19785
Description: Shirling, E.B. and D. Gottlieb. 1968. International Journal of Systematic Bacteriology **18**:69-189.

S. matensis Margalith, Beretta and Timbal 1959
Margalith, P., G. Beretta and M.T. Timbal. 1959. Antiobiotics and Chemotherapy **9**:71-75. **Type strain:** ATCC 23935
Description: Shirling, E.B. and D. Gottlieb. 1968. International Journal of Systematic Bacteriology **18**:279-392.

S. mauvecolor Okami and Umezawa 1960
Okami, Y. and H. Umezawa. 1960 in Murase, M., T. Hikiji, K. Nitta, Y. Okami, T. Takeuchi and H. Umezawa. Journal of Antibiotics (Tokyo) Series A **14**:113-118. **Type strain:** ATCC 29835
Description: Murase, M., T. Hikiji, K. Nitta, Y. Okami, T. Takeuchi and H. Umezawa. 1960. Ibid.

S. mediolani Arcamone, Camerino, Cotta, Franceschi, Grein, Penco and Spalla 1969
Arcamone, F., B. Camerino, E. Cotta, G. Franceschi, A. Grein, S. Penco and C. Spalla. 1969. Experientia **24**:241-242. **Type strain:** NCIB 10969
Description: Bianchi, M.L., A. Grein, P. Julita, M.P. Marnati and C. Spalla. 1970. Zeitschrift fur Allgemeine Mikrobiologie **10**:237-244.

S. melanogenes Sugawara and Onuma 1957

Sugawara, A. and M. Onuma. 1957. Journal of Antibiotics (Tokyo) Series A **10**:138-142. **Type strain:** ATCC 23937
Description: Shirling, E.B. and D. Gottlieb. 1968. International Journal of Systematic Bacteriology **18**:279-392.

S. melanosporofaciens Arcamone, Bertazzoli, Ghione and Scotti 1959
Arcamone, F., C. Bertazzoli, M. Ghione and T. Scotti. 1959. Giornale di Microbiologia **7**:207-216. **Type strain:** ATCC 25473
Description: Shirling, E.B. and D. Gottlieb. 1969. International Journal of Systematic Bacteriology **19**:391-512.

S. michiganensis Corbaz, Ettlinger, Keller-Schierlein and Zahner 1957
Corbaz, R., L. Ettlinger, W. Keller-Schierlein and H. Zahner. 1957. Archiv fur Mikrobiologie **26**:192-208. **Type strain:** ATCC 14970
Description: Shirling, E.B. and D. Gottlieb. 1968. International Journal of Systematic Bacteriology **18**:69-189.

S. microflavus (Krainsky 1914) Waksman and Henrici 1948
Krainsky, A. 1914. Zentralblatt fur Bakteriologie, Parasitenkunde, Infektionskrankheiten und Hygiene. Abteilung II. **41**:649-688; Waksman, S.A. and A.T. Henrici. 1948 in Breed, R.S., E.G.D. Murray and A.P. Hitchens (eds). Bergey's Manual of Determinative Bacteriology, 6th ed. The Williams and Wilkins Co., Baltimore. pp. 929-980. **Type strain:** CBS 124.18
Description: Shirling, E.B. and D. Gottlieb. 1969. International Journal of Systematic Bacteriology **19**:391-512.

S. minutiscleroticus (Thirumalachar 1965) Pridham 1970
Thirumalachar, M.J. 1965 in Thirumalachar, M.J., P.W. Rahalkar, P.V. Deshmukh and R.S. Sukapure. Hindustan Antibiotics Bulletin **8**:6-9; Pridham, T.G. 1970. Bulletin of the United States Department of Agriculture **1424**:1-55. **Type strain:** ATCC 17757
Description: Shirling, E.B. and D. Gottlieb. 1972. International Journal of Systematic Bacteriology **22**:265-394.

S. mirabilis Ruschmann 1952
Ruschmann, G. 1952. Pharmazie **7**:542-550. **Type strain:** ATCC 27447
Description: Shirling, E.B. and D. Gottlieb. 1972. International Journal of Systematic Bacteriology **22**:265-394.

S. misakiensis Nakamura 1961
Nakamura, G. 1961. Journal of Antibiotics (Tokyo) Series A **14**:86-89. **Type strain:** ATCC 23938
Description: Shirling, E.B. and D. Gottlieb. 1968. International Journal of Systematic Bacteriology **18**:279-392.

S. misionensis Cercos, Eilberg, Goyena, Souto, Vautier and Widuezynski 1962
Cercos, A.P., B.L. Eilberg, J.G. Goyena, J. Souto, E.E. Vautier and I. Widuezynski. 1962. Revista de Investigaciones Agricolas **16**:5-27. **Type strain:** ATCC 14991
Description: Shirling, E.B. and D. Gottlieb. 1969. International Journal of Systematic Bacteriology **19**:391-512.

S. murinus Frommer 1959
Frommer, W. 1959. Archiv fur Mikrobiologie **32**:187-206. **Type strain:** ATCC 19788
Description: Shirling, E.B. and D. Gottlieb. 1968. International Journal of Systematic Bacteriology **18**:69-189.

S. mutabilis (Preobrazhenskaya and Ryabova 1957) Pridham, Hesseltine and Benedict 1958
Preobrazhenskaya, T.P. and I.D. Ryabova. 1957 in Gauze, G.F., T.P. Preobrazhenskaya, E.S. Kudrina, N.O. Blinov, I.D. Ryabova and M.A. Sveshnikova. Problems of classification of actinomycetes-antagonists. Government Publishing House of Medical Literature, Medgiz, Moscow. pp. 1-398; Pridham, T.G., C.W. Hesseltine and R.G. Benedict. 1958. Applied Microbiology **6**:52-79. **Type strain:** ATCC 19919
Description: Shirling, E.B. and D. Gottlieb. 1968. International Journal of Systematic Bacteriology **18**:69-189.

S. naganishii Yamaguchi and Saburi 1955
Yamaguchi, H. and Y. Saburi. 1955. Journal of General and Applied Microbiology **1**:201-135. **Type strain:** ATCC 23939
Description: Shirling, E.B. and D. Gottlieb. 1968. International Journal of Systematic Bacteriology **18**:279-392.

S. narbonensis Corbaz, Ettlinger, Gaumann,

Keller-Schierlein, Kradolfer, Kyburz, Neipp, Prelog, Reusser and Zahner 1955
Corbaz, R., L. Ettlinger, E. Gaumann, W. Keller-Schierlein, F. Kradolfer, E. Kyburz, L. Neipp, V. Prelog, R. Reusser and H. Zahner. 1955. Helvetica Chimica Acta **38**:935-942. **Type strain:** ATCC 19790
Description: Shirling, E.B. and D. Gottlieb. 1968. International Journal of Systematic Bacteriology **18**:69-189.

S. nashvillensis McVeigh and Reyes 1961
McVeigh, I. and C.R. Reyes. 1961. Antibiotics and Chemotherapy **11**:312-319. **Type strain:** ATCC 25476
Description: Shirling, E.B. and D. Gottlieb. 1969. International Journal of Systematic Bacteriology **19**:391-512.

S. neyagawaensis Yamamoto, Nakazawa, Horii and Miyake 1960
Yamamoto, H., K. Nakazawa, S. Horii and A. Miyake. 1960. Journal of the Agricultural Chemical Society of Japan **34**: 268-272. **Type strain:** ATCC 27449
Description: Shirling, E.B. and D. Gottlieb. 1972. International Journal of Systematic Bacteriology **22**:265-394.

S. nigrescens (Sveshnikova 1957) Pridham, Hesseltine and Benedict 1958
Sveshnikova, M.H. 1957 in Gauze, G.F., T.P. Preobrazhenskaya, E.S. Kudrina, N.O. Blinov, I.D. Ryabova and M.A. Sveshnikova. Problems in the classification of actinomycetes-antagonists. State Publishing House of Medical Literature, Medgiz, Moscow, USSR; Pridham, T.G., C.W. Hesseltine and R.G. Benedict. 1958. Applied Microbiology **6**:52-79. **Type strain:** ATCC 23941
Description: Shirling, E.B. and D. Gottlieb. 1968. International Journal of Systematic Bacteriology **18**:279-392.

S. nigrifaciens Waksman 1961
Waksman, S.A. 1961. The Actinomycetes. Vol. 2. Classification, identification and descriptions of genera and species. The Williams and Wilkins Co., Baltimore. pp. 1-363. **Type strain:** ATCC 19791
Description: Shirling, E.B. and D. Gottlieb. 1968. International Journal of Systematic Bacteriology **18**:69-189.

S. nitrosporeus Okami 1952
Okami, Y. 1952. Classification of the antagonistic ray fungi of Japan of the family Streptomycetaceae (in Japanese). Doctoral Dissertation, Hokkaido University. **Type strain:** ATCC 12769
Description: Shirling, E.B. and D. Gottlieb. 1968. International Journal of Systematic Bacteriology **18**:69-189.

S. niveoruber Ettlinger, Corbaz and Hutter 1958
Ettlinger, L., R. Corbaz and R. Hutter. 1958. Archiv fur Mikrobiologie **31**:326-358. **Type strain:** ATCC 14971
Description: Ettlinger, L., R. Corbaz and R. Hutter. 1958. Ibid.

S. niveus Smith, Dietz, Sokolski and Savage 1956
Smith, C.G., A. Dietz, W.T. Sokolski and G.M. Savage. 1956. Antibiotics and Chemotherapy **6**:135-142. **Type strain:** NRRL 2466
Description: Hutter, R. 1963. Giornale di Microbiologia **11**:191-246.

S. noboritoensis Isono, Yamashita, Tomiyama, Suzuki and Sakai 1957
Isoni, K., S. Yamashita, Y. Tomiyama, S. Suzuki and H. Sakai. 1957. Journal of Antibiotics (Tokyo) Series A **10**:21-30. **Type strain:** ATCC 25477
Description: Shirling, E.B. and D. Gottlieb. 1969. International Journal of Systematic Bacteriology **19**:391-512.

S. nodosus Trejo 1961
Trejo, W. 1961 in Waksman, S.A. The Actinomycetes. Vol. 2. Classification, identification and descriptions of genera and species. The Williams and Wilkins Co., Baltimore. pp. 1-363. **Type strain:** ATCC 14899
Description: Shirling, E.B. and D. Gottlieb. 1968. International Journal of Systematic Bacteriology **18**:279-392.

S. nogalater Bhuyan and Dietz 1966
Bhuyan, B.K. and A. Dietz. 1966. Antimicrobial Agents and Chemotherapy **1965**: 836-844. **Type strain:** NRRL 3035
Description: Bhuyan, B.K. and A. Dietz. 1966. Ibid.

S. nojiriensis Ishida, Kumagai, Niida, Hamamoto and Shomura 1967
Ishida, N., K. Kumagai, T. Niida, K. Hamamoto and T. Shomura. 1967. Journal of Antibiotics (Tokyo) Series A **20**:62-65. **Type strain:** ATCC 29781

Description: Ishida, N., K. Kumagai, T. Niida, K. Hamamoto and T. Shomura. 1967. Ibid.

S. noursei Brown, Hazen and Mason 1953
Brown, R., E.L. Hazen and A. Mason. 1953. Science **117**:609-610. **Type strain:** ATCC 11455
Description: Ettlinger, L., R. Corbaz and R. Hutter. 1958. Archiv fur Mikrobiologie **31**:326-358.

S. novaecaesareae Waksman and Henrici 1948
Waksman, S.A. and A.T. Henrici. 1948 in Breed, R.S., E.G.D. Murray and A.P. Hitchens (eds). Bergey's Manual of Determinative Bacteriology, 6th ed. The Williams and Wilkins Co., Baltimore. pp. 929-980. **Type strain:** ATCC 27452
Description: Shirling, E.B. and D. Gottlieb. 1972. International Journal of Systematic Bacteriology **22**:265-394.

S. ochraceiscleroticus Pridham 1970
Pridham, T.G. 1970. Bulletin of the United States Department of Agriculture **1424**: 1-55. **Type strain:** ATCC 15814
Description: Shirling, E.B. and D. Gottlieb. 1972. International Journal of Systematic Bacteriology **22**:265-394.

S. odorifer (Rullmann 1895) Waksman 1953
Rullmann, W. 1895. Inaugural Dissertation. Akademische Buchdruckerei von F. Strauv, Munchen, pp. 1-47; Waksman, S.A. 1953 in Waksman, S.A. and H.A. Lechevalier. Guide to the classification and identification of the Actinomycetes and their antibiotics. The Williams and Wilkins Co., Baltimore. pp. 1-246. **Type strain:** ATCC 6246
Description: Shirling, E.B. and D. Gottlieb. 1972. International Journal of Systematic Bacteriology **22**:265-394.

S. olivaceiscleroticus Pridham 1970
Pridham, T.G. 1970. Bulletin of the United States Department of Agriculture **1424**: 1-55. **Type strain:** ATCC 15722
Description: Shirling, E.B. and D. Gottlieb. 1972. International Journal of Systematic Bacteriology **22**:265-394.

S. olivaceoviridis (Preobrazhenskaya and Ryabova 1957) Pridham, Hesseltine and Benedict 1958
Preobrazhenskaya, T.P. and I.D. Ryabova. 1957 in Gauze, G.F., T.P. Preobrazhenskaya, E.S. Kudrina, N.O. Blinov, I.D. Ryabova and M.A. Sveshnikova. Problems of classification of actinomycetes-antagonists. Government Publishing House of Medical Literature, Medgiz, Moscow. pp. 1-398; Pridham, T.G., C.W. Hesseltine and R.G. Benedict. 1958. Applied Microbiology **6**:52-79. **Type strain:** ATCC 23630
Description: Shirling, E.B. and D. Gottlieb. 1969. International Journal of Systematic Bacteriology **19**:391-512.

S. olivaceus (Waksman 1923) Waksman and Henrici 1948
Waksman, S.A. 1923 in Bergey, D.H., F.C. Harrison, R.S. Breed, B.W. Hammer and F.M. Huntoon. Bergey's Manual of Determinative Bacteriology, 1st ed. The Williams and Wilkins Co., Baltimore. pp. 339-371; Waksman, S.A. and A.T. Henrici. 1948 in Breed, R.S., E.G.D. Murray and A.P. Hitchens (eds). Bergey's Manual of Determinative Bacteriology, 6th ed. The Williams and Wilkins Co., Baltimore. pp. 929-980. **Type strain:** ATCC 3335
Description: Shirling, E.B. and D. Gottlieb. 1968. International Journal of Systematic Bacteriology **18**:68-189.

S. olivochromogenes (Waksman 1923) Waksman and Henrici 1948
Waksman, S.A. 1923 in Bergey, D.H., F.C. Harrison, R.S. Breed, B.W. Hammer and F.M. Huntoon. Bergey's Manual of Determinative Bacteriology, 1st ed. The Williams and Wilkins Co., Baltimore. pp. 339-371; Waksman, S.A. and A.T. Henrici. 1948 in Breed, R.S., E.G.D. Murray and A.P. Hitchens (eds). Bergey's Manual of Determinative Bacteriology, 6th ed. The Williams and Wilkins Co., Baltimore. pp. 929-980. **Type strain:** ATCC 3336
Description: Shirling, E.B. and D. Gottlieb. 1969. International Journal of Systematic Bacteriology **19**:391-512.

S. olivoviridis (Kuchaeva, Krassilnikov, Skryabin and Taptykova 1960) Pridham 1970
Kuchaeva, A.G., N.A. Krassilnikov, G.K. Skryabin and S.D. Taptykova. 1960 in Rautenshtein, Ya. I. Transactions of the Institute of Microbiology, Academy of Sciences USSR **8**:1-344; Pridham, T.G. 1970. Bulletin of the United States De-

partment of Agriculture **1424**:1-55. **Type strain:** ATCC 15882
Description: Shirling, E.B. and D. Gottlieb. 1968. International Journal of Systematic Bacteriology **18**:279-392.

S. omiyaensis Umezawa and Okami 1950
Umezawa, H. and Y. Okami. 1950 in Umezawa, H., T. Tazaki, Y. Okami and S. Fukuyama. Journal of Antibiotics (Tokyo) **3**:292-296. **Type strain:** ATCC 27454
Description: Shirling, E.B. and D. Gottlieb. 1972. International Journal of Systematic Bacteriology **22**:265-394.

S. pactum Bhuyan, Dietz and Smith 1962
Bhuyan, B.K., A. Dietz and C.G. Smith. 1962. Antimicrobial Agents and Chemotherapy **1961**:184-190. **Type strain:** NRRL 2939
Description: Bhuyan, B.K., A. Dietz and C.G. Smith. 1962. Ibid.

S. parvulus Waksman and Gregory 1954
Waksman, S.A. and F.J. Gregory. 1954. Antibiotics and Chemotherapy **4**:1050-1056. **Type strain:** ATCC 12434
Description: Shirling, E.B. and D. Gottlieb. 1968. International Journal of Systematic Bacteriology **18**:69-189.

S. parvus (Krainsky 1914) Waksman and Henrici 1948
Krainsky, A. 1914. Zentralblatt fur Bakteriologie, Parasitenkunde, Infektionskrankheiten und Hygiene. Abteilung II. **41**:649-688; Waksman, S.A. and A.T. Henrici. 1948 in Breed, R.S., E.G.D. Murray and A.P. Hitchens (eds). Bergey's Manual of Determinative Bacteriology, 6th ed. The Williams and Wilkins Co., Baltimore. pp. 929-980. **Type strain:** NRRL B-1455
Description: Shirling, E.B. and D. Gottlieb. 1972. International Journal of Systematic Bacteriology **22**:265-394.

S. peucetius Grein, Spalla, Di Marco and Canevazzi 1963
Grein, A., C. Spalla, A. Di Marco and G. Canevazzi. 1963. Giornale de Microbiologia **11**:109-118. **Type strain:** ATCC 29050
Description: Bergey 8.

S. phaeochromogenes (Conn 1917) Waksman 1957
Conn, H.J. 1917. Bulletin of the New York State Agricultural Experiment Station **60**:3-25; Waksman, S.A. 1957 in Breed, R.S., E.G.D. Murray and N.R. Smith (eds). Bergey's Manual of Determinative Bacteriology, 7th ed. The Williams and Wilkins Co., Baltimore. p. 778. **Type strain:** ATCC 3338
Description: Ettlinger, L., R. Corbaz and R. Hutter. 1958. Archiv fur Mikrobiologie **32**:326-358.

S. phaeofaciens Maeda, Okami, Taya and Umezawa 1952
Maeda, K., Y. Okami, O. Taya and H. Umezawa. 1952. Japanese Journal of the Medical Sciences and Biology **5**:327-339. **Type strain:** ATCC 15034
Description: Shirling, E.B. and D. Gottlieb. 1972. International Journal of Systematic Bacteriology **22**:265-394.

S. phaeopurpureus Shinobu 1957
Shinobu, R. 1957. Memoirs of the Osaka University of the Liberal Arts and Education B Natural Science **6**:63-73. **Type strain:** ATCC 23946
Description: Shirling, E.B. and D. Gottlieb. 1968. International Journal of Systematic Bacteriology **18**:279-392.

S. phaeoviridis Shinobu 1957
Shinobu, R. 1957. Memoirs of the Osaka University of the Liberal Arts and Education B Natural Science **6**:63-73. **Type strain:** ATCC 23947
Description: Shirling, E.B. and D. Gottlieb. 1968. International Journal of Systematic Bacteriology **18**:279-392.

S. pilosus Ettlinger, Corbaz and Hutter 1958
Ettlinger, L., R. Corbaz and R. Hutter. 1958. Archiv fur Mikrobiologie **31**:326-358. **Type strain:** ATCC 19797
Description: Shirling, E.B. and D. Gottlieb. 1968. International Journal of Systematic Bacteriology **18**:69-189.

S. platensis Tresner and Backus 1956
Tresner, H.D. and E.J. Backus. 1956. Applied Microbiology **4**:243-250. **Type strain:** NRRL 2364
Description: Shirling, E.B. and D. Gottlieb. 1968. International Journal of Systematic Bacteriology **18**:279-392.

S. plicatus Pridham, Hesseltine and Benedict 1958
Pridham, T.G., C.W. Hesseltine and R.G. Benedict. 1958. Applied Microbiology **6**:52-79. **Type strain:** NRRL 2428
Description: Shirling, E.B. and D.

Gottlieb. 1969. International Journal of Systematic Bacteriology **19**:391-512.

S. pluricolorescens Okami and Umezawa 1961
Okami, Y. and H. Umezawa. 1961 in Waksman, S.A. The Actinomycetes. Vol. 2. Classification, identification and descriptions of genera and species. The Williams and Wilkins Co., Baltimore. pp. 259-260. **Type strain:** ATCC 19798
Description: Shirling, E.B. and D. Gottlieb. 1968. International Journal of Systematic Bacteriology **18**:69-189.

S. polychromogenes Hageman, Penasse and Teillon 1964
Hageman, G., L. Penasse and J. Teillon. 1964 in Hutter, R. Zentralblatt fur Bakteriologie, Parasitenkunde, Infektionskrankheiten und Hygiene. Abteilung II. **117**:603- 661. **Type strain:** ATCC 12595
Description: Shirling, E.B. and D. Gottlieb. 1969. International Journal of Systematic Bacteriology **19**:391-512.

S. poonensis (Thirumalachar 1960) Pridham 1970
Thirumalachar, M.J. 1960 in Kalakoutskii, L.V. and N.A. Krassilnikov in Rautenshtein, Ya. I. Transactions of the Institute of Microbiology, Academy of Sciences USSR **8**:1-344; Pridham, T.G. 1970. Bulletin of the United States Department of Agriculture **1424**:1-55. **Type strain:** ATCC 15723
Description: Shirling, E.B. and D. Gottlieb. 1972. International Journal of Systematic Bacteriology **22**:265-394.

S. praecox (Millard and Burr 1926) Waksman 1953
Millard, W.A. and S. Burr. 1926. Annals of Applied Biology **13**:580-644; Waksman, S.A. 1953 in Waksman, S.A. and H.A. Lechevalier. Guide to the classification and identification of the Actinomycetes and their antibiotics. The Williams and Wilkins Co., Baltimore. pp. 1-246. **Type strain:** ATCC 3374
Description: Shirling, E.B. and D. Gottlieb. 1969. International Journal of Systematic Bacteriology **19**:391-512.

S. prasinopilosus Ettlinger, Corbaz and Hutter 1958
Ettlinger, L., R. Corbaz and R. Hutter. 1958. Archiv fur Mikrobiologie **31**:326-358. **Type strain:** ATCC 19799
Description: Shirling, E.B. and D. Gottlieb. 1968. International Journal of Systematic Bacteriology **18**:69-189.

S. prasinosporus Tresner, Hayes and Backus 1966
Tresner, H., J.A. Hayes and E.J. Backus. 1966. International Journal of Systematic Bacteriology **16**:161-169. **Type strain:** ATCC 17918
Description: Tresner, H., J.A. Hayes and E.J. Backus. 1966. Ibid.

S. prasinus Ettlinger, Corbaz and Hutter 1958
Ettlinger, L., R. Corbaz and R. Hutter. 1958. Archiv fur Mikrobiologie **31**:326-358. **Type strain:** ATCC 19800
Description: Ettlinger, L., R. Corbaz and R. Hutter. 1958. Ibid.

S. prunicolor (Ryabova and Preobrazhenskaya 1957) Pridham, Hesseltine and Benedict 1958
Ryabova, I.D. and T.P. Preobrazhenskaya. 1957 in Gauze, G.F., T.P. Preobrazhenskaya, E.S. Kudrina, N.O. Blinov, I.D. Ryabova and M.A. Sveshnikova. Problems of classification of actinomycetes-antagonists. Government Publishing House of Medical Literature, Medgiz, Moscow. pp. 1-398; Pridham, T.G., C.W. Hesseltine and R.G. Benedict. 1958. Applied Microbiology **6**:52-79. **Type strain:** ATCC 25487
Description: Shirling, E.B. anD D. Gottlieb. 1969. International Journal of Systematic Bacteriology **19**:391-512.

S. psammoticus Virgilio and Hengeller 1960
Virgilio, A. and C. Hengeller. 1960. Farmaco, Edizione Scientifica **15**:164-174. **Type strain:** CBS 175.61
Description: Shirling, E.B. and D. Gottlieb. 1969. International Journal of Systematic Bacteriology **19**:391-512.

S. pseudogriseolus Okami and Umezawa 1955
Okami, Y. and H. Umezawa. 1955 in Okami, Y., R. Utahara, H. Oyagi, S. Makamura, H. Umezawa, K. Yanagisawa and Y. Tsunematsu. Journal of Antibiotics (Tokyo) Series A **8**:126-131. **Type strain:** ATCC 12770
Description: Hutter, R. 1967. Systematik der Streptomyceten unter besonderer Berucksichtigung der von ihnen gebildeten Antibiotica. Bibl. Microbiol. Fasc. 6,5. Karger AG, Basel, Switzerland. pp. 262-

288.

S. pseudovenezuelae (Kuchaeva, Krassilnikov, Taptykova and Gesheva 1961) Pridham 1970
Kuchaeva, A.G., N.A. Krassilnikov, S.D. Taptykova and R.L. Gesheva. 1961. Izvestiya Na Mikrobiologicheskiya Institut, Bulgarian Academy of Sciences, Class of Biological Sciences, Sofia **13**:103-124; Pridham, T.G. 1970. Bulletin of the United States Department of Agriculture **1424**: 1-55. **Type strain:** ATCC 23951
Description: Shirling, E.B. and D. Gottlieb. 1968. International Journal of Systematic Bacteriology **18**:279-392.

S. pulveraceus Shibata, Higashide, Kanzaki, Yamamoto and Nakazawa 1961
Shibata, M., E. Higashide, T. Kanzaki, H. Yamamoto and K. Nakazawa. 1961. Agricultural and Biological Chemistry **25**: 171-175. **Type strain:** ATCC 13875
Description: Shibata, M., E. Higashide, T. Kanzaki, H. Yamamoto and K. Nakazawa. 1961. Ibid.

S. puniceus Patelski 1951
Patelski, R.A. 1951 in Routien, J.B. and A. Hofmann. Antibiotics and Chemotherapy **1**:387-389. **Type strain:** ATCC 19801
Description: Hutter, R. 1963. Giornale di Microbiologia **11**:191-246.

S. purpeofuscus Yamaguchi and Saburi 1955
Yamaguchi, T. and Y. Saburi. 1955. Journal of General and Applied Microbiology **1**:201-235. **Type strain:** ATCC 23952
Description: Shirling, E.B. and D. Gottlieb. 1968. International Journal of Systematic Bacteriology **18**:279-392.

S. purpurascens Lindenbein 1952
Lindenbein, W. 1952. Archiv fur Mikrobiologie **17**:361-383. **Type strain:** ATCC 25489
Description: Shirling, E.B. and D. Gottlieb. 1969. International Journal of Systematic Bacteriology **19**:391-512.

S. purpurogeneiscleroticus Pridham 1970
Pridham, T.G. 1970. Bulletin of the United States Department of Agriculture **1424**: 1-55. **Type strain:** ATCC 19348
Description: Shirling, E.B. and D. Gottlieb. 1972. International Journal of Systematic Bacteriology **22**:265-394.

S. racemochromogenes Sugai 1956
Sugai, T. 1956. Journal of Antibiotics (Tokyo) Series B **9**:170-179. **Type strain:** ATCC 23954
Description: Shirling, E.B. and D. Gottlieb. 1968. International Journal of Systematic Bacteriology **18**:279-392.

S. rameus Shibata 1959
Shibata, M. 1959. Journal of Antibiotics (Tokyo) Series B **12**:398-400. **Type strain:** ATCC 21273
Description: Shibata, M. 1959. Ibid.

S. ramulosus Ettlinger, Gaumann, Hutter, Keller-Schierlein, Kradolfer, Neipp, Prelog and Zahner 1958
Ettlinger, L., E. Gaumann, R. Hutter, W. Keller-Schierlein, F. Kradolfer, L. Neipp, V. Prelog and H. Zahner. 1958. Helvetica Chimica Acta **41**:216-219. **Type strain:** ATCC 19802
Description: Ettlinger, L., E. Gaumann, R. Hutter, W. Keller-Schierlein, F. Kradolfer, L. Neipp, V. Prelog and H. Zahner. 1958. Ibid.

S. rangoon (Erikson 1935) Pridham, Hesseltine and Benedict 1958
Erikson, D. 1935. Medical Research Council (Great Britain) Special Report Series **203**:5-61; Pridham, T.G., C.W. Hesseltine and R.G. Benedict. 1958. Applied Microbiology **6**:52-79. **Type strain:** ATCC 6860
Description: Shirling, E.B. and D. Gottlieb. 1969. International Journal of Systematic Bacteriology **19**:391-512.

S. recifensis (Goncalves de Lima, Machado, Araujo, Falcao de Morais and Biermann 1955) Falcao de Morais, Goncalves de Lima and Dalia Maia 1957
Goncalves de Lima, O., M.P. Machado, L. de A. Araujo, J.O. Falcao de Morais and H. Biermann. 1955. Anais da Sociedade de Biologia de Pernambuco **13**:21-36; Falcao de Morais, J.O., O. Goncalves de Lima and M.H. Dalia Maia. 1957. Anais da Sociedade de Biologia de Pernambuco **15**:239-253. **Type strain:** ATCC 19803
Description: Shirling, E.B. and D. Gottlieb. 1968. International Journal of Systematic Bacteriology **18**:69-189.

S. regensis Gupta, Sobti and Chopra 1963
Gupta, K.C., R.R. Sobti and I.C. Chopra. 1963. Hindustan Antibiotics Bulletin

6:12-16. **Type strain:** ATCC 27461
Description: Shirling, E.B. and D. Gottlieb. 1972. International Journal of Systematic Bacteriology **22**:265-394.

S. resistomycificus Lindenbein 1952
Lindenbein, W. 1952. Archiv fur Mikrobiologie **17**:361-383. **Type strain:** NRRL 2290
Description: Shirling, E.B. and D. Gottlieb. 1968. International Journal of Systematic Bacteriology **18**:69-189.

S. rimosus Sobin, Finlay and Kane 1953
Sobin, B.A., A.C. Finlay and J.H. Kane. 1953 in Waksman, S.A. and H.A. Lechevalier. Guide to the classification and identification of the Actinomycetes and their antibiotics. The Williams and Wilkins Co., Baltimore. pp. 1-246. **Type strain:** NRRL 2234
Description: Shirling, E.B. and D. Gottlieb. 1968. International Journal of Systematic Bacteriology **18**:279-392.

S. rimosus subsp. *paromomycinus* Coffey, Anderson, Fisher, Galbraith, Hillegas, Kohberger, Thompson, Weston and Ehrlich 1959
Coffey, G.L., L.E. Anderson, M.W. Fisher, M.M. Galbraith, A.B. Hillegas, D.L. Kohberger, P.E. Thompson, K.S. Weston and J. Ehrlich. 1959. Antibiotics and Chemotherapy **9**:730-738. **Type strain:** NRRL 2455
Description: Bergey 8.

S. rimosus subsp. *rimosus* Sobin, Finlay and Kane 1953
Sobin, B.A., A.C. Finlay and J.H. Kane. 1953 in Waksman, S.A. and H.A. Lechevalier. Guide to the classification and identification of the Actinomycetes and their antibiotics. The Williams and Wilkins Co., Baltimore. pp. 1-246. **Type strain:** NRRL 2234
Description: Shirling, E.B. and D. Gottlieb. 1968. International Journal of Systematic Bacteriology **18**:279-392.

S. rishiriensis Kawaguchi, Tsukiura, Okanishi, Miyaki, Ohmori, Fujisawa and Koshiyama 1965
Kawaguchi, H., H. Tsukiura, M. Okanishi, T. Miyaki, T. Ohmori, K. Fujisawa and H. Koshiyama. 1965. Journal of Antibiotics (Tokyo) Series A **18**:1-10. **Type strain:** ATCC 14812
Description: Shirling, E.B. and D. Gottlieb. 1972. International Journal of Systematic Bacteriology **22**:265-394.

S. rochei Berger, Jampolsky and Goldberg 1953
Berger, J., L.M. Jampolsky and M.W. Goldberg. 1953 in Waksman, S.A. and H.A. Lechevalier. Guide to the classification and identification of the Actinomycetes and their antibiotics. The Williams and Wilkins Co., Baltimore. pp. 1-246. **Type strain:** ATCC 10739
Description: Shirling, E.B. and D. Gottlieb. 1968. International Journal of Systematic Bacteriology **18**:279-392.

S. roseiscleroticus Pridham 1970
Pridham, T.G. 1970. Bulletin of the United States Department of Agriculture **1424**: 1-55. **Type strain:** ATCC 17755
Description: Shirling, E.B. and D. Gottlieb. 1972. International Journal of Systematic Bacteriology **22**:265-394.

S. roseodiastaticus (Duche 1934) Waksman 1953
Duche, J. 1934. Encyclopedie Mycologique **6**:1-375; Waksman, S.A. 1953 in Waksman, S.A. and H.A. Lechevalier. Guide to the classification and identification of the actinomycetes and their antibiotics. The Williams and Wilkins Co., Baltimore. pp. 1-246. **Type strain:** CBS 102.34
Description: Ettlinger, L., R. Corbaz and R. Hutter. 1958. Archiv fur Mikrobiologie **31**:326-358.

S. roseoflavus Arai 1951
Arai, T. 1951. Journal of Antibiotics (Tokyo) Series A **4**:215-221. **Type strain:** NRRL B-2789
Description: Shirling, E.B. and D. Gottlieb. 1972. International Journal of Systematic Bacteriology **22**:265-394.

S. roseofulvus (Preobrazhenskaya 1957) Pridham, Hesseltine and Benedict 1958
Preobrazhenskaya, T.P. 1957 in Gauze, G.F., T.P. Preobrazhenskaya, E.S. Kudrina, N.O. Blinov, I.D. Ryabova and M.A. Sveshnikova. Problems of classification of actinomycetes-antagonists. Government Publishing House of Medical Literature Medgiz, Moscow. pp. 1-398; Pridham, T.G., C.W. Hesseltine and R.G. Benedict. 1958. Applied Microbiology **6**:52-79. **Type strain:** ATCC 19921

Description: Shirling, E.B. and D. Gottlieb. 1968. International Journal of Systematic Bacteriology **18**:69-189.

S. roseolilacinus (Preobrazhenskaya and Sveshnikova 1957) Pridham, Hesseltine and Benedict 1958
Preobrazhenskaya, T.P. and M.A. Sveshnikova. 1957 in Gauze, G.F., T.P. Preobrazhenskaya, E.S. Kudrina, N.O. Blinov, I.D. Ryabova and M.A. Sveshnikova. Problems of classification of actinomycetes-antagonists. Government Publishing House of Medical Literature, Medgiz, Moscow. pp. 1-398; Pridham, T.G., C.W. Hesseltine and R.G. Benedict. 1958. Applied Microbiology **6**:52-79. **Type strain:** ATCC 19922
Description: Shirling, E.B. and D. Gottlieb. 1968. International Journal of Systematic Bacteriology **18**:69-189.

S. roseolus (Preobrazhenskaya and Sveshnikova 1957) Pridham, Hesseltine and Benedict 1958
Preobrazhenskaya, T.P. and M.A. Sveshnikova. 1957 in Gauze, G.F., T.P. Preobrazhenskaya, E.S. Kudrina, N.O. Blinov, I.D. Ryabova and M.A. Sveshnikova. Problems of classification of actinomycetes-antagonists. Government Publishing House of Medical Literature, Medgiz, Moscow. pp. 1-398; Pridham, T.G., C.W. Hesseltine and R.G. Benedict. 1968. Applied Microbiology **6**:52-79. **Type strain:** ATCC 23210
Description: Shirling, E.B. and D. Gottlieb. 1968. International Journal of Systematic Bacteriology **18**:69-189.

S. roseosporus Falcao de Morais and Dalia Maia 1961
Falcao de Morais, J.O. and M.H. Dalia Maia. 1961. Revista do Instituto de Antibioticos, Universidade do Recife **3**:33-60. **Type strain:** ATCC 23958
Description: Shirling, E.B. and D. Gottlieb. 1968. International Journal of Systematic Bacteriology **18**:279-392.

S. roseoviolaceus (Sveshnikova 1957) Pridham, Hesseltine and Benedict 1958
Sveshnikova, M.A. 1957 in Gauze, G.F., T.P. Preobrazhenskaya, E.S. Kudrina, N.O. Blinov, I.D. Ryabova and M.A. Sveshnikova. Problems of classification of actinomycetes-antagonists. Government Publishing House of Medical Literature, Medgiz, Moscow. pp. 1-398; Pridham, T.G., C.W. Hesseltine and R.G. Benedict. 1958. Applied Microbiology **6**:52-79. **Type strain:** ATCC 25493
Description: Shirling, E.B. and D. Gottlieb. 1969. International Journal of Systematic Bacteriology **19**:391-512.

S. roseoviridis (Preobrazhenskaya 1957) Pridham, Hesseltine and Benedict 1958
Preobrazhenskaya, T.P. 1957 in Gauze, G.F., T.P. Preobrazhenskaya, E.S. Kudrina, N.O. Blinov, I.D. Ryabova and M.A. Sveshnikova. Problems of classification of actinomycetes-antagonists. Government Publishing House of Medical Literature, Medgiz, Moscow. pp. 1-398; Pridham, T.G., C.W. Hesseltine and R.G. Benedict. 1958. Applied Microbiology **6**:52-79. **Type strain:** ATCC 23959
Description: Shirling, E.B. and D. Gottlieb. 1968. International Journal of Systematic Bacteriology **18**:279-392.

S. rubiginosohelvolus (Kudrina 1957) Pridham, Hesseltine and Benedict 1958
Kudrina, E.S. 1957 in Gauze, G.F., T.P. Preobrazhenskaya, E.S. Kudrina, N.O. Blinov, I.D. Ryabova and M.A. Sveshnikova. Problems of classification of actinomycetes-antagonists. Government Publishing House of Medical Literature, Medgiz, Moscow. pp. 1-398; Pridham, T.G., C.W. Hesseltine and R.G. Benedict. 1958. Applied Microbiology **6**:52-79. **Type strain:** ATCC 19926
Description: Shirling, E.B. and D. Gottlieb. 1968.

S. rubiginosus (Preobrazhenskaya, Blinov and Ryabova 1957) Pridham, Hesseltine and Benedict 1958
Preobrazhenskaya, T.P., N.O. Blinov and I.D. Ryabova. 1957 in Gauze, G.F., T.P. Preobrazhenskaya, E.S. Kudrina, N.O. Blinov, I.D. Ryabova and M.A. Sveshnikova. Problems of classification of actinomycetes-antagonists. Government Publishing House of Medical Literature, Medgiz, Moscow. pp. 1-398; Pridham, T.G., C.W. Hesseltine and R.G. Benedict. 1958. Applied Microbiology **6**:52-79. **Type strain:** ATCC 19927
Description: Shirling, E.B. and D. Gottlieb. 1968. International Journal of

Systematic Bacteriology **18**:279-392.

S. rutgersensis (Waksman and Curtis 1916) Waksman and Henrici 1948
Waksman, S.A. and R.E. Curtis. 1916. Soil Science **1**:99-134; Waksman, S.A. and A.T. Henrici. 1948 in Breed, R.S., E.G.D. Murray and A.P. Hitchens (eds). Bergey's Manual of Determinative Bacteriology, 6th ed. The Williams and Wilkins Co., Baltimore. pp. 929-980. **Type strain:** ATCC 3350
Description: Shirling, E.B. and D. Gottlieb. 1968. International Journal of Systematic Bacteriology **18**:69-189.

S. rutgersensis subsp. *castelarensis* Cercos 1954
Cercos, A.P. 1954. Revista de Investigaciones Agricolas **8**:263-283. **Type strain:** ATCC 15191
Description: Bergey 8.

S. rutgersensis subsp. *rutgersensis* (Waksman and Curtis 1916) Waksman and Henrici 1948
Waksman, S.A. and R.E. Curtis. 1916. Soil Science **1**:99-134; Waksman, S.A. and A.T. Henrici. 1948 in Breed, R.S., E.G.D. Murray and A.P. Hitchens (eds). Bergey's Manual of Determinative Bacteriology, 6th ed. The Williams and Wilkins Co., Baltimore. pp. 929-980. **Type strain:** ATCC 3350
Description: Shirling, E.B. and D. Gottlieb. 1968. International Journal of Systematic Bacteriology **18**:69-189.

S. sampsonii (Millard and Burr 1926) Waksman 1953
Millard, W.A. and S. Burr. 1926. Annals of Applied Biology **13**:580-644; Waksman, S.A. 1953 in Waksman, S.A. and H.A. Lechevalier. Guide to the classification and identification of the actinomycetes and their antibiotics. The Williams and Wilkins Co., Baltimore. pp. 1-246. **Type strain:** ATCC 25495
Description: Shirling, E.B. and D. Gottlieb. 1969. International Journal of Systematic Bacteriology **19**:391-512.

S. sclerotialus Pridham 1970
Pridham, T.G. 1970. Bulletin of the United States Department of Agriculture **1424**: 1-55. **Type strain:** ATCC 15721
Description: Shirling, E.B. and D. Gottlieb. 1972. International Journal of Systematic Bacteriology **22**:265-394.

S. setonii (Millard and Burr 1926) Waksman 1953
Millard, W.A. and S. Burr. 1926. Annals of Applied Biology **13**:580-644; Waksman, S.A. 1953 in Waksman, S.A. and H.A. Lechevalier. Guide to the classification and identification of the actinomycetes and their antibiotics. The Williams and Wilkins Co., Baltimore. pp. 1-246. **Type strain:** ATCC 25497
Description: Shirling, E.B. and D. Gottlieb. 1972. International Journal of Systematic Bacteriology **22**:265-394.

S. showdoensis Nishimura, Mayama, Komatsu, Kato, Shimaoka and Tanaka 1964
Nishimura, H., M. Mayama, Y. Komatsu, H. Kato, N. Shimaoka and Y. Tanaka. 1964. Journal of Antibiotics (Tokyo) Series A **17**:148-155. **Type strain:** ATCC 15105
Description: Shirling, E.B. and D. Gottlieb. 1972. International Journal of Systematic Bacteriology **22**:265-394.

S. sindenensis Nakazawa and Fujii 1957
Nakazawa, K. and S. Fujii. 1957. Annual Report Takeda Research Laboratory **16**: 109-110. **Type strain:** ATCC 23963
Description: Shirling, E.B. and D. Gottlieb. 1968. International Journal of Systematic Bacteriology **18**:279-392.

S. sioyaensis Nishimura, Okamoto, Mayama, Ohtsuka, Nakajima, Tawara, Shimohira and Shimaoka 1961
Nishimura, H., S. Okamoto, M. Mayama, H. Ohtsuka, K. Nakajima, K. Tawara, M. Shimohira and N. Shimaoka. 1961. Journal of Antibiotics (Tokyo) Series A **14**:255-263. **Type strain:** ATCC 13989
Description: Shirling, E.B. and D. Gottlieb. 1968. International Journal of Systematic Bacteriology **18**:69-189.

S. somaliensis (Brumpt 1906) Waksman and Henrici 1948
Brumpt, E. 1906. Archives de Parasitologie **10**:489-527; Waksman, S.A. and A.T. Henrici. 1948 in Breed, R.S., E.G.D. Murray and A.P. Hitchens (eds). Bergey's Manual of Determinative Bacteriology, 6th ed. The Williams and Wilkins Co., Baltimore. pp. 929-980. **Type strain:** ATCC 33201
Description: Waksman, S.A. and A.T.

Henrici. 1948. Ibid.

S. sparsogenes Owen, Dietz and Camiener 1963
Owen, S.P., A. Dietz and G.W. Camiener. 1963. Antimicrobial Agents and Chemotherapy **1962**:772-779. **Type strain:** NRRL 2940
Description: Shirling, E.B. and D. Gottlieb. 1969. International Journal of Systematic Bacteriology **19**:391-512.

S. spectabilis Mason, Dietz and Smith 1961
Mason, D.J., A. Dietz and R.M. Smith. 1961. Antibiotics and Chemotherapy **11**: 118-122. **Type strain:** NRRL 2494
Description: Shirling, E.B. and D. Gottlieb. 1972. International Journal of Systematic Bacteriology **22**:265-394.

S. spheroides Wallick, Harris, Reagan, Ruger and Woodruff 1956
Wallick, H., D.A. Harris, M.A. Reagan, M. Ruger and H.B. Woodruff. 1956. Antibiotics Annual **1955-56**:909-917. **Type strain:** NRRL 2449
Description: Shirling, E.B. and D. Gottlieb. 1968. International Journal of Systematic Bacteriology **18**:279-392.

S. spiroverticillatus Shinobu 1958
Shinobu, R. 1958. Botanical Magazine (Tokyo) **71**:87-93. **Type strain:** ATCC 19811
Description: Shirling, E.B. and D. Gottlieb. 1968. International Journal of Systematic Bacteriology **18**:69-189.

S. subrutilus Arai, Kuroda, Yamagishi and Katoh 1964
Arai, T., S. Kuroda, S. Yamagishi and Y. Katoh. 1964. Journal of Antibiotics (Tokyo) Series A **17**:23-28. **Type strain:** ATCC 27467
Description: Shirling, E.B. and D. Gottlieb. 1972. International Journal of Systematic Bacteriology **22**:265-394.

S. sulphureus (Gasperini 1894) Waksman 1953
Gasperini, G. 1894. Processi verbali della Societa Toscana di Scienze naturali di Pisa **9**:64-89; Waksman, S.A. 1953 in Waksman, S.A. and H.A. Lechevalier. Guide to the classification and identification of the actinomycetes and their antibiotics. The Williams and Wilkins Co., Baltimore. pp. 1-246. **Type strain:** ATCC 27468
Description: Shirling, E.B. and D. Gottlieb. 1972. International Journal of Systematic Bacteriology **22**:265-394.

S. tanashiensis Hata, Ohki and Higuchi 1952
Hata, T., N. Ohki and T. Higuchi. 1952. Journal of Antibiotics (Tokyo) Series A **5**: 529-534. **Type strain:** ATCC 23967
Description: Shirling, E.B. and D. Gottlieb. 1968. International Journal of Systematic Bacteriology **18**:279-392.

S. tendae Ettlinger, Corbaz and Hutter 1958
Ettlinger, L., R. Corbaz and R. Hutter. 1958. Archiv fur Mikrobiologie **31**:326-358. **Type strain:** ATCC 19812
Description: Ettlinger, L., R. Corbaz and R. Hutter. 1958. Ibid.

S. termitum Duche, Heim and Laboureur 1951
Duche, J., R. Heim and P. Laboureur. 1951 in Heim, R. Bulletin de la Societe Mycologique France **67**:359-364. **Type strain:** ATCC 25499
Description: Shirling, E.B. and D. Gottlieb. 1969. International Journal of Systematic Bacteriology **19**:391-512.

S. thermodiastaticus (Bergey, Harrison, Breed, Hammer and Huntoon 1923) Waksman 1953
Bergey, D.H., F.C. Harrison, R.S. Breed, B.W. Hammer and F.M. Huntoon. 1923. Bergey's Manual of Determinative Bacteriology, 1st ed. The Williams and Wilkins Co., Baltimore. pp. 1-442; Waksman, S.A. 1953 in Waksman, S.A. and H.A. Lechevalier. Guide to the classification and identification of the actinomycetes and their antibiotics. The Williams and Wilkins Co., Baltimore. pp. 1-246. **Type strain:** ATCC 27472
Description: Shirling, E.B. and D. Gottlieb. 1972. International Journal of Systematic Bacteriology **22**:265-394.

S. thermonitrificans Desai and Dhala 1967
Desai, A.J. and S.A. Dhala. 1967. Antonie van Leeuwenhoek Journal of Microbiology and Serology **33**:137-144. **Type strain:** ATCC 23385
Description: Shirling, E.B. and D. Gottlieb. 1972. International Journal of Systematic Bacteriology **22**:265-394.

S. thermoviolaceus Henssen 1957
Henssen, A. 1957. Archiv fur Mikrobiologie **26**:373-414. **Type strain:** ATCC 19283
Description: Shirling, E.B. and D. Gottlieb. 1972. International Journal of Systematic Bacteriology **22**:265-394.

S. thermoviolaceus subsp. *apingens* Henssen

1957
Henssen, A. 1957. Archiv fur Mikrobiologie **26**:373-414. **Type strain:** ATCC 19994
Description: Bergey 8.

S. thermoviolaceus subsp. *thermoviolaceus* Henssen 1957
Henssen, A. 1957. Archiv fur Mikrobiologie **26**:373-414. **Type strain:** ATCC 19283
Description: Shirling, E.B. and D. Gottlieb. 1972. International Journal of Systematic Bacteriology **22**:265-394.

S. thermovulgaris Henssen 1957
Henssen, A. 1957. Archiv fur Mikrobiologie **26**:373-414. **Type strain:** ATCC 19284
Description: Shirling, E.B. and D. Gottlieb. 1969. International Journal of Systematic Bacteriology **19**:391-512.

S. torulosus Lyons and Pridham 1971
Lyons, A.J. and T.G. Pridham. 1971. Applied Microbiology **22**:190-193. **Type strain:** NRRL B-3889
Description: Lyons, A.J. and T.G. Pridham. 1971. Ibid.

S. toxytricini (Preobrazhenskaya and Sveshnikova 1957) Pridham, Hesseltine and Benedict 1958
Preobrazhenskaya, T.P. and M.A. Sveshnikova. 1957 in Gauze, G.F., T.P. Preobrazhenskaya, E.S. Kudrina, N.O. Blinov, I.D. Ryabova and M.A. Sveshnikova. Problems of classification of actinomycetes-antagonists. Government Publishing House of Medical Literature, Medgiz, Moscow. pp. 1-398; Pridham, T.G., C.W. Hesseltine and R.G. Benedict. 1958. Applied Microbiology **6**:52-79. **Type strain:** ATCC 19813
Description: Shirling, E.B. and D. Gottlieb. 1968. International Journal of Systematic Bacteriology **18**:69-189.

S. tricolor (Wollenweber 1920) Waksman 1961
Wollenweber, H.W. 1920. Arbeiten des Forschungsinstitutes fur Kartoffelbau, Verlagsbuchandlung Paul Parey, Berlin, No. 2, pp. 1-102; Waksman, S.A. 1961. The Actinomycetes. Vol. 2. Classification, identification and descriptions of genera and species. The Williams and Wilkins Co., Baltimore. pp. 1-363. **Type strain:** CBS 103.21
Description: Bergey 8.

S. tubercidicus Nakamura 1961
Nakamura, G. 1961. Journal of Antibiotics (Tokyo) Series A **14**:90-93. **Type strain:** ATCC 25502
Description: Shirling, E.B. and D. Gottlieb. 1969. International Journal of Systematic Bacteriology **19**:391-512.

S. tuirus Albert and Malaquias de Querioz 1963
Albert, C.A. and V.M. Malaquias de Querioz. 1963. Revista do Instituto de Antibioticos, Universidade do Recife **5**:43-51. **Type strain:** ATCC 19007
Description: Shirling, E.B. and D. Gottlieb. 1972. International Journal of Systematic Bacteriology **22**:265-394.

S. umbrinus (Sveshnikova 1957) Pridham, Hesseltine and Benedict 1958
Sveshnikova, M.A. 1957 in Gauze, G.F., T.P. Preobrazhenskaya, E.S. Kudrina, N.O. Blinov, I.D. Ryabova and M.A. Sveshnikova. Problems of classification of actinomycetes-antagonists. Government Publishing House of Medical Literature, Medgiz, Moscow. pp. 1-398; Pridham, T.G., C.W. Hesseltine and R.G. Benedict. 1958. Applied Microbiology **6**:52-79. **Type strain:** ATCC 19929
Description: Shirling, E.B. and D. Gottlieb. 1969. International Journal of Systematic Bacteriology **19**:391-512.

S. variabilis (Preobrazhenskaya, Ryabova and Blinov 1957) Pridham, Hesseltine and Benedict 1958
Preobrazhenskaya, T.P., I.D. Ryabova and N.O. Blinov. 1957 in Gauze, G.F., T.P. Preobrazhenskaya, E.S. Kudrina, N.O. Blinov, I.D. Ryabova and M.A. Sveshnikova. Problems of classification of actinomycetes-antagonists. Government Publishing House of Medical Literature, Medgiz, Moscow. pp. 1-398; Pridham, T.G., C.W. Hesseltine and R.G. Benedict. 1958. Applied Microbiology **6**:52-79. **Type strain:** ATCC 19930
Description: Shirling, E.B. and D. Gottlieb. 1968. International Journal of Systematic Bacteriology **18**:69-189.

S. varsoviensis Kurylowicz and Woznicka 1967
Kurylowicz, W. and W. Woznicka. 1967. Medycyna Doswiadczalna i Mikrobiologia **19**:1-9. **Type strain:** ATCC 25505
Description: Shirling, E.B. and D. Gottlieb. 1969. International Journal of Systematic Bacteriology **19**:391-512.

S. vastus Szabo and Marton 1958

Szabo, I. and M. Marton. 1958. Agrokemia es Talajtan **7**:243-262. **Type strain:** ATCC 25506
Description: Shirling, E.B. and D. Gottlieb. 1969. International Journal of Systematic Bacteriology **19**:391-512.

S. venezuelae Ehrlich, Gottlieb, Burkholder, Anderson and Pridham 1948
Ehrlich, J., D. Gottlieb, P.R. Burkholder, L.E. Anderson and T.G. Pridham. 1948. Journal of Bacteriology **56**:467-477. **Type strain:** NRRL 2277
Description: Shirling, E.B. and D. Gottlieb. 1969. International Journal of Systematic Bacteriology **19**:391-512.

S. vinaceus Jones 1952
Jones, K.L. 1952. Papers from the Michigan Academy of Science, Arts and Letters **37**:47-48. **Type strain:** NRRL 2382
Description: Shirling, E.B. and D. Gottlieb. 1972. International Journal of Systematic Bacteriology **22**:265-394.

S. vinaceusdrappus Pridham, Hesseltine and Benedict 1958
Pridham, T.G., C.W. Hesseltine and R.G. Benedict. 1958. Applied Microbiology **6**:52-79. **Type strain:** NRRL 2363
Description: Shirling, E.B. and D. Gottlieb. 1969. International Journal of Systematic Bacteriology **19**:391-512.

S. violaceochromogenes (Ryabova and Preobrazhenskaya 1957) Pridham 1970
Ryabova, I.D. and T.P. Preobrazhenskaya. 1957 in Gauze, G.F., T.P. Preobrazhenskaya, E.S. Kudrina, N.O. Blinov, I.D. Ryabova and M.A. Sveshnikova. Problems of classification of actinomycetes-antagonists. Government Publishing House of Medical Literature, Medgiz, Moscow. pp. 1-398; Pridham, T.G. 1970. Bulletin of the United States Department of Agriculture **1424**:1-55. **Type strain:** ATCC 19932
Description: Shirling, E.B. and D. Gottlieb. 1969. International Journal of Systematic Bacteriology **19**:391-512.

S. violaceolatus (Krassilnikov, Sorokina, Alferova and Bezzubenkova 1965) Pridham 1970
Krassilnikov, N.A., E.J. Sorokina, V.A. Alferova and A.P. Bezzubenkova. 1965 in Krassilnikov, N.A. (ed). Biology of selected groups of actinomycetes (in Russian). Institute of Microbiology, Academy of Science, Publishing Firm Nauka, Moscow, USSR. pp. 1-372; Pridham, T.G. 1970. Bulletin of the United States Department of Agriculture **1424**:1-55. **Type strain:** ATCC 19847
Description: Shirling, E.B. and D. Gottlieb. 1969. International Journal of Systematic Bacteriology **19**:391-512.

S. violaceorectus (Ryabova and Preobrazhenskaya 1957) Pridham, Hesseltine and Benedict 1958
Ryabova, I.D. and T.P. Preobrazhenskaya. 1957 in Gauze, G.F., T.P. Preobrazhenskaya, E.S. Kudrina, N.O. Blinov, I.D. Ryabova and M.A. Sveshnikova. Problems of classification of actinomycetes-antagonists. Government Publishing House of Medical Literature, Medgiz, Moscow. pp. 1-398; Pridham, T.G., C.W. Hesseltine and R.G. Benedict. 1958. Applied Microbiology **6**:52-79. **Type strain:** ATCC 25514
Description: Shirling, E.B. and D. Gottlieb. 1969. International Journal of Systematic Bacteriology **19**:391-512.

S. violaceoruber (Waksman and Curtis 1916) Pridham 1970
Waksman, S.A. and R.E. Cutris. 1916. Soil Science **1**:99-134; Pridham, T.G. 1970. Bulletin of the United States Department of agriculture **1424**:1-55. **Type strain:** ATCC 14980
Description: Shirling, E.B. and D. Gottlieb. 1968. International Journal of Systematic Bacteriology **18**:69-189.

S. violaceus (Rossi Doria 1891) Waksman 1953
Rossi Doria, T. 1891. Annali dell' Istituto d'Igiene Sperimentale della Rendiconti Universita di Roma **1**:399-438; Waksman, S.A. 1953 in Waksman, S.A. and H.A. Lechevalier. Guide to the classification and identification of the actinomycetes and their antibiotics. The Williams and Wilkins Co., Baltimore. pp. 1-246. **Type strain:** ATCC 15888
Description: Shirling, E.B. and D. Gottlieb. 1969. International Journal of Systematic Bacteriology **19**:391-512.

S. violaceusniger (Waksman and Curtis 1916) Pridham, Hesseltine and Benedict 1958
Waksman, S.A. and R.E. Curtis. 1916. Soil Science **1**:99-134; Pridham, T.G.,

C.W. Hesseltine and R.G. Benedict. 1958. Applied Microbiology **6**:52-79. **Type strain:** NRRL B-1476
Description: Shirling, E.B. and D. Gottlieb. 1972. International Journal of Systematic Bacteriology **22**:265-394.

S. violarus (Artamonova and Krassilnikov 1960) Pridham 1970
Artamonova, O.I. and N.A. Krassilnikov. 1960 in Rautenshtein, Ya. I. Transactions of the Institute of Microbiology, Academy of Science USSR **8**:1-344; Pridham, T.G. 1970. Bulletin of the United States Department of Agriculture **1424**:1-55. **Type strain:** ATCC 15891
Description: Shirling, E.B. and D. Gottlieb. 1969. International Journal of Systematic Bacteriology **19**:391-512.

S. violascens (Preobrazhenskaya and Sveshnikova 1957) Pridham, Hesseltine and Benedict 1958
Preobrazhenskaya, T.P. and M.A. Sveshnikova. 1957 in Gauze, G.F., T.P. Preobrazhenskaya, E.S. Kudrina, N.O. Blinov, I.D. Ryabova and M.A. Sveshnikova. Problems of classification of actinomycetes-antagonists. Government Publishing House of Medical Literature, Medgiz, Moscow. pp. 1-398; Pridham, T.G., C.W. Hesseltine and R.G. Benedict. 1958. Applied Microbiology **6**:52-79. **Type strain:** ATCC 23968
Description: Shirling, E.B. and D. Gottlieb. 1968. International Journal of Systematic Bacteriology **18**:279-392.

S. violatus (Artamonova and Krassilnikov 1960) Pridham 1970
Artamonova, O.I. and N.A. Krassilnikov. 1960 in Rautenshtein, Ya. I. Transactions of the Institute of Microbiology, Academy of Science USSR **8**:1-344; Pridham, T.G. 1970. Bulletin of the United States Department of Agriculture **1424**:1-55. **Type strain:** ATCC 15892
Description: Shirling, E.B. and D. Gottlieb. 1972. International Journal of Systematic Bacteriology **22**:265-394.

S. virginiae Grundy, Whitman, Rdzok, Rdzok, Hanes and Sylvester 1952
Grundy, W.E., A.L. Whitman, E.G. Rdzok, E.J. Rdzok, M.E. Hanes and J.C. Sylvester. 1952. Antibiotics and Chemotherapy **2**:399-408. **Type strain:** ATCC 19817
Description: Shirling, E.B. and D. Gottlieb. 1968. International Journal of Systematic Bacteriology **18**:69-189.

S. viridiviolaceus (Ryabova and Preobrazhenskaya 1957) Pridham, Hesseltine and Benedict 1958
Ryabova, I.D. and T.P. Preobrazhenskaya. 1957 in Gauze, G.F., T.P. Preobrazhenskaya, E.S. Kudrina, N.O. Blinov, I.D. Ryabova and M.A. Sveshnikova. Problems of classification of actinomycetes-antagonists. Government Publishing House of Medical Literature, Medgiz, Moscow. pp. 1-398; Pridham, T.G., C.W. Hesseltine and R.G. Benedict. 1958. Applied Microbiology **6**:52-79. **Type strain:** ATCC 27478
Description: Shirling, E.B. and D. Gottlieb. 1972. International Journal of Systematic Bacteriology **22**:265-394.

S. viridochromogenes (Krainsky 1914) Waksman and Henrici 1948
Krainsky, A. 1914. Zentralblatt fur Bakteriologie, Parasitenkunde, Infektionskrankheiten und Hygiene. Abteilung II. **41**:649-688; Waksman, S.A. and A.T. Henrici. 1948 in Breed, R.S., E.G.D. Murray and A.P. Hitchens (eds). Bergey's Manual of Determinative Bacteriology, 6th ed. The Williams and Wilkins Co., Baltimore. pp. 929-980. **Type strain:** ATCC 14920
Description: Shirling, E.B. and D. Gottlieb. 1972. International Journal of Systematic Bacteriology **22**:265-394.

S. viridodiastaticus (Baldacci, Grein and Spalla 1955) Pridham, Hesseltine and Benedict 1958
Baldacci, E., A. Grein and C. Spalla. 1955. Giornale di Microbiologia **1**:127-143; Pridham, T.G., C.W. Hesseltine and R.G. Benedict. 1958. Applied Microbiology **6**:52-79. **Type strain:** ATCC 25518
Description: Shirling, E.B. and D. Gottlieb. 1969. International Journal of Systematic Bacteriology **19**:391-512.

S. viridosporus Pridham, Hesseltine and Benedict 1958
Pridham, T.G., C.W. Hesseltine and R.G. Benedict. 1958. Applied Microbiology **6**:52-79. **Type strain:** NRRL 2414
Description: Shirling, E.B. and D.

Gottlieb. 1972. International Journal of Systematic Bacteriology **22**:265-394.

S. werraensis Wallhausser, Huber, Nesemann, Prave and Zepf 1964
Wallhausser, K.H., G. Huber, G. Nesemann, P. Prave and K. Zepf. 1964. Arzneimittel Forschung **14**:356-360. **Type strain:** ATCC 14424
Description: Shirling, E.B. and D. Gottlieb. 1972. International Journal of Systematic Bacteriology **22**:265-394.

S. willmorei (Erikson 1935) Waksman and Henrici 1948
Erikson, D. 1935. Medical Research Council (Great Britain) Special Report Series **203**:5-61; Waksman, S.A. and A.T. Henrici. 1948 in Breed, R.S., E.G.D. Murray and A.P. Hitchens (eds). Bergey's Manual of Determinative Bacteriology, 6th ed. The Williams and Wilkins Co., Baltimore. pp. 929-980. **Type strain:** ATCC 6867
Description: Shirling, E.B. and D. Gottlieb. 1972. International Journal of Systematic Bacteriology **22**:265-394.

S. xanthochromogenes Arishima, Sakamoto and Sato 1956
Arishima, M., J.M. Sakamoto and T. Sato. 1956. Journal of the Agricultural Chemical Society of Japan **30**:469-471. **Type strain:** ATCC 19818
Description: Shirling, E.B. and D. Gottlieb. 1968. International Journal of Systematic Bacteriology **18**:69-189.

S. xanthocidicus Asahi, Nagatsu and Suzuki 1966
Asahi, K., J. Nagatsu and S. Suzuki. 1966. Journal of Antibiotics (Tokyo) Series A **19**:195-199. **Type strain:** ATCC 27480
Description: Shirling, E.B. and D. Gottlieb. 1972. International Journal of Systematic Bacteriology **22**:265-394.

S. xantholiticus (Konev and Tsyganov 1962) Pridham 1970
Konev, I.E. and V.A. Tsyganov. 1962. Mikrobiologiya **31**:1023-1028; Pridham, T.G. 1970. Bulletin of the United States Department of Agriculture **1424**:1-55. **Type strain:** ATCC 27481
Description: Shirling, E.B. and D. Gottlieb. 1972. International Journal of Systematic Bacteriology **22**:265-394.

S. xanthophaeus Lindenbein 1952
Lindenbein, W. 1952. Archiv fur Mikrobiologie **17**:361-383. **Type strain:** ATCC 19819
Description: Shirling, E.B. and D. Gottlieb. 1968. International Journal of Systematic Bacteriology **18**:69-189.

S. yokosukanensis Nakamura 1961
Nakamura, G. 1961. Journal of Antibiotics (Tokyo) Series A **14**:94-97. **Type strain:** ATCC 25520
Description: Shirling, E.B. and D. Gottlieb. 1969. International Journal of Systematic Bacteriology **19**:391-512.

S. zaomyceticus Hinuma 1954
Hinuma, Y. 1954. Journal of Antibiotics (Tokyo) Series A **7**:134-136. **Type strain:** ATCC 27482
Description: Shirling, E.B. and D. Gottlieb. 1972. International Journal of Systematic Bacteriology **22**:265-394.

Streptosporangium Couch 1955
Couch, J.N. 1955. Journal of the Elisha Mitchell Scientific Society **71**:148-155. **Type species:** *S. roseum* Couch 1955
Description: Bergey 8.

S. albidum Furumai, Ogawa and Okuda 1968
Furumai, T., H. Ogawa and T. Okuda. 1968. Journal of Antibiotics (Tokyo) **21**: 179-181. **Type strain:** ATCC 25243
Description: Bergey 8.

S. album Nonomura and Ohara 1960
Nonomura, H. and Y. Ohara. 1960. Journal of Fermentation Technology **38**:405-409. **Type strain:** DSM 43023
Description: Bergey 8.

S. amethystogenes Nonomura and Ohara 1960
Nonomura, H. and Y. Ohara. 1960. Journal of Fermentation Technology **38**:405-409. **Type strain:** KCC A-0026
Description: Bergey 8.

S. corrugatum Williams and Sharples 1976
Williams, S.T. and G.P. Sharples. 1976. International Journal of Systematic Bacteriology **26**:45-52. **Type strain:** ATCC 29331
Description: Williams, S.T. and G.P. Sharples. 1976. Ibid.

S. indianense Gupta 1965
Gupta, K.C. 1965. Journal of Antibiotics (Tokyo) Series A **18**:125-127. **Type strain:** KCC A-0053

Description: Gupta, K.C. 1965. Ibid.

S. longisporum Schafer 1969
Schafer, D. 1969. Archiv fur Mikrobiologie **66**:365-373. **Type strain:** ATCC 25212
Description: Bergey 8.

S. nondiastaticum Nonomura and Ohara 1969
Nonomura, H. and Y. Ohara. 1969. Journal of Fermentation Technology **47**:701-709. **Type strain:** ATCC 27101
Description: Bergey 8.

S. pseudovulgare Nonomura and Ohara 1969
Nonomura, H. and Y. Ohara. 1969. Journal of Fermentation Technology **47**:701-709. **Type strain:** ATCC 27100
Description: Bergey 8.

S. roseum Couch 1955
Couch, J.N. 1955. Journal of the Elisha Mitchell Scientific Society **71**:148-155. **Type strain:** ATCC 12428
Description: Bergey 8.

S. violaceochromogenes Kawamoto, Takasawa, Okachi, Kohakura, Takahashi and Nara 1975
Kawamoto, I., S. Takasawa, R. Okachi, M. Kohakura, I. Takahashi and T. Nara. 1975. Journal of Antibiotics (Tokyo) **28**: 358-365. **Type strain:** ATCC 21807
Description: Kawamoto, I., S. Takasawa, R. Okachi, M. Kohakura, I. Takahashi and T. Nara. 1975. Ibid.

S. viridialbum Nonomura and Ohara 1960
Nonomura, H. and Y. Ohara. 1960. Journal of Fermentation Technology **38**:405-409. **Type strain:** KCC A-027
Description: Bergey 8.

S. viridogriseum Okuda, Furumai, Watanabe, Okugawa and Kimura 1966
Okuda, T., T. Furumai, E. Watanabe, Y. Okugawa and S. Kimura. 1966. Journal of Antibiotics (Tokyo) **19**:121-127. **Type strain:** ATCC 25242
Description: Bergey 8.

S. viridogriseum subsp. *kofuense* Nonomura and Ohara 1969
Nonomura, H. and Y. Ohara. 1969. Journal of Fermentation Technology **47**:701-709. **Type strain:** ATCC 27102
Description: Bergey 8.

S. viridogriseum subsp. *viridogriseum* Okuda, Furumai, Watanabe, Okugawa and Kimura 1966
Okuda, T., T. Furumai, E. Watanabe, Y. Okugawa and S. Kimura. 1966. Journal of Antibiotics (Tokyo) **19**:121-127. **Type strain:** ATCC 25242
Description: Bergey 8.

S. vulgare Nonomura and Ohara 1960
Nonomura, H. and Y. Ohara. 1960. Journal of Fermentation Technology **38**:405-409. **Type strain:** KCC A-0028
Description: Bergey 8.

Streptoverticillium Baldacci 1958
Baldacci, E. 1958. Giornale di Microbiologia **6**:10-27. **Type species:** *S. baldaccii* Farina and Locci 1966
Description: Bergey 8.

S. abikoense (Umezawa, Tazaki and Fukuyama 1951) Locci, Baldacci and Petrolini Baldan 1969
Umezawa, H., T. Tazaki and S. Fukuyama. 1951. Japanese Medical Journal **4**: 331-346; Locci, R., E. Baldacci and B. Petrolini Baldan. 1969. Giornale di Microbiologia **17**:1-60. **Type strain:** ATCC 12766
Description: Locci, R., E. Baldacci and B. Petrolini Baldan. 1969. Ibid.

S. albireticuli (Nakazawa 1955) Locci, Baldacci and Petrolini Baldan 1969
Nakazawa, K. 1955. Journal of the Agricultural Chemical Society of Japan **29**: 647-649; Locci, R., E. Baldacci and B. Petrolini Baldan. 1969. Giornale di Microbiologia **17**:1-60. **Type strain:** ATCC 19721
Description: Shirling, E.B. and D. Gottlieb. 1968. International Journal of Systematic Bacteriology **18**:69-189.

S. album Locci, Baldacci and Petrolini Baldan 1969
Locci, R., E. Baldacci and B. Petrolini Baldan. 1969. Giornale di Microbiologia **17**:1-60. **Type strain:** NRRL 2401
Description: Locci, R., E. Baldacci and B. Petrolini Baldan. 1969. Ibid.

S. ardum (De Boer, Dietz, Lummis and Savage 1961) Locci, Baldacci and Petrolini Baldan 1969
De Boer, C., A. Dietz, N.E. Lummis and G.M. Savage. 1961. Antimicrobial Agents Annual pp. 17-22; Locci, R., E. Baldacci and B. Petrolini Baldan. 1969. Giornale di Microbiologia **17**:1-60. **Type strain:** NRRL 2817
Description: Locci, R., E. Baldacci and

B. Petrolini Baldan. 1969. Ibid.

S. aureoversile Locci, Baldacci and Petrolini Baldan 1969
Locci, R., E. Baldacci and B. Petrolini Baldan. 1969. Giornale di Microbiologia **17**:1-60. **Type strain:** ATCC 15853
Description: Shirling, E.B. and D. Gottlieb. 1969. International Journal of Systematic Bacteriology **19**:391-512.

S. baldaccii Farina and Locci 1966
Farina, G. and R. Locci. 1966. Giornale di Microbiologia **14**:33-52. **Type strain:** ATCC 23654
Description: Farina, G. and R. Locci. 1966. Ibid.

S. biverticillatum (Preobrazhenskaya 1957) Farina and Locci 1966
Preobrazhenskaya, T.P. 1957 in Gauze, G.F., T.P. Preobrazhenskaya, E.S. Kudrina, N.O. Blinov, I.D. Ryabova and M.A. Sveshnikova. Problems in the classification of actinomycetes-antagonists. Publishing House of Medical Literature, Medgiz, Moscow. pp. 1-398; Farina, G. and R. Locci. 1966. Giornale di Microbiologia **14**:33-52. **Type strain:** ATCC 23615
Description: Shirling, E.B. and D. Gottlieb. 1968. International Journal of Systematic Bacteriology **18**:279-392.

S. blastmyceticum (Watanabe, Tanaka, Fukuhara, Miyairi, Yonehara and Umezawa 1957) Locci, Baldacci and Petrolini Baldan 1969
Watanabe, K., T. Tanaka, K. Fukuhara, N. Miyairi, H. Yonehara and H. Umezawa. 1957. Journal of Antibiotics (Tokyo) Series A **10**:39-45; Locci, R., E. Baldacci and B. Petrolini Baldan. 1969. Giornale di Microbiologia **17**:1-60. **Type strain:** ATCC 19731
Description: Shirling, E.B. and D. Gottlieb. 1968. International Journal of Systematic Bacteriology **18**:69-189.

S. cinnamoneum (Benedict, Dvonch, Shotwell, Pridham and Lindenfelser 1952) Baldacci, Farina and Locci 1966
Benedict, R.G., W. Dvonch, O.L. Shotwell, T.G. Pridham and L.A. Lindenfelser. 1952. Antibiotics and Chemotherapy **2**:591-594; Baldacci, E., G. Farina and R. Locci. 1966. Giornale di Microbiologia **14**:153-171. **Type strain:** NRRL B-1285
Description: Shirling, E.B. and D. Gottlieb. 1968. International Journal of Systematic Bacteriology **18**:279-392.

S. cinnamoneum subsp. *albosporum* Thirumalachar 1968
Thirumalachar, M.J. 1968 in Rahalkar, P.W. and M.J. Thirumalachar. Hindustan Antibiotics Bulletin **11**:90-96. **Type strain:** ATCC 25186
Description: Locci, R., E. Baldacci and B. Petrolini Baldan. 1969. Giornale di Microbiologia **17**:1-60.

S. cinnamoneum subsp. *cinnamoneum* (Benedict, Dvonch, Shotwell, Pridham and Lindenfelser 1952) Baldacci, Farina and Locci 1966
Benedict, R.G., W. Dvonch, O.L. Shotwell, T.G. Pridham and L.A. Lindenfelser. 1952. Antibiotics and Chemotherapy **2**:591-594; Baldacci, E., G. Farina and R. Locci. 1966. Giornale di Microbiologia **14**:153-171. **Type strain:** NRRL B-1285
Description: Shirling, E.B. and D. Gottlieb. 1968. International Journal of Systematic Bacteriology **18**:279-392.

S. cinnamoneum subsp. *lanosum* Thirumalachar 1968
Thirumalachar, M.J. 1968 in Rahalkar, P.W. and M.J. Thirumalachar. Hindustan Antibiotics Bulletin **11**:90-96. **Type strain:** ATCC 25187
Description: Locci, R., E. Baldacci and B. Petrolini Baldan. 1969. Giornale di Microbiologia **17**:1-60.

S. cinnamoneum subsp. *sparsum* Thirumalachar 1968
Thirumalachar, M.J. 1968 in Rahalkar, P.W. and M.J. Thirumalachar. Hindustan Antibiotics Bulletin **11**:90-96. **Type strain:** ATCC 25185
Description: Locci, R., E. Baldacci and B. Petrolini Baldan. 1969. Giornale di Microbiologie **17**:1-60.

S. distallicum Locci, Baldacci and Petrolini Baldan 1969
Locci, R., E. Baldacci and B. Petrolini Baldan. 1969. Giornale di Microbiologia **17**:1-60. **Type strain:** NRRL 2886
Description: Locci, R., E. Baldacci and B. Petrolini Baldan. 1969. Ibid.

S. ehimense (Shibata, Honjo, Tokui and Nakazawa 1954) Locci, Baldacci and Petrolini Baldan 1969
Shibata, M., M. Honjo, Y. Tokui and N.

Nakazawa. 1954. Journal of Antibiotics (Tokyo) Series B **7**:168; Locci, R., E. Baldacci and B. Petrolini Baldan. 1969. Giornale di Microbiologia **17**:1-60. **Type strain:** ATCC 23903
Description: Locci, R., E. Baldacci and B. Petrolini Baldan. 1969. Ibid.

S. **eurocidicum** (Okami, Utahara, Nakamura and Umezawa 1954) Locci, Baldacci and Petrolini Baldan 1969
Okami, Y., R. Utahara, S. Nakamura and H. Umezawa. 1954. Journal of Antibiotics (Tokyo) Series A **7**:98-103; Locci, R., E. Baldacci and B. Petrolini Baldan. 1969. Giornale di Microbiologia **17**:1-60. **Type strain:** ATCC 27428
Description: Shirling, E.B. and D. Gottlieb. 1972. International Journal of Systematic Bacteriology **22**:265-394.

S. **fervens** (De Boer, Dietz, Evans and Michaels 1960) Locci, Baldacci and Petrolini Baldan 1969
De Boer, C., A. Dietz, J.S. Evans and R.M. Michaels. 1960. Antibiotics Annual pp. 220-226; Locci, R., E. Baldacci and B. Petrolini Baldan. 1969. Giornale di Microbiologia **17**:1-60. **Type strain:** NRRL 2755
Description: Shirling, E.B. and D. Gottlieb. 1972. International Journal of Systematic Bacteriology **22**:265-394.

S. **fervens** subsp. *fervens* Baldacci and Locci 1974
Baldacci, E. and R. Locci. 1974. Bergey 8. **Type strain:** NRRL 2755
Description: Shirling, E.B. and D. Gottlieb. 1972. International Journal of Systematic Bacteriology **22**:265-394.

S. **fervens** subsp. *melrosporus* Mason, Lummis and Dietz 1965
Mason, D.J., W.L. Lummis and A. Dietz. 1965. Antimicrobial Agents and Chemotherapy **1964**:110-113. **Type strain:** NRRL 3117
Description: Locci, R., E. Baldacci and B. Petrolini Baldan. 1969. Giornale di Microbiologia **17**:1-60.

S. **flavopersicum** (Oliver, Goldstein, Bower, Holper and Otto 1961) Locci, Baldacci and Petrolini Baldan 1969
Oliver, T.J., A. Goldstein, R.R. Bower, J.C. Holper and R.H. Otto. 1961. Antimicrobial Agents and Chemotherapy **1960**: 495-502; Locci, R., E. Baldacci and B. Petrolini Baldan. 1969. Giornale di Microbiologia **17**:1-60. **Type strain:** NRRL 2820
Description: Locci, R., E. Baldacci and B. Petrolini Baldan. 1969. Ibid.

S. **griseocarneum** (Benedict, Stodola, Shotwell, Borud and Lindenfelser 1950) Baldacci, Farina and Locci 1966
Benedict, R.G., F.H. Stodola, O.L. Shotwell, A.M. Borud and L.A. Lindenfelser. 1950. Science (Washington) **112**:77-78; Baldacci, E., G. Farina and R. Locci. 1966. Giornale di Microbiologia **14**:153-171. **Type strain:** NRRL B-1068
Description: Shirling, E.B. and D. Gottlieb. 1968. International Journal of Systematic Bacteriology **18**:69-189.

S. **griseoverticillatum** (Shinobu and Shimada 1962) Locci, Baldacci and Petrolini Baldan 1969
Shinobu, R. and Y. Shimada. 1962. Botanical Magazine (Tokyo) **75**:170-175; Locci, R., E. Baldacci and B. Petrolini Baldan. 1969. Giornale di Microbiologia **17**:1-60. **Type strain:** ATCC 27436
Description: Shirling, E.B. and D. Gottlieb. 1972. International Journal of Systematic Bacteriology **22**:265-394.

S. **hachijoense** (Hosoya, Komatsu, Soeda and Sonoda 1952) Locci, Baldacci and Petrolini Baldan 1969
Hosoya, S., N. Komatsu, M. Soeda and Y. Sonoda. 1952. Japanese Journal of Experimental Medicine **22**:505-509; Locci, R., E. Baldacci and B. Petrolini Baldan. 1969. Giornale di Microbiologia **17**:1-60. **Type strain:** ATCC 19769
Description: Locci, R., E. Baldacci and B. Petrolini Baldan. 1969. Ibid.

S. **hiroshimense** (Shinobu 1955) Farina and Locci 1966
Shinobu, R. 1955. Seibutsugakkaishi **6**:43-46; Farina, G. and R. Locci. 1966. Giornale di Microbiologia **14**:33-52. **Type strain:** ATCC 19772
Description: Shirling, E.B. and D. Gottlieb. 1968. International Journal of Systematic Bacteriology **18**:69-189.

S. **kashmirense** (Gupta and Chopra 1963) Locci, Baldacci and Petrolini Baldan 1969
Gupta, K.C. and I.C. Chopra. 1963. Hindustan Antibiotics Bulletin **5**:110-112;

Locci, R., E. Baldacci and B. Petrolini Baldan. 1969. Giornale di Microbiologia **17**:1-60. **Type strain:** ATCC 27439
Description: Shirling, E.B. and D. Gottlieb. 1972. International Journal of Systematic Bacteriology **22**:265-394.

S. kentuckense (Barr and Carman 1956) Baldacci, Farina and Locci 1966
Barr, F.S. and P.E. Carman. 1956. Antibiotics and Chemotherapy **6**:286-289; Baldacci, E., G. Farina and R. Locci. 1966. Giornale di Microbiologia **14**:153-171. **Type strain:** ATCC 12691
Description: Shirling, E.B. and D. Gottlieb. 1968. International Journal of Systematic Bacteriology **18**:279-392.

S. kishiwadense (Shinobu and Kayamura 1964) Locci, Baldacci and Petrolini Baldan 1969
Shinobu, R. and Y. Kayamura. 1964. Botanical Magazine (Tokyo) **77**:176-180; Locci, R., E. Baldacci and B. Petrolini Baldan. 1969. Giornale di Microbiologia **17**:1-60. **Type strain:** ATCC 25464
Description: Locci, R., E. Baldacci and B. Petrolini Baldan. 1969. Ibid.

S. ladakanum Hanka, Evans, Mason and Dietz 1966
Hanka, L.J., J.S. Evans, D.J. Mason and A. Dietz. 1967. Antimicrobial Agents and Chemotherapy **1966**:619-624. **Type strain:** NRRL 3191
Description: Shirling, E.B. and D. Gottlieb. 1972. International Journal of Systematic Bacteriology **22**:265-394.

S. lavenduligriseum Locci, Baldacci and Petrolini Baldan 1969
Locci, R., E. Baldacci and B. Petrolini Baldan. 1969. Giornale di Microbiologia **17**:1-60. **Type strain:** ATCC 13306
Description: Locci, R., E. Baldacci and B. Petrolini Baldan. 1969. Ibid.

S. lilacinum (Nakazawa, Tanabe, Shibata, Miyake and Takewaka 1956) Locci, Baldacci and Petrolini Baldan 1969
Nakazawa, K., K. Tanabe, M. Shibata, A. Miyake and T. Takewaka. 1956. Journal of Antibiotics (Tokyo) Series B **9**:81; Locci, R., E. Baldacci and B. Petrolini Baldan. 1969. Giornale di Microbiologia **17**:1-60. **Type strain:** ATCC 23930
Description: Locci, R., E. Baldacci and B. Petrolini Baldan. 1969. Ibid.

S. luteoverticillatum (Shinobu 1956) Locci, Baldacci and Petrolini Baldan 1969
Shinobu, R. 1956. Memoirs of the Osaka University of the Liberal Arts and Education B **5**:84-93; Locci, R., E. Baldacci and B. Petrolini Baldan. 1969. Giornale di Microbiologia **17**:1-60. **Type strain:** ATCC 23933
Description: Locci, R., E. Baldacci and B. Petrolini Baldan. 1969. Ibid.

S. mashuense (Sawazaki, Suzuki, Nakamura, Kawasaki, Yamashita, Isono, Anzai, Serizawa and Sekiyama 1955) Locci, Baldacci and Petrolini Baldan 1969
Sawazaki, T., S. Suzuki, G. Nakamura, M. Kawasaki, S. Yamashita, K. Isono, K. Anzai, Y. Serizawa and Y. Sekiyama. 1955. Journal of Antibiotics (Tokyo) Series A **8**:44-47; Locci, R., E. Baldacci and B. Petrolini Baldan. 1969. Giornale di Microbiologia **17**:1-60. **Type strain:** ATCC 23934
Description: Locci, R., E. Baldacci and B. Petrolini Baldan. 1969. Ibid.

S. mobaraense (Nagatsu and Suzuki 1963) Locci, Baldacci and Petrolini Baldan 1969
Nagatsu, J. and S. Suzuki. 1963 in Tamura, S., N. Takahashi, S. Miyamoto, R. Mori, S. Suzuki and J. Nagatsu. 1963. Agricultural and Biological Chemistry **27**: 576-582; Locci, R., E. Baldacci and B. Petrolini Baldan. 1969. Giornale di Microbiologia **17**:1-60. **Type strain:** ATCC 29032
Description: Locci, R., E. Baldacci and B. Petrolini Baldan. 1969. Ibid.

S. netropsis (Finlay, Hochstein, Sobin and Murphy 1951) Baldacci, Farina and Locci 1966
Finlay, A.C., F.A. Hochstein, B.A. Sobin and F.X. Murphy. 1951. Journal of the American Chemical Society **73**:341-343; Baldacci, E., G. Farina and R. Locci. 1966. Giornale di Microbiologia **14**:153-171. **Type strain:** NRRL 2268
Description: Shirling, E.B. and D. Gottlieb. 1968. International Journal of Systematic Bacteriology **18**:279-392.

S. olivoreticuli (Arai, Nakada and Suzuki 1957) Baldacci, Farina and Locci 1966
Arai, M., T. Nakada and M. Suzuki. 1957. Antibiotics and Chemotherapy **7**: 435-442; Baldacci, E., G. Farina and R. Locci. 1966. Giornale di Microbiologia

14:153-171. **Type strain:** ATCC 23943
Description: Shirling, E.B. and D. Gottlieb. 1968. International Journal of Systematic Bacteriology **18**:279-392.

S. olivoverticillatum (Shinobu 1956) Baldacci, Farina and Locci 1966
Shinobu, R. 1956. Memoirs of the Osaka University of the Liberal Arts and Education B **5**:84-93; Baldacci, E., G. Farina and R. Locci. 1966. Giornale di Microbiologia **14**:153-171. **Type strain:** NRRL B-1994
Description: Baldacci, E., G. Farina and R. Locci. 1966. Ibid.

S. orinoci Cassinelli, Grein, Orezzi, Pennella and Sanfilippo 1967
Cassinelli, G., A. Grein, P. Orezzi, P. Pennella and A. Sanfilippo. 1967. Archiv fur Mikrobiologie **55**:358-368. **Type strain:** ATCC 23202
Description: Shirling, E.B. and D. Gottlieb. 1972. International Journal of Systematic Bacteriology **22**:265-394.

S. parvisporogenes Locci, Baldacci and Petrolini Baldan 1969
Locci, R., E. Baldacci and B. Petrolini Baldan. 1969. Giornale di Microbiologia **17**:1-60. **Type strain:** ATCC 12568
Description: Shirling, E.B. and D. Gottlieb. 1972. International Journal of Systematic Bacteriology **22**:265-394.

S. rectiverticillatum (Krassilnikov and Yuan 1965) Locci, Baldacci and Petrolini Baldan 1969
Krassilnikov, N.A. and T. Yuan. 1965 in Krassilnikov, N.A. (ed). Biology of individual groups of Actinomycetes (in Russian). Institute of Microbiology, Academy of Science, Publishing Firm Nauka, Moscow, USSR. pp. 28-57; Locci, R., E. Baldacci and B. Petrolini Baldan. 1969. Giornale di Microbiologia **17**:1-60. **Type strain:** ATCC 19845
Description: Shirling, E.B. and D. Gottlieb. 1969. International Journal of Systematic Bacteriology **19**:391-512.

S. roseoverticillatum (Shinobu 1956) Farina and Locci 1966
Shinobu, R. 1956. Memoirs of the Osaka University of the Liberal Arts and Education B **5**:84-93; Farina, G. and R. Locci. 1966. Giornale di Microbiologia **14**:33-52. **Type strain:** ATCC 19807
Description: Shirling, E.B. and D. Gottlieb. 1968. International Journal of Systematic Bacteriology **18**:69-189.

S. salmonis (Baldacci, Farina and Locci 1966) Locci, Baldacci and Petrolini Baldan 1969
Baldacci, E., G. Farina and R. Locci. 1966. Giornale di Microbiologia **14**:153-171; Locci, R., E. Baldacci and B. Petrolini Baldan. 1969. Giornale di Microbiologia **17**:1-60. **Type strain:** NRRL B-1472
Description: Locci, R., E. Baldacci and B. Petrolini Baldan. 1969. Ibid.

S. septatum Locci, Baldacci and Petrolini Baldan 1969
Locci, R., E. Baldacci and B. Petrolini Baldan. 1969. Gionale di Microbiologia **17**:1-60. **Type strain:** NRRL 2974
Description: Shirling, E.B. and D. Gottlieb. 1972. International Journal of Systematic Bacteriology **22**:265-394.

S. thioluteum (Okami 1952) Baldacci, Farina and Locci 1966
Okami, Y. 1952. Doctoral Dissertation, Hokkaido University; Baldacci, E., G. Farina and R. Locci. 1966. Giornale di Microbiologia **14**:153-171. **Type strain:** CBS 642.72
Description: Shirling, E.B. and D. Gottlieb. 1972. International Journal of Systematic Bacteriology **22**:265-394.

Succinimonas Bryant, Small, Bouma and Chu 1958
Bryant, M.P., N. Small, C. Bouma and H. Chu. 1958. Journal of Bacteriology **76**:15-23. **Type species:** *S. amylolytica* Bryant, Small, Bouma and Chu 1958
Description: Bergey 8.

S. amylolytica Bryant, Small, Bouma and Chu 1958
Bryant, M.P., N. Small, C. Bouma and H. Chu. 1958. Journal of Bacteriology **76**:15-23. **Type strain:** ATCC 19206
Description: Holdeman, L.V., E.P. Cato and W.E.C. Moore (eds). 1977. Anaerobe Laboratory Manual, 4th ed. Virginia Polytechnic Institute and State University, Blacksburg, Virginia. pp. 1-156.

Succinivibrio Bryant and Small 1956
Bryant, M.P. and N. Small. 1956. Journal of Bacteriology **72**:22-26. **Type species:** *S. dextrinosolvens* Bryant and Small 1956

Description: Bergey 8.

S. dextrinosolvens Bryant and Small 1956
Bryant, M.P. and N. Small. 1956. Journal of Bacteriology **72**:22-26. **Type strain:** ATCC 19716
Description: Holdeman, L.V., E.P. Cato and W.E.C. Moore (eds). 1977. Anaerobe Laboratory Manual, 4th ed. Virginia Polytechnic Institute and State University, Blacksburg, Virginia. pp. 1-156.

Sulfolobus Brock, Brock, Belly and Weiss 1972
Brock, T.D., K.M. Brock, R.T. Belly and R.L. Weiss. 1972. Archiv fur Mikrobiologie **84**:54-68. **Type species:** *S. acidocaldarius* Brock, Brock, Belly and Weiss 1972
Description: Brock, T.D., K.M. Brock, R.T. Belly and R.L. Weiss. 1972. Ibid.

S. acidocaldarius Brock, Brock, Belly and Weiss 1972
Brock, T.D., K.M. Brock, R.T. Belly and R.L. Weiss. 1972. Archiv fur Mikrobiologie **84**:54-68. **Type strain:** DSM 639
Description: Brock, T.D., K.M. Brock, R.T. Belly and R.L. Weiss. 1972. Ibid.

Symbiotes Philip 1956
Philip, C.B. 1956. Canadian Journal of Microbiology **2**:261-270. **Type species:** *S. lectularius* (Arkwright, Atkin and Bacot 1921) Philip 1956
Description: Bergey 8.

S. lectularius (Arkwright, Atkin and Bacot 1921) Philip 1956
Arkwright, J.A., E.E. Atkin and A. Bacot. 1921. Parasitology **13**:27-36; Philip, C.B. 1956. Canadian Journal of Microbiology **2**:261-270. **Type strain:** no culture isolated.
Description: Bergey 8.

Thermoactinomyces Tsilinsky 1899
Tsilinsky, P. 1899. Annales de l'Institut Pasteur (Paris) **13**:500-505. **Type species:** *T. vulgaris* Tsilinsky 1899
Description: Bergey 8.

T. candidus Kurup, Barboriak, Fink and Lechevalier 1975
Kurup, V.P., J.J. Barboriak, J.N. Fink and M.P. Lechevalier. 1975. International Journal of Systematic Bacteriology **25**: 150-154. **Type strain:** ATCC 27868
Description: Kurup, V.P., J.J. Barboriak, J.N. Fink and M.P. Lechevalier. 1975. Ibid.

T. dichotomicus (Krassilnikov and Agre 1964) Cross and Goodfellow 1973
Krassilnikov, N.A. and N.S. Agre. 1964. Mikrobiologiya **33**:935-943; Cross, T. and M. Goodfellow. 1973 in Sykes, G. and F.A. Skinner (eds). Actinomycetales, characteristics and practical importance. Academic Press, London. pp. 11-112. **Type strain:** strain 114 (Department Soil Biology, Moscow University)
Description: Krassilnikov, N.A. and N.S. Agre. 1964. Ibid.

T. peptonophilus Nonomura and Ohara 1971
Nonomura, H. and Y. Ohara. 1971. Journal of Fermentation Technology **49**:895-903. **Type strain:** ATCC 27302
Description: Nonomura, H. and Y. Ohara. 1971. Ibid.

T. sacchari Lacey 1971
Lacey, J. 1971. Journal of General Microbiology **66**:327-338. **Type strain:** ATCC 27375
Description: Lacey, J. 1971. Ibid.

T. vulgaris Tsilinsky 1899
Tsilinsky, P. 1899. Annales de l'Institut Pasteur (Paris) **13**:500-505. **Type strain:** KCC A-0162
Description: Bergey 8.

Thermomicrobium Jackson, Ramaley and Meinschein 1973
Jackson, T.J., R.F. Ramaley and W.G. Meinschein. 1973. International Journal of Systematic Bacteriology **23**:28-36. **Type species:** *T. roseum* Jackson, Ramaley and Meinschein 1973
Description: Jackson, T.J., R.F. Ramaley and W.G. Meinschein. 1973. Ibid.

T. fosteri Phillips and Perry 1976
Phillips Jr., W.E. and J.J. Perry. 1976. International Journal of Systematic Bacteriology **26**:220-225. **Type strain:** ATCC 29033
Description: Phillips Jr., W.E. and J.J. Perry. 1976. Ibid.

T. roseum Jackson, Ramaley and Meinschein 1973
Jackson, T.J., R.F. Ramaley and W.G. Meinschein. 1973. International Journal of Systematic Bacteriology **23**:28-36. **Type strain:** ATCC 27502
Description: Jackson, T.J., R.F. Ramaley

and W.G. Meinschein. 1973. Ibid.

Thermomonospora Henssen 1957
Henssen, A. 1957. Archiv fur Mikrobiologie **26**:373-414. **Type species:** *T. curvata* Henssen 1957
Description: Bergey 8.

T. alba (Locci, Baldacci and Petrolini Baldan 1967) Cross and Goodfellow 1973
Locci, R., E. Baldacci and B. Petrolini Baldan. 1967. Giornale di Microbiologia **15**:79-91; Cross, T. and M. Goodfellow. 1973 in Sykes, G. and F.A. Skinner. Actinomycetales, characteristics and practical importance. Academic Press. London. pp. 11-112. **Type strain:** IPV 1900
Description: Cross, T. and M. Goodfellow. 1973. Ibid.

T. curvata Henssen 1957
Henssen, A. 1957. Archiv fur Mikrobiologie **26**:373-414. **Type strain:** ATCC 19995
Description: Bergey 8.

T. mesophila Nonomura and Ohara 1971
Nonomura, H. and Y. Ohara. 1971. Journal of Fermentation Technology **49**:895-903. **Type strain:** ATCC 27303
Description: Nonomura, H. and Y. Ohara. 1971. Ibid.

T. mesouviformis Nonomura and Ohara 1974
Nonomura, H. and Y. Ohara. 1974. Journal of Fermentation Technology **52**:10-13. **Type strain:** ATCC 27644
Description: Nonomura, H. and Y. Ohara. 1974. Ibid.

Thermoplasma Darland, Brock, Samsonoff and Conti 1970
Darland, G., T.D. Brock, W. Samsonoff and S.F. Conti. 1970. Science (Washington) **170**:1416-1418. **Type species:** *T. acidophilum* Darland, Brock, Samsonoff and Conti 1970
Description: Bergey 8.

T. acidophilum Darland, Brock, Samsonoff and Conti 1970
Darland, G., T.D. Brock, W. Samsonoff and S.F. Conti. 1970. Science (Washington) **170**:1416-1418. **Type strain:** AMRC-C165
Description: Brock, T.D. 1978. Thermophilic microorganisms and life at high temperatures. Springer-Verlag, New York. pp. 92-116.

Thermus Brock and Freeze 1969
Brock, T.D. and H. Freeze. 1969. Journal of Bacteriology **98**:289-297. **Type species:** *T. aquaticus* Brock and Freeze 1969
Description: Brock, T.D. and H. Freeze. 1969. Ibid.

T. aquaticus Brock and Freeze 1969
Brock, T.D. and H. Freeze. 1969. Journal of Bacteriology **98**:289-297. **Type strain:** ATCC 25104
Description: Brock, T.D. and H. Freeze. 1969. Ibid.

Thiobacillus Beijerinck 1904
Beijerinck, M.W. 1904. Archives Neerlandaises des Sciences Exactes et Naturelles (Section 2) **9**:131-157. **Type species:** *T. thioparus* Beijerinck 1904
Description: Bergey 8.

T. concretivorus Parker 1945
Parker, C.D. 1945. Australian Journal of Experimental Biology and Medical Science **23**:81-90. **Type strain:** NCIB 8345
Description: Parker, C.D. 1945. Ibid.

T. ferrooxidans Temple and Colmer 1951
Temple, K.L. and A.R. Colmer. 1951. Journal of Bacteriology **62**:605-611. **Type strain:** ATCC 23270
Description: Temple, K.L. and A.R. Colmer. 1951. Ibid.

T. intermedius London 1963
London, J.P. 1963. Archiv fur Mikrobiologie **46**:329-337. **Type strain:** ATCC 15466
Description: London, J.P. 1963. Ibid.

T. neapolitanus Parker 1957
Parker, C.D. 1957 in Breed, R.S., E.G.D. Murray and N.R. Smith (eds). Bergey's Manual of Determinative Bacteriology, 7th ed. The Williams and Wilkins Co., Baltimore. pp. 83-88. **Type strain:** NCIB 8539
Description: Bergey 8.

T. novellus Starkey 1934
Starkey, R.L. 1934. Journal of Bacteriology **28**:365-386. **Type strain:** ATCC 8093
Description: Bergey 8.

T. perometabolis London and Rittenberg 1967
London, J.P. and S.C. Rittenberg. 1967. Archiv fur Mikrobiologie **59**:218-225. **Type strain:** ATCC 23370
Description: London, J.P. and S.C. Rittenberg. 1967. Ibid.

T. thiooxidans Waksman and Joffe 1922
Waksman, S.A. and J.S. Joffe. 1922. Journal of Bacteriology **7**:239-256. **Type strain:** ATCC 19377
Description: Bergey 8.

T. thioparus Beijerinck 1904
Beijerinck, M.W. 1904. Archives Neerlandaises des Sciences Exactes et Naturelles (Section 2) **9**:131-157. **Type strain:** ATCC 8158
Description: Bergey 8.

Thiocapsa Winogradsky 1888
Winogradsky, S. 1888. Beitrage zur Morphologie und Physiologie der Bacterien. Heft I. Zur Morphologie und Physiologie der Schwefelbacterien. Arthur Felix, Leipzig. pp. 1-120. **Type species:** *T. roseopersicina* Winogradsky 1888
Description: Bergey 8.

T. pfennigii Eimhjellen 1970
Eimhjellen, K.E. 1970. Archiv fur Mikrobiologie **73**:193-194. **Type strain:** DSM 1375
Description: Bergey 8.

T. roseopersicina Winogradsky 1888
Winogradsky, S. 1888. Beitrage zur Morphologie und Physiologie der Bacterien. Heft I. Zur Morphologie und Physiologie der Schwefelbacterien. Arthur Felix, Leipzig. pp. 1-120. **Type strain:** DSM 217
Description: Bergey 8.

Thiocystis Winogradsky 1888
Winogradsky, S. 1888. Beitrage zur Morphologie und Physiologie der Bacterien. Heft I. Zur Morphologie und Physiologie der Schwefelbacterien. Arthur Felix, Leipzig. pp. 1-120. **Type species:** *T. violacea* Winogradsky 1888
Description: Bergey 8.

T. gelatinosa (Winogradsky 1888) Pfennig and Truper 1971
Winogradsky, S. 1888. Beitrage zur Morphologie und Physiologie der Bacterien. Heft I. Zur Morphologie und Physiologie der Schwefelbacterien. Arthur Felix, Leipzig. pp. 1-120; Pfennig, N. and H.G. Truper. 1971. International Journal of Systematic Bacteriology **21**:11-14. **Type strain:** DSM 215
Description: Bergey 8.

T. violacea Winogradsky 1888
Winogradsky, S. 1888. Beitrage zur Morphologie und Physiologie der Bacterien. Heft I. Zur Morphologie und Physiologie der Schwefelbacterien. Arthur Felix, Leipzig. pp. 1-120. **Type strain:** DSM 207
Description: Bergey 8.

Thiodictyon Winogradsky 1888
Winogradsky, S. 1888. Beitrage zur Morphologie und Physiologie der Bacterien. Heft I. Zur Morphologie und Physiologie der Schwefelbacterien. Arthur Felix, Leipzig. pp. 1-120. **Type species:** *T. elegans* Winogradsky 1888
Description: Bergey 8.

T. bacillosum (Winogradsky 1888) Pfennig and Truper 1971
Winogradsky, S. 1888. Beitrage zur Morphologie und Physiologie der Bacterien. Heft I. Zur Morphologie und Physiologie der Schwefelbacterien. Arthur Felix, Leipzig. pp. 1-120; Pfennig, N. and H.G. Truper. 1971. International Journal of Systematic Bacteriology **21**:11-14. **Type strain:** DSM 234
Description: Bergey 8.

T. elegans Winogradsky 1888
Winogradsky, S. 1888. Beitrage zur Morphologie und Physiologie der Bacterien. Heft I. Zur Morphologie und Physiologie der Schwefelbacterien. Arthur Felix, Leipzig. pp. 1-120. **Type strain:** DSM 232
Description: Bergey 8.

Thiomicrospira Kuenen and Veldkamp 1972
Kuenen, J.G. and H. Veldkamp. 1972. Antonie van Leeuwenhoek Journal of Microbiology and Serology **38**:241-256. **Type species:** *T. pelophila* Kuenen and Veldkamp 1972
Description: Kuenen, J.G. and H. Veldkamp. 1972. Ibid.

T. denitrificans Timmer-ten Hoor 1975
Timmer-ten Hoor, A. 1975. Netherlands Journal of Sea Research **9**:343-351. **Type strain:** DSM 1251
Description: Timmer-ten Hoor, A. 1975. Ibid.

T. pelophila Kuenen and Veldkamp 1972
Kuenen, J.G. and H. Veldkamp. 1972. Antonie van Leeuwenhoek Journal of Microbiology and Serology **38**:241-256. **Type strain:** ATCC 27801

Description: Kuenen, J.G. and H. Veldkamp. 1972. Ibid.

Thiopedia Winogradsky 1888
Winogradsky, S. 1888. Beitrage zur Morphologie und Physiologie der Bacterien. Heft I. Zur Morphologie und Physiologie der Schwefelbacterien. Arthur Felix, Leipzig. pp. 1-120. **Type species:** *T. rosea* Winogradsky 1888
Description: Bergey 8.

T. rosea Winogradsky 1888
Winogradsky, S. 1888. Beitrage zur Morphologie und Physiologie der Bacterien. Heft I. Zur Morphologie und Physiologie der Schwefelbacterien. Arthur Felix, Leipzig. pp. 1-120. **Type strain:** no pure culture.
Description: Bergey 8.

Thioploca Lauterborn 1907
Lauterborn, R. 1907. Berichte der Deutschen Botanischen Gesellschaft **25**:238-242. **Type species:** *T. schmidlei* Lauterborn 1907
Description: Bergey 8.

T. schmidlei Lauterborn 1907
Lauterborn, R. 1907. Berichte der Deutschen Botanischen Gesellschaft **25**:238-242. **Type strain:** no culture isolated.
Description: Lauterborn, R. 1907. Ibid.

Thiospira Visloukh 1914
Visloukh, S.M. 1914. Zhurnal Mikrobiologii **1**:42-51. **Type species:** *T. winogradskyi* (Omelianski 1905) Visloukh 1914
Description: Bergey 8.

T. winogradskyi (Omelianski 1905) Visloukh 1914
Omelianski, W. 1905. Zentralblatt fur Bakteriologie, Parasitenkunde, Infektionskrankheiten und Hygiene. Abteilung II. **14**:769-772; Visloukh, S.M. 1914. Zhurnal Mikrobiologii **1**:42-51. **Type strain:** no culture available.
Description: Bergey 8.

Thiospirillum Winogradsky 1888
Winogradsky, S. 1888. Beitrage zur Morphologie und Physiologie der Bacterien. Heft I. Zur Morphologie und Physiologie der Schwefelbacterien. Arthur Felix, Leipzig. pp. 1-120. **Type species:** *T. jenense* (Ehrenberg 1838) Migula 1900
Description: Bergey 8.

T. jenense (Ehrenberg 1838) Migula 1900
Ehrenberg, C.G. 1838. Die Infusionsthierchen als vollkommene Organismen. L. Voss, Leipzig; Migula, W. 1900. System der Bakterien, Vol. 2. Gustav Fischer, Jena. **Type strain:** DSM 216
Description: Bergey 8.

Thiothrix Winogradsky 1888
Winogradsky, S. 1888. Beitrage zur Morphologie und Physiologie der Bacterien. Heft 1. Zur Morphologie und Physiologie der Schwefelbacterien. Arthur Felix, Leipzig. pp. 1-120. **Type species:** *T. nivea* (Rabenhorst 1865) Winogradsky 1888
Description: Bergey 8.

T. nivea (Rabenhorst 1865) Winogradsky 1888
Rabenhorst, L. 1865. Flora Europaea Algarum aquae dulcis et submarinae. E. Kummer, Leipzig, Section II. pp. 1-319; Winogradsky, S. 1888. Beitrage zur Morphologie und Physiologie der Bacterien. Heft 1. Zur Morphologie und Physiologie der Schwefelbacterien. Arthur Felix, Leipzig. pp. 1-120. **Type strain:** no culture isolated.
Description: Bland, J.A. and J.T. Staley. 1978. Archives of Microbiology **117**:79-87.

Thiovulum Hinze 1913
Hinze, G. 1913. Berichte der Deutschen Botanischen Gesellschaft **31**:189-202. **Type species:** *T. majus* Hinze 1913
Description: Bergey 8.

T. majus Hinze 1913
Hinze, G. 1913. Berichte der Deutschen Botanischen Gesellschaft **31**:189-202. **Type strain:** no culture isolated.
Description: de Boer, W.E., J.W.M. La Riviere and A.L. Houwink. 1961. Antonie van Leeuwenhoek Journal of Microbiology and Serology **27**:447-456.

Toxothrix Molisch 1925
Molisch, H. 1925. Zentralblatt fur Bakteriologie, Parasitenkunde, Infektionskrankheiten und Hygiene. Abteilung II. **65**: 130-139. **Type species:** *T. trichogenes* (Cholodny 1924) Beger 1953
Description: Bergey 8.

T. trichogenes (Cholodny 1924) Beger 1953
Cholodny, N. 1924. Zentralblatt fur Bakteriologie, Parasitenkunde, Infektionskrankheiten und Hygiene. Abteilung II. **61**:292-298; Beger, H. 1953 in Beger, H. and G. Bringmann. Zentralblatt fur Bakteriologie, Parasitenkunde, Infektionskrankheiten und Hygiene. Abteilung II. **107**:318-334. **Type strain:** no culture isolated.
Description: Krul, J.M., P. Hirsch and J.T. Staley. 1970. Antonie van Leeuwenhoek Journal of Microbiology and Serology **36**:409-420.

Treponema Schaudinn 1905
Schaudinn, F. 1905. Deutsche Medizinische Wochenschrift **31**:1728. **Type species:** *T. pallidum* (Schaudinn and Hoffmann 1905) Schaudinn 1905
Description: Bergey 8.

T. hyodysenteriae Harris, Glock, Christensen and Kinyon 1972
Harris, D.L., R.D. Glock, C.R. Christensen and J.M. Kinyon. 1972. Veterinary Medicine - Small Animal Clinician **67**:61-64. **Type strain:** ATCC 27164
Description: Kinyon, J.M. and D.L. Harris. 1979. International Journal of Systematic Bacteriology **29**:102-109.

T. innocens Kinyon and Harris 1979
Kinyon, J.M. and D.L. Harris. 1979. International Journal of Systematic Bacteriology **29**:102-109. **Type strain:** ATCC 29796
Description: Kinyon, J.M. and D.L. Harris. 1979. Ibid.

T. minutum Dobell 1912
Dobell, C.C. 1912. Archiv fur Protistenkunde **26**:117-240. **Type strain:** CIPP 5162
Description: Bergey 8.

T. pallidum (Schaudinn and Hoffmann 1905) Schaudinn 1905
Schaudinn, F. and E. Hoffmann. 1905. Arbeiten aus dem Kaiserlichen Gesundheitsamte (Berlin) **22**:528-534; Schaudinn, F. 1905. Deutsche Medizinische Wochenschrift **31**:1728. **Type strain:** no culture available.
Description: Bergey 8.

T. paraluiscuniculi (Jacobsthal 1920) Smibert 1974
Jacobsthal, E. 1920. Dermatologische Wochenschrift **71**:569-571; Smibert, R.M. 1974. Bergey 8. **Type strain:** no culture available.
Description: Bergey 8.

T. pertenue (Castellani 1905) Castellani and Chalmers 1910
Castellani, A. 1905. Journal of the Ceylon Branch of the British Medical Association **2**:54; Castellani, A. and A.J. Chalmers. 1910. Manual of Tropical Medicine, 1st ed. Bailliere, Tindall and Cox, London. **Type strain:** no culture available.
Description: Bergey 8.

Ureaplasma Shepard, Lunceford, Ford, Purcell, Taylor-Robinson, Razin and Black 1974
Shepard, M.C., C.D. Lunceford, D.K. Ford, R.H. Purcell, D. Taylor-Robinson, S. Razin and F.T. Black. 1974. International Journal of Systematic Bacteriology **24**:160-171. **Type species:** *U. urealyticum* Shepard, Lunceford, Ford, Purcell, Taylor-Robinson, Razin and Black 1974
Description: Shepard, M.C., C.D. Lunceford, D.K. Ford, R.H. Purcell, D. Taylor-Robinson, S. Razin and F.T. Black. 1974. Ibid.

U. urealyticum Shepard, Lunceford, Ford, Purcell, Taylor-Robinson, Razin and Black 1974
Shepard, M.C., C.D. Lunceford, D.K. Ford, R.H. Purcell, D. Taylor-Robinson, S. Razin and F.T. Black. 1974. International Journal of Systematic Bacteriology **24**:160-171. **Type strain:** ATCC 27618
Description: Shepard, M.C., C.D. Lunceford, D.K. Ford, R.H. Purcell, D. Taylor-Robinson, S. Razin and F.T. Black. 1974. Ibid.

Veillonella Prevot 1933
Prevot, A.R. 1933. Annales Societe Nature **15**:23-260. **Type species:** *V. parvula* (Veillon and Zuber 1898) Prevot 1933
Description: Bergey 8.

V. alcalescens Prevot 1933
Prevot, A.R. 1933. Annales Societe Nature **15**:23-260. **Type strain:** ATCC 17745
Description: Bergey 8.

V. alcalescens subsp. *alcalescens* Prevot 1933
Prevot, A.R. 1933. Annales Societe Nature **15**:23-260. **Type strain:** ATCC 17745

Description: Bergey 8.

V. alcalescens subsp. *criceti* Rogosa 1965
Rogosa, M. 1965. Journal of Bacteriology **90**:704-709. **Type strain:** ATCC 17747
Description: Bergey 8.

V. alcalescens subsp. *dispar* Rogosa 1965
Rogosa, M. 1965. Journal of Bacteriology **90**:704-709; **Type strain:** ATCC 17748
Description: Bergey 8.

V. alcalescens subsp. *ratti* Rogosa 1965
Rosgosa, M. 1965. Journal of Bacteriology **90**:704-709. **Type strain:** ATCC 17746
Description: Bergey 8.

V. parvula (Veillon and Zuber 1898) Prevot 1933
Veillon, A. and A. Zuber. 1898. Archives de Medicine Experimentale et d'Anatomie Pathologique **10**:517-545; Prevot, A.R. 1933. Annales Societe Naturelles **15**:23-260. **Type strain:** ATCC 10790
Description: Holdeman, L.V., E.P. Cato and W.E.C. Moore (eds). 1977. Anaerobe Laboratory Manual, 4th ed. Virginia Polytechnic Institute and State University, Blacksburg, Virginia. pp. 1-156.

V. parvula subsp. *atypica* Rogosa 1965
Rogosa, M. 1965. Journal of Bacteriology **90**:704-709; **Type strain:** ATCC 17744
Description: Bergey 8.

V. parvula subsp. *parvula* Rogosa 1965
Rogosa, M. 1965. Journal of Bacteriology **90**:704-709. **Type strain:** ATCC 10790
Description: Holdeman, L.V., E.P. Cato and W.E.C. Moore (eds). 1977. Anaerobe Laboratory Manual, 4th ed. Virginia Polytechnic Institute and State University, Blacksburg, Virginia. pp. 1-156.

V. parvula subsp. *rodentium* Rogosa 1965
Rogosa, M. 1965. Journal of Bacteriology **90**:704-709; **Type strain:** ATCC 17743
Description: Bergey 8.

Vibrio Pacini 1854
Pacini, F. 1854. Gazette Medicale de Italiana Toscano Firenze **6**:405-412. **Type species:** *V. cholerae* Pacini 1854
Description: Bergey 8.

V. albensis Lehmann and Neumann 1896
Lehmann, K.B. and R. Neumann. 1896. Atlas und Grundriss der Bakteriologie und Lehrbuch der speziellen bakteriologischen Diagnostik. 1st ed. J.F. Lehmann, Munchen. **Type strain:** ATCC 14547
Description: Hendrie, M.S., W. Hodgkiss and J.M. Shewan. 1970. Journal of General Microbiology **64**:151-169.

V. alginolyticus (Miyamoto, Nakamura and Takizawa 1961) Sakazaki 1968 (See also Beneckea alginolytica)
Miyamoto, Y., K. Nakamura and K. Takizawa. 1961. Japanese Journal of Microbiology **5**:477-486; Sakazaki, R. 1968. Japanese Journal of Medical Science and Biology **21**:359-362 **Type strain:** ATCC 17749
Description: Sakazaki, R. 1968. Ibid.

V. anguillarum Bergeman 1909
Bergeman, A.M. 1909. Berichte der Bayerischen Biologischen Versuchs-Station **2**: 10-54. **Type strain:** ATCC 19264
Description: Hastein, T. and J.E. Smith. 1977. Journal of Fish Biology **11**:69-75.

V. cholerae Pacini 1854
Pacini, F. 1854. Gazette Medicale de Italiana Toscano Firenze **6**:405-412. **Type strain:** ATCC 14035
Description: Hugh, R. 1965. International Bulletin of Bacteriological Nomenclature and Taxonomy **15**:13-24.

V. costicola Smith 1938
Smith, F.B. 1938. Proceedings of the Royal Society Queensland **49**:29-52. **Type strain:** NCMB 701
Description: Bergey 8.

V. fischeri (Beijerinck 1889) Lehmann and Neumann 1896 (See also Photobacterium fischeri)
Beijerinck, M.W. 1889. Archives neerlandaises des Sciences Exactes et Naturelles **23**:401-427; Lehmann, K.B. and R. Neumann. 1896. Atlas und Grundriss der Bakteriologie und Lehrbuch der speziellen bakteriologischen Diagnostik. 1st ed. J.F. Lehmann, Munchen. **Type strain:** ATCC 7744
Description: Hendrie, M.S., W. Hodgkiss and J.M. Shewan. 1971. International Journal of Systematic Bacteriology **21**: 217-221.

V. metschnikovii Gamaleia 1888
Gamaleia, M.N. 1888. Annales de l'Institut Pasteur (Paris) **2**:482-488. **Type strain:** NCTC 8443
Description: Lee, J.V., T.J. Donovan and A.L. Furniss. 1978. International Journal of Systematic Bacteriology **28**:99-111.

V. parahaemolyticus (Fujino, Okuno, Nakada, Aoyama, Fukai, Mukai and Ueho 1951) Sakazaki, Iwanami and Fukumi 1963 (See also Beneckea parahaemolytica)
Fujino, T., Y. Okuno, D. Nakada, A. Aoyama, K. Fukai, T. Mukai and T. Ueho. 1951. Journal of the Japanese Association of Infectious Diseases **25**:11; Sakazaki, R., S. Iwanami and H. Fukumi. 1963. Japanese Journal of Medical Sciences and Biology **16**:161-188. **Type strain:** ATCC 17802
Description: Fujino, T., R. Sakazaki and K. Tamura. 1974. International Journal of Systematic Bacteriology **24**:447-449.

V. succinogenes Wolin, Wolin and Jacobs 1961
Wolin, M.J., E.A. Wolin and N.J. Jacobs. 1961. Journal of Bacteriology **81**: 911-917. **Type strain:** ATCC 29543
Description: Holdeman, L.V., E.P. Cato and W.E.C. Moore (eds). 1977. Anaerobe Laboratory Manual, 4th ed. Virginia Polytechnic Institute and State University, Blacksburg, Virginia. pp. 1-156.

Vitreoscilla Pringsheim 1949
Pringsheim, E.G. 1949. Bacteriological Reviews **13**:47-98. **Type species:** *V. beggiatoides* Pringsheim 1949
Description: Bergey 8.

V. beggiatoides Pringsheim 1949
Pringsheim, E.G. 1949. Bacteriological Reviews **13**:47-98. **Type strain:** no culture available.
Description: Bergey 8.

V. stercoraria Pringsheim 1951
Pringsheim, E.G. 1951. Journal of General Microbiology **5**:124-149. **Type strain:** ATCC 15218
Description: Bergey 8.

Wolbachia Hertig 1936
Hertig, M. 1936. Parasitology **28**:453-486. **Type species:** *W. pipientis* Hertig 1936
Description: Bergey 8.

W. melophagi (Noller 1917) Philip 1956
Noller, W. 1917. Archiv fur Schiffs und Tropenhygiene **21**:53-94; Philip, C.B. 1956. Canadian Journal of Microbiology **2**:261-270. **Type strain:** no culture isolated.
Description: Bergey 8.

W. persica Suitor and Weiss 1961
Suitor Jr., E.C. and E. Weiss. 1961. Journal of Infectious Diseases **108**:95-106. **Type strain:** ATCC VR 331
Description: Bergey 8.

W. pipientis Hertig 1936
Hertig, M. 1936. Parasitology **28**:453-486. **Type strain:** no culture isolated.
Description: Bergey 8.

Xanthobacter Wiegel, Wilke, Baumgarten, Opitz and Schlegel 1978
Wiegel, J., D. Wilke, J. Baumgarten, R. Opitz and H.G. Schlegel. 1978. International Journal of Systematic Bacteriology **28**:573-581. **Type species:** *X. autotrophicus* (Baumgarten, Reh and Schlegel 1974) Wiegel, Wilke, Baumgarten, Opitz and Schlegel 1978
Description: Wiegel, J., D. Wilke, J. Baumgarten, R. Opitz and H.G. Schlegel. 1978. Ibid.

X. autotrophicus (Baumgarten, Reh and Schlegel 1974) Wiegel, Wilke, Baumgarten, Opitz and Schlegel 1978
Baumgarten, J., M. Reh and H.G. Schlegel. 1974. Archives of Microbiology **100**: 207-217. Wiegel, J., D. Wilke, J. Baumgarten, R. Opitz and H.G. Schlegel. 1978. International Journal of Systematic Bacteriology **28**:573-581. **Type strain:** DSM 432
Description: Wiegel, J., D. Wilkie, J. Baumgarten, R. Opitz and H.G. Schlegel. 1978. Ibid.

X. flavus Malik and Claus 1979
Malik, K.A. and D. Claus. 1979. International Journal of Systematic Bacteriology **29**:283-287. **Type strain:** DSM 338
Description: Malik, K.A. and D. Claus. 1979. Ibid.

Xanthomonas Dowson 1939
Dowson, W.J. 1939. Zentralblatt fur Bakteriologie, Parasitenkunde, Infektionskrankheiten und Hygiene. Abteilung II. **100**:177-193. **Type species:** *X. campestris* (Pammel 1895) Dowson 1939
Description: Bergey 8.

X. albilineans (Ashby 1929) Dowson 1943
Ashby, S.F. 1929. Tropical Agriculture, Trinidad **6**:135-138; Dowson, W.J. 1943. Transactions of the British Mycological Society **26**:4-14. **Type strain:** NCPPB

2969
Description: Bergey 8.

X. ampelina Panagopoulos 1969
Panagopoulos, C.G. 1969. Annales de l'Institut Phytopathologique Benaki NS **9**: 59-81. **Type strain:** NCPPB 2217
Description: Panagopoulos, C.G. 1969. Ibid.

X. axonopodis Starr and Garces 1950
Starr, M.P. and C. Garces. 1950. Revista Facultad Nacional de Agronomia (Medellin) Colombia **12**:73-83. **Type strain:** ATCC 19312
Description: Starr, M.P. and C. Garces. 1950. Ibid.

X. campestris (Pammel 1895) Dowson 1939
Pammel, L.H. 1895. Bulletin of the Iowa State College Agricultural Experiment Station **27**:130-134; Dowson, W.J. 1939. Zentralblatt fur Bakteriologie, Parasitenkunde und Infektionskrankheiten. Abteilung II. **100**:177-193. **Type strain:** NCPPB 528
Description: Bergey 8.

X. fragariae Kennedy and King 1962
Kennedy, B.W. and T.H. King. 1962. Phytopathology **52**:873-875. **Type strain:** NCPPB 1469
Description: Kennedy, B.W. and T.H. King. 1962. Ibid.

Xenorhabdus Thomas and Poinar 1979
Thomas, G.M. and G.O. Poinar Jr. 1979. International Journal of Systematic Bacteriology **29**:352-360. **Type species:** *X. nematophilus* (Poinar and Thomas 1965) Thomas and Poinar 1979
Description: Thomas, G.M. and G.O. Poinar Jr. 1979. Ibid.

X. luminescens Thomas and Poinar 1979
Thomas, G.M. and G.O. Poinar Jr. 1979. International Journal of Systematic Bacteriology **29**:352-360. **Type strain:** ATCC 29999
Description: Thomas, G.M. and G.O. Poinar Jr. 1979. Ibid.

X. nematophilus (Poinar and Thomas 1965) Thomas and Poinar 1979
Poinar Jr., G.O. and G.M. Thomas. 1965. International Bulletin of Bacteriological Nomenclature and Taxonomy **15**: 249-252; Thomas, G.M. and G.O. Poinar Jr. 1979. International Journal of Systematic Bacteriology **29**:352-360. **Type strain:** ATCC 19061
Description: Thomas, G.M. and G.O. Poinar Jr. 1979. Ibid.

Yersinia van Loghem 1944
van Loghem, J.J. 1944. Antonie van Leeuwenhoek Journal of Microbiology and Serology **10**:15-16. **Type species:** *Y. pestis* (Lehmann and Neumann 1896) van Loghem 1944
Description: Bergey 8.

Y. enterocolitica (Schleifstein and Coleman 1939) Frederiksen 1964
Schleifstein, J.I. and M.B. Coleman. 1939. New York State Journal of Medicine **39**:1749-1753; Frederiksen, W. 1964. Proceedings of the XIV Scandinavian Congress of Pathology and Microbiology, Oslo 1964. Norwegian Universities Press, Oslo. pp. 103-104. **Type strain:** ATCC 9610
Description: Bottone, E.J. 1977. CRC Critical Review in Microbiology **5**:211.

Y. pestis (Lehmann and Neumann 1896) van Loghem 1944
Lehmann, K.B. and R. Neumann. 1896. Atlas und Grundriss der Bakteriologie und Lehrbuch der speziellen bacteriologischen Diagnostik. 1st ed. J.F. Lehmann, Munchen. p. 194; van Loghem, J.J. 1944. Antonie van Leeuwenhoek Journal of Microbiology and Serology **10**:15-16. **Type strain:** ATCC 19428
Description: Bergey 8.

Y. philomiragia Jensen, Owen and Jellison 1969
Jensen, W.I., C.R. Owen and W.L. Jellison. 1969. Journal of Bacteriology **100**: 1237-1241. **Type strain:** ATCC 25015
Description: Jensen, W.I., C.R. Owen and W.L. Jellison. 1969. Ibid.

Y. pseudotuberculosis (Pfeiffer 1889) Smith and Thal 1965
Pfeiffer, A. 1889. Uber die bacillare Pseudotuberkulose bei Nagethieren. Verlag von George Thieme, Leipzig. pp. 1-42; Smith, J.E. and E. Thal. 1965. Acta Pathologica and Microbiologica Scandinavica **64**:213-223. **Type strain:** ATCC 29833
Description: Bergey 8.

Y. ruckeri Ewing, Ross, Brenner and Fanning 1978
Ewing, W.H., A.J. Ross, D.J. Brenner

and G.R. Fanning. 1978. International Journal of Systematic Bacteriology **28**:37-44. **Type strain:** ATCC 29473
Description: Ewing, W.H., A.J. Ross, D.J. Brenner and G.R. Fanning. 1978. Ibid.

Zoogloea Itzigsohn 1868
Itzigsohn, H. 1868. Sitzungs-Berichte der Gesellschaft naturforschender Freunde zu Berlin, 19 November 1867, pp. 30-31. **Type species:** *Z. ramigera* Itzigsohn 1868
Description: Bergey 8.

Z. ramigera Itzigsohn 1868
Itzigsohn, H. 1868. Sitzungs-Berichte der Gesellschaft naturforschender Freunde zu Berlin, 19 November 1867, pp. 30-31. **Type strain:** ATCC 19544
Description: Unz, R.F. 1971. International Journal of Systematic Bacteriology **21**: 91-99.

Zymomonas Kluyver and van Niel 1936
Kluyver, A.J. and C.B. van Niel. 1936. Zentralblatt fur Bakteriologie, Parasitenkunde, Infektionskrankheiten und Hygiene. Abteilung II. **94**:369-403. **Type species:** *Z. mobilis* (Lindner 1928) Kluyver and van Niel 1936
Description: Bergey 8.

Z. mobilis (Lindner 1928) Kluyver and van Niel 1936
Lindner, P. 1928. Bericht des Westpreussischen Botanisch-Zoologischen Vereins, 50. pp. 253-255; Kluyver, A.J. and C.B. van Niel. 1936. Zentralblatt fur Bakteriologie, Parasitenkunde, Infektionskrankheiten und Hygiene. Abteilung II. **94**: 369-403. **Type strain:** ATCC 10988
Description: De Ley, J. and J. Swings. 1976. International Journal of Systematic Bacteriology **26**:146-157.

Z. mobilis subsp. *mobilis* (Lindner 1928) De Ley and Swings 1976
Lindner, P. 1928. Bericht des Westpreussischen Botanisch- Zoologischen Vereins, 50. pp. 253-255; De Ley, J. and J. Swings. 1976. International Journal of Systematic Bacteriology **26**:146-157. **Type strain:** ATCC 10988
Description: De Ley, J. and J. Swings. 1976. Ibid.

Z. mobilis subsp. *pomaceae* (Millis 1956) De Ley and Swings 1976
Millis, N.F. 1956. Journal of General Microbiology **15**:521-528; De Ley, J. and J. Swings. 1976. International Journal of Systematic Bacteriology **26**:146-157. **Type strain:** ATCC 29192
Description: De Ley, J. and J. Swings. 1976. Ibid.